NUCLEAR PHYSICS

NUCLEAR PHYSICS

ALI A. ABDULLA

Professor of Physics

Physics Department

College of Science, University of Baghdad, Iraq

Library of Congress Control Number: 2015912134
        ISBN: Hardcover   978-1-5035-9007-6
                 Softcover   978-1-5035-9006-9
                 eBook   978-1-5035-9005-2

Print information available on the last page.

Rev. date: 10/15/2015

To order additional copies of this book, contact:
Xlibris
1-888-795-4274
www.Xlibris.com
Orders@Xlibris.com
716605

Contents

Preface

Nuclear physics is an important branch of physics. It was developed after the discovery of the nucleus of the atom in 1911, as a result of a scattering of alpha particle (α) by a gold foil target, performed by the famous physicist Rutherford. This experiment defined the status of an atom as a nucleus of a positive charge followed by the discovery of the proton and the neutron, as constituents of this nucleus. The mass of the nucleus was found to be almost more than 99% of the mass of the atom. Alsoits size in terms of its radius was determined by Rutherford scattering about 1 fm (fermion), later scattering experiments and other different methods confirmed the radius of the nucleus is related to its mass number (A) as $R = r_o A^{1/3}$, where r_o is the basic radius of the nucleus; the lowest radius is the radius of the lowest electron orbit in the hydrogen ground state. It is called Bohr radius; it is about $0.53A^o$. The average of r_o was measured or calculated by these different methods to be 1.2 fm. The force holding the nucleons bound inside the nucleus size is called a nuclear force which is not quite well-known as the electromagnetic force or the gravitational force, which are very well-known. But the main features of the nuclear force are almost known. These features were observed from the nuclear reactions and the nucleon- nucleon interactions and observed properties of nuclei such as the electromagnetic properties of a nucleus and the configuration of nucleons inside the nucleus volume. These nucleons are distributed under the effect of the exclusion principle. The experiment data plays the main role in developing the theory of nuclear physics. From this information, the nuclear force, which is related to the potential as ($F = -\nabla V$), takes many possible forms. These forms are affected by the spins and orbital angular momenta of different nucleons and also by their isotopic spins, where for the motion of nucleons system (nucleus) there are a spacial coordinates, spin-space coordinates, independent of ($\vec{r}$), and charge or isospace, which depends only on the isotopic spin (τ). Therefore, there are different forms of the Hamiltonian of the nuclear system, which might be of scalar form, vector form, tensor form, or a mixture of two or more of these forms. That depends on the required explanation to the results obtained experimentally or predicted theoretically. However, the nuclear force is quite stronger than the gravitational, the electromagnetic, and the nuclear weak interaction (β-decay, μ-decay) forces. It is claimed that the nuclear force is about 10^{39} times the gravitational force, 10^{13} times the nuclear weak interaction force, and 10^2 times the electromagnetic force. Therefore, gravitational force can be neglected inside the nucleus compared to nuclear force. The nucleus is a microscopic system cannot be described by the classical mechanics. This fact activated scientists to develop a new mechanics to able to deal or to tackle the new physical phenomenon of atomic, nuclear, and molecular system in view of the proposed wave property of a moving system developed by de-Broglie in1924. So during the period 1926–1927, a new mechanics was introduced considering the wave and quantal properties. The wave mechanics was developed by Schrödinger, and the quantum mechanics was developed by Heisenberg, who based his quantum mechanics on the matrix element theory in mathematics, considering the quanta as numbers. But both developments lead to the result its self . Nuclear theory is greatly advanced experimentally and theoretically after the discovery of the nuclear energy in 1939, which was directed to the unhumanitarian uses. The detailed structures of the nucleus were studied for nuclei; the light and the heavy ones during the period 1939–2013 brought out huge information and helped a lot in developing other branches of science. Finally, this book is based on lectures in nuclear physics taught by the author for the last twenty five years at the graduate level. A brief and clear demonstration to the subjects was considered, because the detailed story of the development of nuclear theory is tackled by tens of books. The references considered for these lectures are so many therefore, some of the important one will be listed at the end of the book. The book contains the following,

1-Chapter One

Introduction

2- Chapter Two

Representation of Three-Dimensional Rotation Group

Theory and Applications to Nuclear Physics

3- Chapter Three

Electromagnetic Properties of Nucleus

4- Chapter Four

Emission of Electromagnetic Radiation

(Electromagnetic Transitions)

5- Chapter Five

Emission of Multipole Radiation (EMR)

(Electromagnetic Radiations)

6- Chapter Six

Nuclear β-Decay

7-Chapter Seven

Nuclear Reactions

8- Chapter Eight

High-Energy Physics: Nuclear Particles

Chapter One

Introduction (Brief Necessary Review)

I-1—A General Look at Some Features of the Nucleus

To start with, one might give a comparison between the different classes of matter according to their characteristic forces, where these forces determine in general the class of the matter (atoms, molecules, solids, nuclei, or hadrons). These classes of matter differ actually in the force of interactions, range of this force, the relativity importance in their dynamicalal characters, energy excitation, and the number of particles constituting it. These features are shown in table 1 below

Table 1: Classification of Matter

(de-Shalit and Fastback Vol. I)

Class of Matter	Strength of Interaction	Range of Force	Importance of Relativity	Energy Excitation	Number of Particles
1. Atoms	Weak	Large	Little	$1 - 10^5$ eV	1 to many
Molecules	Weak	Large	Little	$10^{-3} - 10^{-1}$ eV	2 to many
Solids	Weak	Large	Little	$10^{-4} - 1$ eV	Very many
2. Nuclei	Moderate	Short	Some	$10^5 - 10^7$ eV	2 to many
3. Hadrons (baryon + mesons)	Strong	Short	Great	$10^7 - 10^{>9}$ eV	?

From table 1, it is clear that nuclear physics is concerned with a matter of moderate force compared with that of hadrons (high-energy physics). But compared with atoms, molecules, and solids, it is of strong force. As it is well-known to physicists and students of physics, the nuclear forces responsible for binding protons with neutrons, and neutrons with neutrons, and protons with protons are quite strong compared with electromagnetic force, gravitational force, and the weak interaction force of β-decay. Nuclear forces equated to these forces are represented respectively as 10^2, 10^{39}–10^{40}, and 10^{13} since the energy is related physically to the force, whether this force is related to the structure of the nucleus, to the nuclear reaction, or to the creation of a new nuclear particle. Therefore, the field of nuclear physics can be classified (according to the energy) as follows

(1) 0 ~ 1 MeV—the study will be concerned with nuclear structure such as the spin I, magnetic dipole moment μ, quadrupole moment Q, and parity (π).

(2) 1 ~ 10 MeV—this deals with the nuclear reactions of all types under 10 MeV.

(3) > 1000 MeV—this is known as high-energy physics, which deals with nuclear particles, creation, or specifications. This book is specially designed for the first two (1, 2). Although the nuclear force is not well defined, but due to the fact that nuclear physics is an open field, one can deduce in brief the following features of the nuclear force and nuclear system:

1. The force range is about 10^{-13} cm $\cong$ 10^{-15} m.

2. Force law is not quite well-known.

3. The system is of many body problems.

4. It is a microscopic system.

5. Quantum mechanics is the good tool to describe the nuclear system.

6. Hamiltonian ($H = KE + PE$) is not quite well defined for the nucleus, due to the fact that the force is not very well-known, which implies that potential $(V) \cong F_n \cdot d \Rightarrow f \equiv -\nabla V$ is not well defined as shape and quantity.

7. There are two important distinguished models in describing the nuclear system; they are as follows

   a. Model-independent

   b. Model-dependent

   As an example, take an excited state for a nucleus such as

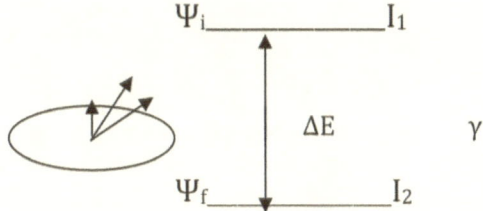

It is known (as it will be shown) that the transition probability[1] from state ψ_i to state ψ_f is given by the well-known golden rule of transition $\tau = \frac{2\pi}{\hbar} |H_{if}|^2 \, d\rho$ (E) or in general $\frac{dN}{dT}$ = B (model) f (θ, I_1, I_2) where B is model dependent factor and f (θ, I_1, I_2) is shape dependent.

For a brief clarification, consider the nucleus; here, τ is a model-dependent (lifetime) that can be found by using Wigner-Eckart theorem (to be shown in the next chapters).

After this brief demonstration to the general description of the nucleus, the question is , what is the real nucleus after ninety-four years since its discovery by Rutherford's α-particle experiment? After the formulations of quantum mechanics in its nonrelativistic and relativistic pictures and then quantum the field theory, the physics of the twentieth century has been concerned with the quantum structure of matter. Each system, such as an atom, a nucleus, or a hadron, studied by high-energy physics, has a ground state and a spectrum of excited states, depending on the energy of the excitation, that are specified by a set of internal quantum numbers, such as spins, in addition to their energies. This claim can be represented by the following diagram:

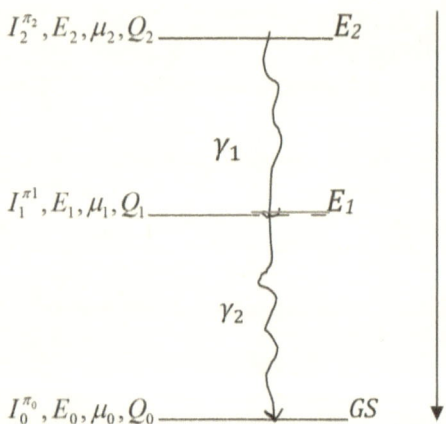

This diagram shows a nucleus of ground state ($I_0^{\pi_0}, E_0, \mu_0, Q_0$), a first excited state, ($I_1^{\pi_1}, E_1, \mu_1, Q_1$), and a second excited state ($I_2^{\pi_2}, E_2, \mu_2, Q_2$). It is a simple system of excited nucleus, because some cases may be of so many excited states that might be of hundred excited states with transitions between these excited states and also between them and the ground state of the excited nucleus. These transitions may be a type of electromagnetic radiations (γ-transitions) or a type of particles transitions (α, β transitions) or a mixture of these types transitions such as $\alpha \rightarrow \beta \rightarrow \gamma$ or $\alpha \rightarrow \beta$, $\alpha \rightarrow \gamma$ or $\beta \rightarrow \gamma$. All these will be clarified in the chapter of α, β, γ decays related to the properties of nuclei in different status.

So at the present status of a nucleus, one claims with confidence that the nucleus entity is composed of two types of particles: a proton of known mass and charge ($m_p \cong 1.67 \times 10^{-24}$ gm, charge = e^+) and a neutron of $m_n > m_p$ and a neutral charge. From table 1, the nucleus is under a force effect of a moderate strength and a short range. The effect of relativity is of some importance and the energies of excitation are in the range (0.5–10) MeV. The number of the particles as constituents of the nucleus starts from two (nucleus of deuteron) to many. Therefore, a nucleus is a many-body system, where any dealing with it is dealing with a many-body problem, which is not so easy to tackle without the approach of approximations. Hence, the well known approximation methods play an important role here. These approximation methods such as WKB, perturbation, vibrational, are well treated by quantum mechanics books (I. Schiff, Merzbacher, Messiah, Saxon, Alonso and Valk and the author. The constituents of the nucleus, the protons and neutrons are treated as one particle of different states, proton state and neutron state, within a defined space called charge space, or isotopic space, isobaric space, or isospace. This will be clarified later. The mass difference between m_n and m_p is about 0.78 MeV, but this does not affect the physical conception of a nucleon proposal for both proton and neutron, because the physical description of the nucleon as proton or neutron is obeying the intrinsic isobaric or isotopic spin defined as

$$\tau = \frac{1}{2}(Z - N)$$ or in some books $$\tau = \frac{1}{2}(N - Z)$$, where N is the number of the neutrons in the nucleus and Z is the number of the protons in the nucleus. (It is equivalent to the electrons' number, and it represents the atomic number in the periodic table of elements. So here, there is a nucleus of a defined size ($R = 10^{-15}$ m) and particles' constituents

situated in definite states (levels), according to the Pauli exclusion principle. Then the nucleon, in the state of the proton, is called a proton, and when it is in the neutron state, it is called a neutron. So this nucleon has two degrees of freedom in the so-called isospace, i.e. charge (or isospin space) is of two dimensions such as

Now: $A = Z + N =$ the mass number   q = 0_____N-state

$Z =$ Atomic number (protonsnumber)   --↑----------------nucleon

Nucleon is either neutron or proton.   q=+1 ──↓────── P-state

$N =$ Neutron number

So an element X is written as $^{A}_{Z}X_{N}$. These particles are nuclear main particles that can be studied by nuclear physics and quantum mechanics. The hadrons are strongly interacting particles,it includes the baryons (nucleons and strange baryons), such as Λ, Σ, Ξ, and Ω, and the mesons, such as π, ρ, κ etc., so the concept of nuclei can be generalized by including in its domain all systems with two baryons (deuteron) or more. When these baryons are nucleons, the systems are the ordinary nuclei (such as H, He, B, Cu, etc.). If some baryons are strange baryons as well as nucleons, the systems are called hypernuclei. It is important to notice from table 1 that the nuclei are the only system that consists of a limited number of particles. The most massive known nucleus consists of 259 nucleons, where the least massive nucleus is the deuteron that consists of two nucleons, with moderately strong forces acting between the particles.

I-2—Nuclear Size

It is well established that the atom is no longer an undividable entity since 1897, where J.J.Thompson discovered the electron as a particle with specified characteristics. Then in 1911, Rutherford had discovered that the atom has a positively charged center with a mass of about 99.9% of the total mass of the atom. This central point of the atom is called a nucleus of the atom. Rutherford has discovered this by performing a scattering experiment of α-particle $(2p + 2n)$ as an incident particle on a target of a gold (Au) foil. The impact parameter (b) was calculated to be about b $\approx 10^{-12}$ cm, which, physically, is the least distance between the points, where α-particle deviates from its straight path and the position where the nucleus is located. The scattering experiments were developed greatly to go more deeply inside the nucleus to study the charge and the current distributions within the nucleus. For this, high-energy (≈ 100 MeV) electrons were used as probes, by the scattering of these electrons, as projectiles, and the nucleus under study, as a target.

This nucleus is surrounded by a "cloud" of electrons. At this time, the atom shape is well established as the nucleus constitutes of protons and neutrons and a "cloud" of electrons moving in their orbits around the nucleus. Also, the size of the atom is 10^{-8} cm (in terms of radius). The radius (R) of a nucleus is given by $R = r_\circ A^{1/3}$, where $r_\circ \cong 1.2 \times 10^{-13}$ cm = 1.2 fermi. So R depends on the mass number (A), which implies that the radius of each nucleus differs according to its mass number. The density of the electrons in the atoms was found with a little change over the 10^{-8} cm dimension. Therefore, in many cases, the atomic spectroscopic data are accurate enough to trace even the effects of the shape of the nuclear change distribution on the dynamic of its surrounding electrons. Due to this fact, it is quite possible to determine the electric quadruple moment (Q) of the nucleus, where Q measures the extent to which the charge distributions in the nucleus deviate from the spherically symmetrical shape and acquire an ellipsoidal shape. Also, the magnetic dipole moments (μ), which reflect the current and spin distributions in nuclei, have generally dramatic effects on atomic spectra and, with further refined measurement, yielded information on nuclear magnetic octupole moments too. The coupling of atomic spins with nuclear spins gives the hyperfine structures for the nuclear spectra as well. The importance of measuring Q is represented by the following

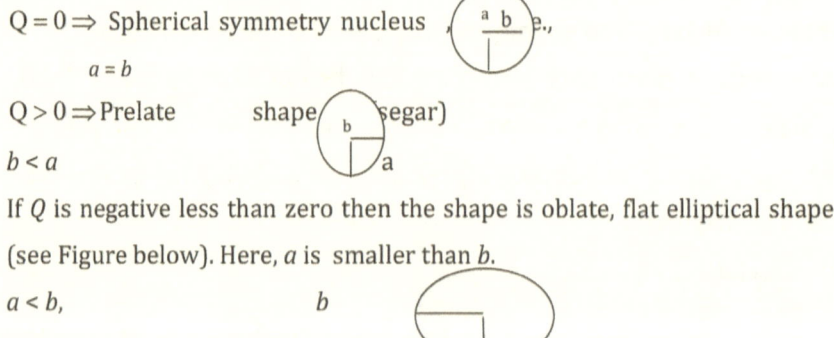

$Q = 0 \Rightarrow$ Spherical symmetry nucleus
$a = b$

$Q > 0 \Rightarrow$ Prelate shape (segar)
$b < a$

If Q is negative less than zero then the shape is oblate, flat elliptical shape (see Figure below). Here, a is smaller than b.

$a < b,$

All the features mentioned before were confirmed by the experimental measurements or studies for the last one hundred years. As mentioned previously, the scattering experiments were of important results concerning the charge, current, and spin distributions in the nucleus. In addition to that, the spectra, magnetic moments, and quadruple moments were measured for many nuclei. Using the electrons as powerful probes with high energy ($\approx$ 100 MeV), where the wavelengths of the electrons' waves are so close to the dimensions of the investigated nuclei, helps a lot to study the charge and current distributions in the nuclei. This is due to the fact that the forces acting on the electrons penetrating the nucleus depend on the details of the charge distributions in the nucleus; in addition to this, the electromagnetic interactions are quite well understood. Therefore, the data obtained from electron scattering experiments provide us with fairly accurate information about electromagnetic properties of the nucleus from all kinds of measurements, such as electron-scattering, atomic spectroscopy data, mu-mesic X-rays (muonic atoms, where $\bar{u}$-meson is captured in an atomic orbit that is even much better than $\bar{e}$, for probing the nucleus for charge distribution study). The mass of $\bar{u}$ is about 105.66 MeV. This indicates that $m_{\bar{u}} \cong 200\, m_e$, so the Bohr radius for this muonic atom is two hundred times less than that of $\bar{e}$, whence $\bar{u}$-meson probes the nuclear charge distribution from a much closer distance.

Mu-mesic X-rays have greatly helped in clarifying the charge distribution and the current distribution in nuclei. All these experiments and related others conforming the mass of the nucleus (A = mass number) are independent of nuclear density, which imply that the radius of the nucleus in general is given by the following

$R = r_o A^{1/3}$, where $r_o \approx 1.12$–1.2 fm. And A is the mass number of the nucleus ($Z + N$). Figure 1 gives the general charge distribution inside the nucleus.

Figure1 ,typical charge distributio

ρ is the average density, which is about 1.72×10^{38} nucleons/cm³.

τ is the skin depth, which is about 2.4 fm (fm = 10^{-13} cm).

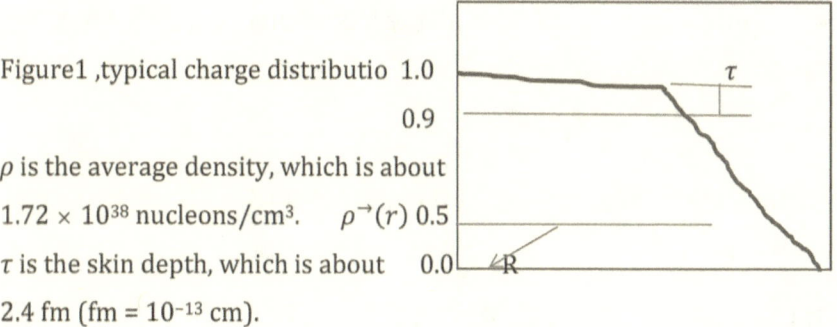

I-3—Nuclear Mass

The mass of a nucleus can be determined from so many experiments such as nuclear reactions. But it was found that this measured mass is quite different from either the carbon mass C^{12} ($A = 12$) or the oxygen mass O^{16} ($A = 16$). Here, C^{12} is chosen as a reference of scale, where the atomic mass is 12, and one atomic unit is $(1.66043 \pm 0.00002) \times 10^{-24}$ gm, which is equivalent to (931.478 ± 0.005) MeV/C^2. On this base, the following important data found

$M_H = 1.00782522$ atomic mass units, hydrogen atom mass

$M_O = 15.99491494 \pm 0.00000028$ atomic mass units, oxygen atom mass

Therefore, the bare nuclear mass is given by the following

$$M_{nuc} = M_{at.} - (zm_e + B_e(z)) \tag{1}$$

where m_e is the electron mass and B_e is the binding energy of the Z-electrons in a neutral atom, where it is estimated to be about $15.73\ z^{7/3}$ eV (from Fermi-Thomas model). The mass of an electron is measured to be $(5.48597 \pm 0.00003) \times 10^{-4}$ atomic mass units, which is equivalent to 0.511 MeV. Using equation (1) and the estimated value of $B_e(Z)$, one can get the following:

M_p (proton mass) = $(1.00727663 \pm 0.00000008)$ amu $\equiv (938.256 \pm 0.005)$ MeV

M_N (neutron mass) = $(1.0086654 \pm 0.0000004)$ amu $\equiv (939.550 \pm 0.005)$ MeV

Now the binding energy of the nucleus $B(Z, N)$ is defined as the following:

$$M(Z,N) = ZM_H + NM_N - B(Z,N) \tag{2}$$

In (2), a small correction, due to the electronic binding energy, is neglected.

$M(Z, N)$ is the atomic mass of the nucleus, with Z = protons number, and $N = A - Z$ is the neutrons number. So the binding energy $B(Z, N) \equiv B(Z, A - Z)$ is the energy required to break up the nucleus into its nucleons: $B > 0$.

The binding energy $B(Z, N)$ is an increasing function, generally speaking, of Z and N, i.e., $\dfrac{B(Z,N)}{A(Z+N)} \cong$ constant for the range $A = 12$ and up. For a good approximation, one empirically finds that

$$\frac{B(Z,N)}{A} \approx 8.5\,MeV/nucleon - \quad \text{(crude result)} \tag{3}$$

Figure 2 represents B/A versus A

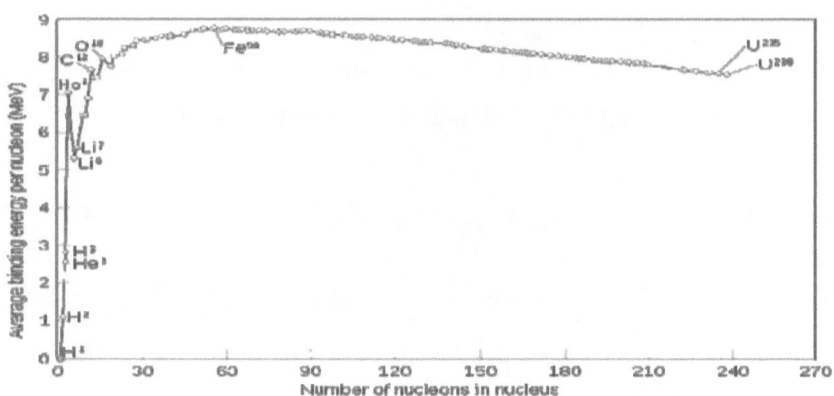

As a comparison with the electronic binding energy per electron, it is noticed that

$$\frac{B_e(Z)}{Z} = 15.73 \times Z^{4/3} = 15.73 Z^{4/3}$$

$$(4)$$

Equation (4) shows that $\frac{B_e}{Z}$ increases as $Z^{4/3}$ increases, where for a nucleus, $\frac{B}{A} \cong 8.5\, MeV/nucleon$ is about constant. If $\frac{B_e}{Z}$ is drawn versus Z (atomic number), figure 3 will be obtained.

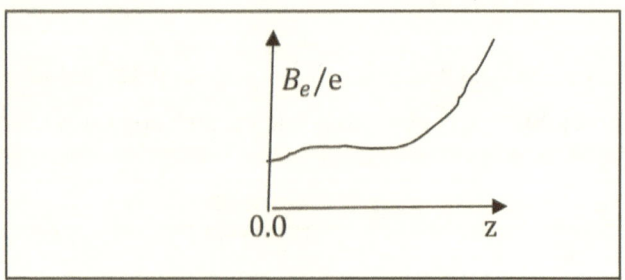

Figure 3: B_e/e versus Z

I-3-1—Semiempirical Mass Formula (Weisăker Formula)

This semiempirical formula for the nuclear mass of any nucleus was developed by Weisăker, which is given as follows:

$$M_{nuc.} = ZM_p + NM_n - a_1 A + a_2 A^{2/3} + a_3 \frac{Z^2}{A^{1/3}} + a_4 \frac{(Z-N)^2}{A} + \delta(A) \quad (5)$$

where a_i (I = 1, 2, 3, 4) are coefficients of measured values.

Now the interoperation of the terms of (5) is given as follows:

$ZM_p + NM_n$ are the total mass of the protons and neutrons, which represents, apparently, the mass of the nucleus as such. $a_1 A$ represents the volume energy; $a_2A^{2/3}$ is the energy term that is proportional to the nuclear surface area, and because the nucleons at the surface of the nucleus contribute less to the total binding energy, then $a_2A^{2/3}$ has to be subtracted from the volume energy, i.e., (a_1A) has to be reduced by

$a_2A^{2/3}$. The term $a_3\dfrac{Z^2}{A^{1/3}}$ represents the coulomb energy that is due to the coulomb force, which actually increases the mass of the nucleus.

The term $a_4\dfrac{(Z-N)^2}{A}$ represents the symmetry energy, which is zero for $Z = N$ nuclei. $\delta(A)$ represents the pairing energy coefficients that have different values according to whether the nuclei are odd-odd or odd-even or even-even. The empirical values of a_i and $\delta(A)$ are found to be the following:

$a_1 = 15.6\text{MeV}$

$a_2 = 18.56\text{MeV}$

$a_3 = 0.717\text{MeV}$

$a_4 = 28.100\text{MeV}$

$$\delta(A) = \begin{cases} 34A^{-3/4}\,MeV & \text{for odd-odd nuclei} \\ 0 & \text{for odd-even nuclei} \\ -34A^{-3/4}\,MeV & \text{for even-even nuclei} \end{cases}$$

I-3-2—Nuclear Forces: Entity and Features

The entity of nuclear force is not quite well-known, as it is the case with the electromagnetic force and the gravitational force. But in general, most of its features are recognized. It is the dominant force inside the nucleus that holds and binds together the protons, neutrons, and protons-neutrons. Therefore, nuclear binding energy is very much different from that of electromagnetic and gravitational binding energy, i.e., $B_n > B_g$; $B_n > B_e$. As has been shown previously, the nuclear force is about one hundred times of f_e, 10^{39-40} times of f_G, and 10^{13} times of f_{nw}, where f_e is the electromagnetic force, f_G is the gravitational force, and f_{nw} is nuclear weak force (nuclei-decay). The nuclear force (f_n) is well studied since the mid of the twentieth century, through many experiments in nuclear reactions, especially those experiments dealing with nucleon-nucleon scattering. The experimental data have shown the following features of the nuclear force:

1. Inside the nucleus, the $f_n \gg f_e$; otherwise, most nuclei that are stable would not exist, because of the repulsion force between the protons (+ charge).

2. The boundness of nuclei means f_n is attractive or, at least, has dominant attractive components in it, within a certain range of the radius (r).

3. f_n is a short range force ($r \sim 10^{-13}$ cm = 1fm).

4. From figure 1, the binding energy (BE) is almost increased linearly with A, which implies that f_n is a saturated force.

5. f_n strength can be measured, in addition to BE, by another energy to be compared with two nucleon system such as deuteron $(p + n)$. This nuclear system can be graphically represented as

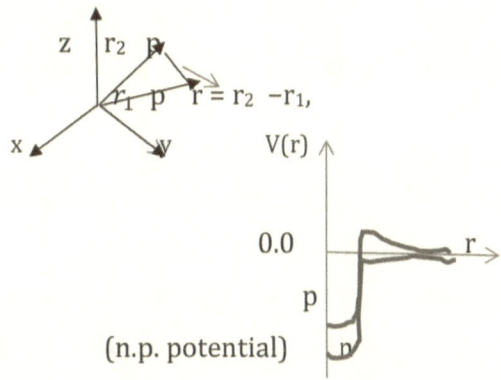

(n.p. potential)

This energy is the one forcing these two nucleons to be within the range (r) of their mutual interaction, which implies an increase to their relative momentum, because $\Delta p \Delta x \approx \hbar$ leads to an increase in their kinetic energy in the center mass system, i.e.,

$$E_k = \frac{p^2}{M} = \frac{(\Delta p)^2}{M} = \frac{\hbar^2}{M(\Delta x)^2}$$

; M is reduced mass.

This kinetic (relative) energy has its effect on the nucleons' interaction. It opposes their binding within their range of interaction and consequently forbids a formation of a bound state. Therefore, the nuclear interaction has to overcome this kinetic energy (E_k). Also, it is important to know whether BE is large or small compared to (E_k). Since deuteron has only one bound state that represents two nucleons $(p + n)$, the nuclear interaction is barely strong enough to overcome the kinetic energy (relative) E_k, which comes about when two nucleons come within each other's range. This indicates that nuclear interaction is basically a weak interaction with the task of overcoming E_k. In the case of atoms, the situation is quite different because the coulomb force

is inversely proportional to the square of relative distance, i.e, $F_c \propto \dfrac{1}{r^2}$

. Hence, $E_k \approx \dfrac{1}{r}$, so at a large distance, $E_k \sim$ small, compared with E_c (coulomb energy). Therefore, a hydrogen atom has many bound states.

6. f_n is not only depending on the relative distance $(r_1 - r_2)$ that separates between the two nucleons; it is also depends on the nucleons' intrinsic degree of freedom related to their spins and their isotopic spins, i.e., in addition to the spacial space (x, y, z), there are the spin space (↑↓) and the charge space (isospace). For the spin dependence, it is known for deuteron (as the simplest nucleus). There are two nucleons, the proton and the neutron, where the possible freedom for their coupling is

1. Bound state indicates parallel spins such as $\underset{p\ n}{\uparrow\uparrow} \Rightarrow J = 1$ where J is the total spin ($S_p + S_n = 1/2 + 1/2 = 1$).

2. Unbound state indicates antiparallel spins such as $\underset{p\ n}{\downarrow\uparrow} or \underset{p\ n}{\uparrow\downarrow} \Rightarrow J = 0 = \left(\dfrac{1}{2} + \left(-\dfrac{1}{2}\right)\right) = 0$.

The spin dependency of the nuclear force is represented by three different ways:

a. A direct spin—spin interaction of the form $(\sigma_1 \cdot \sigma_2) V(|r_1 - r_2|)$, where σ_i is Pauli matrices related to the spin as $S_i = \dfrac{1}{2}\sigma_i$, and $V\left(|r_1 - r_2|\right)$ is the potential under which the nucleons are behaving (see figure 4).

b. Spin-orbit interaction of the type $\left[(J_1 + J_2)\cdot(r_1 - r_2)(p_1 - p_2)\times V(r)\right]$, where $\left[(r_1 - r_2)(p_1 - p_2) = r \times p \propto L_{re.}\right], r_1 - r_2 =$ relative distance, $L_{re.}$ is relative angular momentum, $p = p_1 - p_2 =$ relative momentum. Therefore, spin-orbit interaction is $(\sigma_1 + \sigma_2)L_{re}V(r)$.

c. A   tensor   interaction   type   given   by

$$\left\{3(\sigma_1 \cdot (r_1 - r_2))[\sigma_2 \cdot (r_1 - r_2)] - (r_1 - r_2)^2 \sigma_1 \cdot \sigma_2\right\} V(r_1 - r_2) \quad (7)$$

where $\sigma_i \equiv$ Pauli spin operator (Pauli matrices) and $\sigma_a\sigma_{b+}\sigma_b\sigma_a = 2\delta_{ab}$, $\sigma_a\sigma_b = i\sigma_c$, a, b, c cyclical.

Now the question that comes about is, How are all these types of nuclear force dependencies proposed by physicists working in the field of nuclear physics theory? These come from the following physically observed phenomena:

a. Nucleon-Nucleon scattering experimental data.

b. The measured quadruple moment (Q) for the simplest nucleon (deuteron) suggested a tensor-force type, where Q for the $_1^2D$ is found to be about $2.78\times10^{-27} cm^2 = 2.78\times10^{-3} bn$, which means that $Q_D > 0$ and is accountable (non-negligible); hence, this suggests a prelate (segar) shape to the deuteron nucleus such as

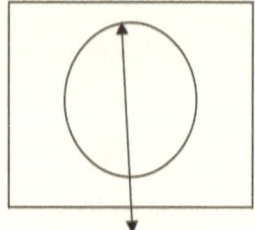

$(a - b) > 0, Q > 0, J = 1 \Rightarrow$ bound state

If Q was a negative quantity, i.e., $Q < 0$, then the shape of the nucleus is oblate (dough nut), as in the following diagram:

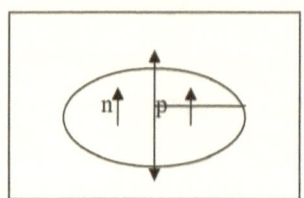 $(a - b) < 0, Q < 0$

The preference between a prelate shape and an oblate shape comes only through an interaction that couples the spins to the relative vector $\vec{r} = \vec{r_1} - \vec{r_2}$, which represents the distance between the interacting nucleons. In the case of the deuteron ($_1^2D$), this can be achieved only via a tensor force.

7.   (a) Charge-symmetry type of force, i.e., p.p $\cong$ n.n, neglecting f_c

(b) Charge-independent type of force, i.e., p.p $\cong$ p.n $\cong$ n.n.

(a) and (b) must carefully treat in (a) coulomb force, f_c is very small compared to nuclear force $\left[\dfrac{f_n}{f_c} \approx 10^2\right]$, so f_c can be neglected. And in (b), where p.p $\cong$ n.n $\cong$ p.n, the p.n system that is symmetrical under spin and space variables' exchange (Pauli principle operates for protons or neutrons but does not exclude states for the p.n system). The p.n pair can be formed also in states that have no counterpart in the p.p or n.n system. In such states, its interaction is different, and in fact, it is stronger than p.p or n.n interaction; i.e., (p.n) > p.p and n.n. But in similar space-spin states, it turns out that p.n $\cong$ p.n $\cong$ n.n. (charge independent). Practically, charge symmetry and charge independency are not quite well satisfied. The deviation from charge independency is due to electromagnetic effects. The failure of the charge symmetry is in the order of a few percent.

8. Interaction between two nucleons involves an exchange of momentum as it is another interaction, but in addition to that, nucleon-nucleon interaction exchanges their charges at the same time. This feature dramatically appears in p.n scattering experiments, where it shows a peaking of the differential scattering cross section $\left(\dfrac{d\sigma(\theta)}{d\Omega} \right)$ at backward angles in the center of mass system (c.m.s.) (figure 5). This occurs due to a charge exchange (figure 6)

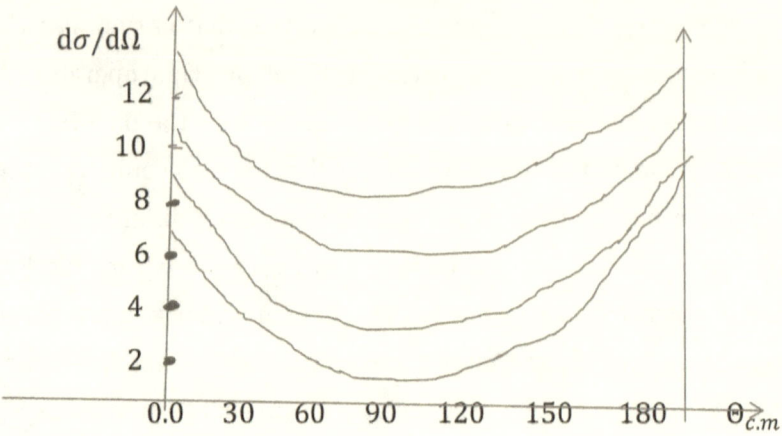

Figure 5:Backward angles scattering at 90–580 MeV energy (de-Shalit, vol. 1.)

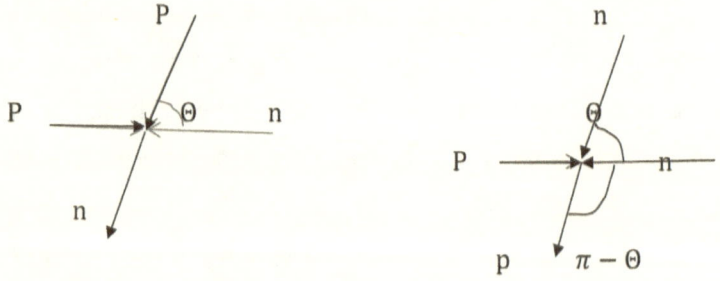

Figure 6: (a) no charge exchange　　　(b) charge exchange

The component of the potential V (r), which is responsible for the exchange of charge in such reaction is called the exchange potential V_{ex} (r).

9. The nuclear force becomes a repulsive one at a range that is about 0.5 fm, where the potential at this point is called a repulsive potential, and then the nucleus has a hard core. This physical fact

was discovered during the performing of a scattering experiment with high energy (> 200 MeV) in the laboratory system. It was found that at the range r = 0.5 fm, the interaction between nucleons becomes repulsive, which is called hard core.

But still, our understanding to nuclear forces is rather limited. Most probable, these forces may arise from the strong coupling of the nucleons to various mesons that are already discovered.The works leading to this discovery show that when a nucleon with high energy (E) is decelerated, a prolific production of mesons is quite noticed. Also, when nucleons exchange these mesons, they exchange a momentum that physically means a force is raised between the nucleons. This can be expressed graphically as follows:

$$P \leftarrow -\pi^+ - \rightarrow P, N \leftarrow -\pi^0 - \rightarrow N, P \rightarrow --\pi^- - \rightarrow N$$

Now let us take the case of the electromagnetic force where its important features can easily be derived from the analogous exchange of vertical photons. Consider two points, 1 and 2, where a photon is emitted from 1 and is absorbed by 2, then we have

$\Delta p = p_2 - p_1 = E/C - (-E/C) = 2E/$, which violates the law of conservation of energy by an amount E/C, so this process can last only for time Δt, where by uncertainty principles, $\Delta E \Delta t \approx \hbar$; hence, $\Delta t = \hbar /\Delta E = c \hbar /E$, and $\Delta t = r/c$, r = distance, and c is the light velocity. Therefore, the range of the force is $r = c^2 \hbar /E$, which is the separation between the two particles, when they exchange momentum ΔP during interval time Δt. The force, hence, between these two particles at $r = c^2 \hbar /E$ is given by $f = \Delta p/\Delta t = 2 \hbar^2 /E^2. (E^2/r^2)$ (8)

Equation (8) can be written as

$$F = (1/E^2/\hbar c) \cdot (2E^2/r^2) = 1/\alpha \cdot 2E^2/r^2, \text{ so } f_c = f \cdot \alpha = 2e^2/r^2$$

By investigation of this, one can notice that coulomb force can be obtained by multiplying equation (8) by the probability of the absorbing and emitting the photon, where this probability must be of the order of the fine structure constant ($\alpha = E^2/\hbar c$), then coulomb force law is obtained.

From (8) it is clear that the force at large distances is communicated by low-energy photons. These photons violate the conservation of energy by a little bit, so they can afford to stay outside the sources for a long time, transferring their momentum over greater distances. However, because their energy is low, they also carry small momenta; this is why the force is decreasing as the distances are increasing. Now let us apply a similar argument to the nuclear force (F_n), where the virtual particles (π^{+-}, π^0) are exchanged particles between nucleons; they have masses that have to be taken in consideration. Also, there is a minimum amount by which energy conservation has to be considered: $\Delta E > m_\pi c^2$, where m_π is the mass of the lightest known meson that is strongly coupled to the nucleon:

$$m_\pi c^2 = E \approx 139.57 \pm 0.011 \text{ MeV} \qquad \text{(for charged pions)}$$

$$\approx 134.972 \pm 0.012 \text{Me} \qquad \text{(for neutral pions)}$$

Therefore, there is an upper limit on the time (t). That is a virtual meson (π meson $\equiv$ pion) can stay away from its source; i.e., Δt is less than ($\hbar/2\pi m_\pi c^2$). This indicates that there is a maximum distance over which a meson can carry a momentum; i.e., Δr is less than $c\Delta t = (\hbar/2\pi)/m_\pi c = 1.4$ fm. This leads to the claim that nuclear force range is &$1/m_\pi$. Here, m_π is the mass of the lightest observed meson.

From the above arguments, one concludes the following:

1. Heavier mesons, or virtual transitions that involve the emission of more than one meson, will be effective only at shorter distance.

2. The complexity of possible mesons that can be exchanged at such shorter distances has thus far prevented a derivation of any reliable force from such fundamental processes.

3. In fact, the situation is even more complex since, although the range at which heavy mesons' contributions to nuclear forces is somehow smaller, the strength of this contributions might be much greater.

4. As an example, for rho (ρ) meson that has a mass $m_\rho = 756.0$ MeV, its concentration still accounts for nearly 10% of F_n at 1.4 fm.

5. The experimental (empirical) data of F_n at distances of 0.7 to 1.4 fm are not completed.

6. It is found experimentally that F_n that differs considerably from each other in the region range 0.7–1.4 fm fits the data equally well.

7. The tail of the nucleon-nucleon potential from distances

$d^{-1} = \hbar / m_\pi c = 1.4$fm and up can be calculated on the basis of one-pion exchange at that part of the potential, which is called one-pion-exchange potential (OPEP). The potential is found to be

$$V_{12} = (1/3)\ (g^2/\hbar c)\ m_\pi c^2\ (\tau_1, \tau_2)\ (\sigma_1, \sigma_2) + (1 + 3/\mu r + 3/(\mu r)^2)S_{12})\ e^{-r/\mu}\ /\ \mu r) \qquad (9)$$

where τ is the isospin operator and S_{12} is a tensor operator given by

$$S_{12} = 1/r^2\ (3\ (\sigma_1 \cdot r)\ (\sigma_2 \cdot r) - (\sigma_1 \cdot \sigma_2)\ r^2) \qquad (10)$$

where r is the mutual separation of the two nucleons and $(g^2 / \hbar c)$ is the coupling constant ≈ 0.081, which is determined by π-meson-nucleon coupling constant.

By these important remarks, we have concluded the arguments about the available information about the important features of the nuclear forces, which is the predominant one inside the nuclear domain; the em and gr forces are neglected against it. Here, the discussion and the demonstrations so far are related to the force that combined the nucleons inside a domain of a dimension proportional to about 1.4 fm or represented by a radius $R = r \, A^{1/3}$, where $r \approx 1.2$–1.4 fm and A is the mass number. The question that may be asked is, How is the nucleon itself consisted. Is it an elementary particle (point particle), or it is of a structure? The answer is, it is of a structure (made of quarks), where a strong nuclear force is needed to bined these quarks to form a nucleon. This will be discussed later in chapters 8 and 9.

1-4 —Nuclear Separation Energy of Neutron $\equiv S_n$

The nucleons (neutron or proton) are greatly combined in the domain of the nucleus. They are distributed among the available states in the domain governed Pauli exclusion principles. So to separate a nucleon from this configuration, it requires energy at least equivalent to the energy that is binding it to its physical position.

Let us define the energy that is required to separate the neutron or a group of neutrons from their original nuclei as S_n, where S_n is a very important physical quantity. Now, consider M (Z, N) as the mass of a nucleus before separating the neutron from it, and M (Z, N − 1) is the mass of this nucleus after removing a neutron from it. The M (Z, N) is given by

$$M (Z, N) = M (Z, N - 1) + Mn - Sn \tag{11}$$

One can easily show that

M (Z, N) < M (Z, N - 1) + Mn, therefore, from (11)

$$S_n = (M (Z, N - 1) + M_n - M (Z, N)) > 0 \tag{12}$$

from the fact that M (Z, N) = ZM_p + NM_n - Be (Z, N)

The S_n is given by

$$S_n = Be (Z, N) - Be (Z, N - 1) \tag{13}$$

Here, S_n is a neutron energy separation required to separate it from the nucleus m (Z, N) >. If a pair of neutrons is to be separated, then S_{2n} is given by

$$S_{2n} = Be (Z, N) - Be (Z, N - 2) \tag{14}$$

$$S_g = Be (Z, N) - Be (Z, N - g), \text{g is 1, 2, 3 group} \tag{15}$$

By the same method, one can find that the separation energy for a proton or group of protons is given by

$$S_p = Be (Z, N) - Be (Z - 1, N)$$

$$S_{2p} = Be (Z, N) - Be (Z - 2, N)$$

$$S_{gp} = Be (Z, N) - Be (Z - g, N), g = 1, 2, 3 \tag{16}$$

As has been shown before, the average binding energy per nucleon (Be (Z, N) / A) is about 8.5 MeV (see figure 2). Also, it can be seen that Sn and Sp increase as the number of N and P increase, where this regularity represents a refinement over the crude constant Be (Z, N) / A ≈ 8.5 MeV/ nucleon mentioned above. If this is precisely valid, then Sn would be 8.5, but as a matter of fact, Sn is not constant; it is fluctuating, which ensures the 8.5 MeV is of approximate nature. This fact implies that the regular behavior of Sn as a function of both Z and N points out at some regularity in the deviations of the actual (BE) from the average value (8.5 MeV). According to these remarks, it was found that some numbers of P and N show that nuclei with these numbers are more stable and their shells are closed. These numbers are 2, 8, 20, 28, 50, 82, and 126; they are called magic numbers. The nuclei with Z and N are magic numbers called double magic nuclei. The interpretation of these magic numbers is based on the coupling of the orbital angular momentum (L) with the spin (S), i.e., the coupling between the orbital motion field and the spinning motion field, i.e., the (L . S) coupling producing a total angular momentum J, where $J = L + S$, vector sum. This interpretation is very well demonstrated in the undergraduate nuclear physics courses (see Meyerhof, Enge, and others).

I-5—Coulomb Forces and Mirror Nuclei

Mirror nuclei are isobaric nuclei (same A), but they exchange protons and neutrons such as $^{27}Al_{13, 14}$, $^{27}Si_{14, 13}$, $^{17}O_{8, 9}$, $^{17}F_{9, 8}$, $^{13}C_{6, 7}$, $^{13}N_{7, 6}$, and $^{27}Mg_{12, 15}$ is an isobaric nucleus as these nuclei but is not a mirror to any of them. These nuclei are mostly strongly bound. But if a comparison is made between the excited states of Al and Si, one finds a great similarity between them (figure 7). If these two nuclei compared with Mg (which is isobaric but not mirror of them), no similarity is there (figure 7).

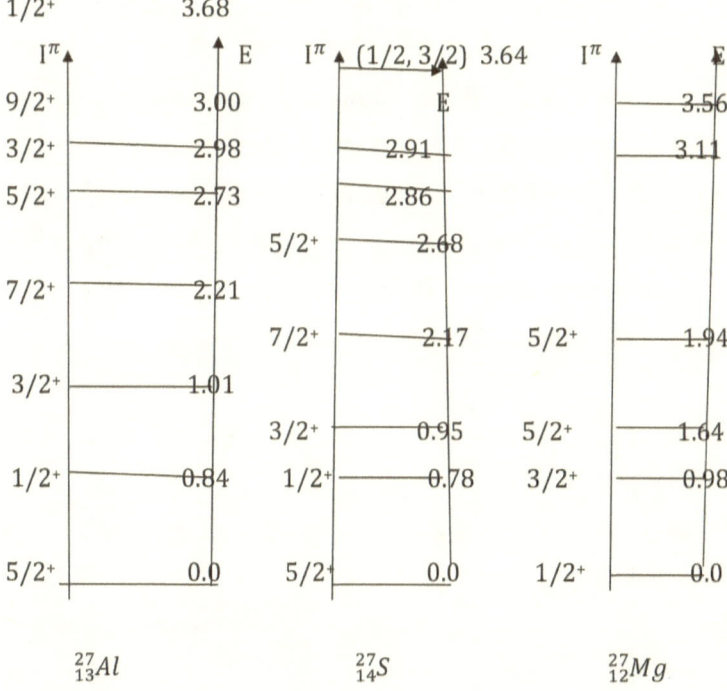

Figure 7 shows a similarity between the two mirror nuclei, Al and Si, and their difference with the isobaric non-mirror with them, the nucleus Mg. By investigating the above spectra of Al, Si, and Mg, there is a great similarity between the excited states' energy (E_{ex}) and angular momentum J and the parity (π) for the mirror nuclei Al and Si. But if both nuclei, Al and Si, are compared with Mg, one finds a striking difference in the sequence of the energy (Eex) and the total angular momentum J, as it is clearly shown in figure 7. There are many pairs of mirror-nuclei in the periodic table such as those mentioned above. This argument indicates that mirror nuclei have similar intrinsic structures. The difference is the nuclei with small (z) is more strongly bound due to the fact that Be (for small z) – Be (for large z) $\cong \Delta E_c \cong \dfrac{Zee}{R}$ (17)

ALI A. ABDULLA

ΔE_c is the change in coulomb energy (Ec) that can be calculated

to be $E_c \propto ze^2/R \cong \frac{3}{5} ze^2/R$ (for uniformly charged sphere), and R is the radius of the nucleus. (As an exercise, show that Ec, then find ΔEc, where $Z^- = Z-1, Z-2$.) If the change in the coulomb energy for mirror nuclei is taken in consideration, one can show the similarity in ΔE_{ex} and J^π for the mirror nuclei pairs $^{17}O_9$, $^{17}F_8$, and $^{13}C_7$, $^{13}N_6$ as shown in figure 8.

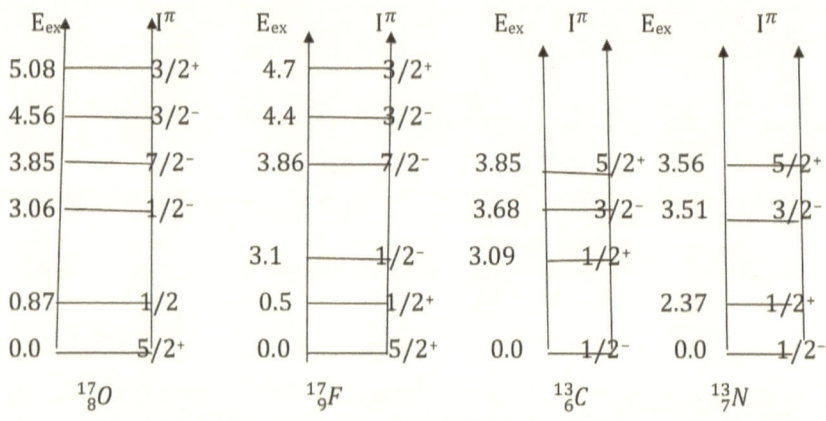

Figure 8 shows the similarity in ΔE_c and J_π

(Reference: de-Shalit, Feshbach, vol. 1)

Note that the existence of two distinct nucleons, proton and neutron, leads to three distinct classes of nuclear interaction: p.p, p.n, and n.n, which imply a comparison of two mirror nuclei with each other, which means comparing p-p and n-n interactions. The p.n interactions involve the same number of n-p pairs for both mirror nuclei. This is generally not the case for non-mirror nuclei, as can be seen clearly from table 2, where number of pairs $\sum n = 1$ for identical particles and n.p for non-identical particles.

Table 2: Pair—number of p.p, n.n, and p.n

Nucleus	^{27}Mg	^{27}Al	^{27}Si
Protons	12	13	14
Neutrons	15	14	13
p.p pairs	66	78	91
N.N pairs	105	182	182
P.N pairs	180	182	182

It can be seen that the difference between the structure of ^{27}Mg and ^{27}Al comes from the fact that p-n pair can exist in states that are not allowed to p-p or n-n pairs and characteristically have a strong interaction in these states. The similarity between the structures of mirror nuclei can be explained in terms of charge symmetry of a nuclear force, which is going to be dealt with in the following section.

I-6—Charge Symmetry and Charge Independence of Nuclear Forces

These features of nuclear forces are well understood through studying the two nucleons, the proton and the neutron. If the two nucleons (p, n) put in the range of the nuclear force, they will be bound to form a simplest nucleus; it is the deuteron (^{2}D$_1$). The proton is a charged nucleon, while the neutron is a neutral nucleon. They have different masses; the neutron mass is heavier than that of the proton with about 0.78 MeV, which is more than the mass of the electron (= 0.511 MeV). The nucleon (p or n) has an intrinsic spin given by $1/2\,\hbar$

Therefore, $S_{p=1/2}$, $S_{n}=1/2$ in the unit of $\hbar$. The experimental data also showed that each nucleon has a magnetic moment given by

$u_p \approx 2.792782 \pm 0.000017$ nuclear magneton

$u_n \approx -1.913148 \pm 0.000066$ nuclear magneton

It is quite clear that both p and n are fermion that obey Fermi-Dirac statistic. Both p and n have intrinsic magnetic momenta that are $u_{p(0)} =$ 1nm and $u_{n(0)=0\,nm}$ (nm = nuclear magneton).

Therefore, $u_n - u_{p(0)} = 1.79$ nm, $u_n - u_{n(0)} = -1.91$ nm

According to the mirror nuclei theory, n.n interaction is charge symmetry, but p.p interaction consists of nuclear and coulomb interactions due to the charge, so if coulomb force is subtracted, then the nuclear force is charge symmetry too.

By definition of the nucleon, the proton and the neutron are different states for the nucleon entity. To understand this idea, one has to understand, in the same sense, the phenomenal idea of the positively and negatively polarized electron, where these represent two states belonging to the same entity, the electron. This analogy is fairly important to be understood. The electron in this case has either spin down (−1/2) or spin up (1/2), so its wave function in this spin space is given by (1/2, −1/2 = (ψ+, ψ−), by the same understanding that the nucleon has a two-dimensional space called charge space, isotopic space, or isospin space. It is related to the charge that is related to the isotopic spin, as it will be clarified later. This is quite clear by the definition of the nucleon stated before. Hence, it is described by the proton state when it is there and by the state of the neutron when it is there, so the wave function of the nucleon is given by ψ_{nuc} (r) = (ψ+, ψ−)   (18)

Where $\psi+$ is the proton wave function and $\psi-$ is the neutron wave function; $+$ and $-$ signs due to spinning up or down. Also, both wave functions in equation (18) have spin 1/2; therefore, each has two components in spin space, either up or down. So by analogy, we can write ($\psi+$ (r) and $\psi-$ (r)) as two components of the wave function of the nucleon in the charge or isospace. This isospace is a purely formulated idea, because in the actuality, one can conceive only the states of the form (18), where one of the components vanishes. Equation (18) describes the nucleon with the probability amplitude of it as being a proton and the probability of being a neutron. However, the actual similarity between p and n makes it sensible to talk of a nucleon state that has mixed properties of proton and neutron.

In quantum physics, systems are described by wave functions, of spacial and time coordinates, and each dynamic variable of a certain physical system has to be associated with an operator consistent with the type of space chosen, coordinate or momentum space. Also, in this revolutionary quantum physics, there are different kinds of spaces of different dimensions, and some of these spaces are abstract spaces such as charges—spaces. But since these spaces are independent of each other (such as coordinate space, spin space, charge, or isospin space), the wave function describing a physical system, such as a nucleus or its constituents, is to be written as

$$\Psi\ (r, s, \tau) = \psi(r)\ \phi(s)\ \chi(\tau) \ —(19)$$

where $\psi(r)$ is describing a system in special space (x, y, z)

and $\phi(s)$ is describing a system in spin space (s),

$\chi(\tau)$ is describing a system in charge space,

τ is an operator related to charge space, to its isotopic spin.

Hence, τ is a proposed operator to operate in the isospin space (charge space) helping in demonstrating isospin-space formalism. There are only four independent operators in the space of dichotomy (space of two branches) variables, i.e., in space of two components' wave function as in (18).

These four operators can be chosen in different ways, but it is found as most convenient to use Pauli-Hermitian operators that are three with an identity operator (which are usually used in spin space). On these bases, the following operators can be introduced:

1) $I = \begin{pmatrix} 1 & 0 \\ 0 & 1 \end{pmatrix}$ the identity operator

2) $\tau_1 = \begin{pmatrix} 0 & 1 \\ 1 & 0 \end{pmatrix}, \tau_2 = \begin{pmatrix} 0 & i \\ i & 0 \end{pmatrix}, \tau_3 = \begin{pmatrix} 1 & 0 \\ 0 & -1 \end{pmatrix}$ (20)

where τ_i (i = 1, 2, 3) satisfies the commutation relations of the Pauli spin matrices; i.e.,

$$\tau_1 \tau_2 = i\tau_3, \tau2 \tau_3 = i\tau_1, \tau_3\tau_1 = i\tau_2 \tag{21}$$

The third component (τ_3) tells whether a given wave function describes a proton or neutron or whether the nucleon is in the proton or neutron state. Now let τ_3 operate on ψ_p and on ψ_n (see 18 to see what the yielding results are).

$$\tau_3 \psi_p = +\psi_p$$

$$\tau_n \psi_n = -\psi_n \tag{22}$$

Equations in (22) are eigenvalue equations that show that the eigenvalue (+1) belongs to the proton eigenstate, and the (−1) belongs to the neutron eigenstate for the operator τ_3, the third component of τ_i. So if τ_{i3} is the τ_3 operator that operates on i^{th} nucleon, then it follows from (20) that the charge of this i^{th} nucleon is given by

$$E\psi_i = e \left(\frac{1+\tau_{i3}}{2}\right)\psi_i \qquad (23)$$

where (e) is the charge unit.

The charge of the nucleon in e unit is thus the eigenvalue of the operator.

If the eigenvalue of τ_{i3} in equation (24) is (-1), $q_i = 0$, then the nucleon is a neutron. But if its eigenvalue is $(+1)$, then $qi = 1$, so the nucleon is a proton. q_i is the charge operator. Now the total charge (Q) of a system of A (mass number) nucleons in unit of (e) is given by the operator:

$$Q_0 = \frac{1}{2}\sum (1 + \tau_{i3}) \qquad (i = 1, 2\text{—}A) \qquad (25)$$

whence coulomb interaction operator of (A) nucleons is given by

$$^\wedge H_c = e^2/4\sum(1 + \tau_{i3})(1 + \tau_{i3})/ r_i r_j, i < j \text{ —}A, \qquad (26)$$

This takes place, formally, between all nucleons including the neutrons, i.e., as an interaction between nucleons according to the definition of a nucleon. But the operator τ_3 in equation (23) shows that the actual contribution to coulomb interaction comes about from proton-proton interaction only. It is well-known that the charge of a system of nucleons is conserved. Thus, if ψ is a function of (A) nucleons with energy eigenvalue (E) that is also an eigenfunction of Hc (as in eq. 26), then

$$Q\psi = 1/2 \left(\sum^A_i (1 + \tau_{i3})\right) \psi = z\psi \qquad (26a)$$

is also an eigenfunction of H_c, belonging to the same eigenvalue, which implies that every nuclear wave function must satisfy

$$H_0 (Q_0 \, \psi) = Q_0 (H_0 \psi) = Q_0 \, H_0 \, \psi \tag{27}$$

where Q_0 is given by (25). This indicates that every nuclear Hamiltonian (H) has to satisfy the commutation relations,

$$H_0 \, Q_0 - Q_0 H_0 = \lfloor H, Q \rfloor = 0 \tag{28}$$

Or since $Q_0 \equiv Q = 1/2 \, (A) + T_3$, where $T_3 = 1/2 \sum_{i=1}^{A} \tau_{i3}$, then, $\lceil H, T \rceil = 0$

Here, $H = H_0$ and $T = T_3$

Also, one can show that H_c and T_3 commute.

(Do it as an exercise.).

Problem 1: Show that OPEP given by equation (9) commutes with (Q), i.e., (OPEP, Q) = 0

One can also show that the operator $\tau_1 . \tau_2$ is given by

$$\tau_1 \cdot \tau_2 = \tau_{i1} \, \tau_{k1} + \tau_{i2} \, \tau_{k2} + \tau_{i3} \, \tau_{k3}$$

which does not commute separately with either τ_{i3} or τ_{k3}, i.e.

$$\lfloor \tau_1 . \tau_2 , \tau_{i3} \rfloor = -2i\tau_{i2}\tau_{k1} + i\tau_{i1}\tau_{k2}$$

$$\tag{30}$$

$$\lfloor \tau_1 . \tau_2 , \tau_{k3} \rfloor = -i\tau_{i1} \, \tau_{k2} + i \, \tau_{i2} \, \tau_{k1}$$

However, it is still true that

$$(\tau_i \, \tau_k, \, \tau_{i3} + \tau_k) = 0 \text{ commutes} \tag{31}$$

Problem 2: Verify (30) and (31), then one finds Q does commute with V_{12} (eq. 9).

By studying or investigating in details the equation (30), one can conclude the following important remarks:

1. It indicates that the individual charge of each nucleon is not conserved separately by the one-pion-exchange potential (OPEP). But the total charge is conserved. Hence, OPEP may lead to a charge exchange between nucleons. In fact, OPEP exchanges $\pi^{\pm}$ (pi-mesons = pions) between the proton and the neutron as in charge exchange scattering (figure 6).

1. Isospin operators τ_i and τ_k that behave as vectors in isospace (charge space) appear in equation (9) for OPEP (V12) as scalar product $(\tau_1 . \tau_2)$; thus, it commutes with the operator, where it represents a rotation around the third axis in isospace.

2. The charge symmetry feature of Fn (nuclear force) can also be formulated in terms of the isospin operators.

3. It is well noticed that τ_1 (operator) has the property of converting (changing) the neutron state into the proton state and vice versa:

$$\left. \begin{array}{l} \tau_1 \psi_p = \psi_n \\ \\ \\ \tau_1 \psi_n = \psi_p \end{array} \right\} \tag{32}$$

Let us define an operator $A_r (1, \text{--------}, A)$ that takes the form

$$A_r(1, \ldots, A) \equiv \tau_1(1)\, \tau_1(2)\, \tau_1(3) \ldots \tau_1(A) \tag{33}$$

With the properties that if $A(1, \ldots A)$ operates on $\psi(1\text{—}A)$, it will yield a wave function, in which the energy of the neutron is converted into proton and conversely. If the nuclear force is charge symmetry, the energy (E) of the system is unchanged by this operation, whence

$$H(A_{r_1} \psi_{(r)}) = (A_r H\psi) = E(A_r, \psi) = A_r H\psi \rightarrow (H A_r)\psi = (A_r H)\psi \rightarrow HA_r - A_r H = 0$$

so the operator A_r commutes with the Hamilton H, since it is valid for any eigenfunction of H.

Therefore A_r operator commutes with Hamiltonian as

$$[H, A_r] = 0 \tag{34}$$

{As an exercise, verify that $[(\tau_1 \cdot \tau_2, A_r] = 0$} —(35)

which implies that the one-pion-exchange potential V12 in equation (9) is charge symmetry. But H_c in equation (26) does not commute with A_r, which means $[H_c, Ar] \neq 0$. Why? Find out! (36)

The interchange between protons and neutrons by the operator A_r leads to a change in coulomb energy Ec, except for the cases where (N = Z). Why? Therefore, equation (36) is an expectable result. Now if F_n is charge independent, i.e., p.p = n.n = p.n (neglecting coulomb force compared to nuclear force), that requires the Hamiltonian of the system (H) commutes with T_1, T_2 and T_3, i.e., with the whole isovector T as such:

$$[H, T] = 0 \tag{37}$$

The question now is, What does it mean to say the nuclear force is charge independent? It means, as shown before, the n.p and p.p and n.n interactions are the same (neglecting, as said before, fc), but this is possible if both pairs are in the same states ("space spin")! From the identical particles theory, it is well-known that p.p and n.n pairs are subjected to the Pauli exclusion principle restriction and the permutation rule, where such pairs have to be in antisymmetric states only, where

$$\psi_{nn}(r_1\,s_1,\,r_2\,s_2) = -\psi_{nn}(r_2\,s_{2,}\,r_1\,s_1)$$

$$\left.\rule{0pt}{60pt}\right\}\qquad(38)$$

$$\psi_{pp}(r_1\,s_1,\,r_2\,s_2) = -\psi_{pp}(r_2\,s_2,\,r_1\,s_1)$$

where S_i spin coordinates of the ith particle

and $r_i(i = 1, 2, 3)$ are the spacial coordinate.

p.n pair is not the identical particles system; hence, there is no restriction on their occupying their states, but it is possible to break up any p.n wave function into a part, which behaves like (38), and a part that is symmetrical with respect to the exchange of particles 1 and 2.

Let χ_i be the wave function of the p.n. system. One might simplify the notations by removing the charge labels (p, n) and talk about isospin space such as

$$\chi_p = (1, 0),\, \chi_n = (0, 1) \qquad\qquad (39)$$

Then, the total wave function of the system is

$$\psi(r_p\,s_{p,}\,r_n\,s_n) \equiv \psi(r_1\,s_1,\,r_2\,s_2)\chi_p(1)\,\chi_n(2) \qquad (40)$$

The isospin wave function $\chi_p(1)\chi_n(2)$ indicates that particle 1 is a proton and particle 2 is a neutron. It is well-known from the principle of symmetrization that the wave function of a physical system is either symmetrical or antisymmetric. Therefore, one might break up ψ_{rs} into two parts,symmetrical and antisymmetric,against space and spin exchanges, respectively, such as

$$\psi \equiv \frac{1}{\sqrt{2}}\,[\psi^o(r_1\,s_1,\,r_2\,s_2) + \psi^1\,(r_2\,s_2,\,r_1\,s_1)] \qquad (41)$$

where

$$^{0}\psi_{sy}(r_1\,s_1,\,r_2\,s_2) = \frac{1}{\sqrt{2}}\,[\psi(r_1\,s_1,\,r_2s_2) + \psi(r_2\,s_2,\,r_1\,s_1)] \tag{42}$$

$$^{1}\psi_{as}(r_1\,s_1,\,r_2\,s_2) = \frac{1}{\sqrt{2}}\,[\psi(r_1\,s_1,\,r_2\,s_2) - \psi(r_2\,s_2,\,r_1\,s_1)] \tag{43}$$

Now if p-n wave function (40) depends specially only on the relative separation of p and n (and not, say, on their separate coordinates, which would have been the case for a p-n pair in an external coulomb field), then it really makes no difference whether particle 1 is a proton and particle 2 is neutron and conversely; hence, it would be possible to use a linear combination of both situations. An important remark one can state with evidential causes, nature has chosen to have an antisymmetric combination of isospin wave functions with a symmetrical combination of space-spin wave function, while the symmetrical combination of the spin wave function is combined with antisymmetrical space-spin wave function. We would have just as well chosen, at this stage of this discussion, the opposite combination, but it can be seen that charge exchange forces introduce a physical difference between the two choices. But still, both combinations are eigenvectors of the operator $A_r(1, 2)$ but with eigenvalues of opposite signs. Therefore, one writes

$$\psi^{0}_{pn}(r_1\,s_1,\,r_2\,s_2) = 1/2[\psi_{pn}(1,\,2) + \psi_{pn}(2,\,1)][\chi_p(1)\,\chi_n(2) - \chi_n(1)\,\chi_p(2)] \tag{44}$$

and,

$$^{1}\psi_{pn}(r\,s,\,r\,s) = \frac{1}{\sqrt{2}}\,[\psi_{pn}(r_1\,s_1,\,r_2\,s_2) + \psi_{pn}(r_2\,s_2,\,r_1\,s_2)][\chi_p(1)\,\chi_n(2) - \chi_n(1)\,\chi_p(1)] \tag{45}$$

Equation (45) can be supplemented with the p.p and n.n wave functions written in the same formalism

$$^1\psi_{pp}(r_1\, s_1,\, r_2\, s_2) = \psi_{pp}(1,\, 2)\, \chi_p(1)\, \chi_p(2) \tag{46}$$

$$^1\psi_{nn}(r_1\, s_1,\, r_2\, s_2) = \psi_{nn}(1,\, 2)\chi_n(1)\, \chi_n(2) \tag{47}$$

In general equations (44-47), the 1 and 2 stand for the space and the spin coordinates of particles 1 and 2, where ψ_{pp} and ψ_{nn} are antisymmetric [refer to equation (38)]. By investigation of these equations, the generalized Pauli exclusion principle is stated as follows:

(Two nucleons' wave functions have to be antisymmetric with respect to the simultaneous exchange of their spaces' spin and isospin coordinates. It can be further noticed that the isospin wave functions that associated with the three wave function $^1\psi_{np}$, $^1\psi_{nn}$, and $^1\psi_{pp}$ are analogous to the ordinary spin combinations for S = 1 (triplet case) wave functions of two particles of one-half spin (s = 1/2). Therefore, one might introduce the total isospin operator

$$T = t_1 + t_2 \tag{48}$$

where $t = (1/2\, \tau)$, τ is an operator similar to σ (Pauli matrix operator).

Accordingly, equations (44–47) take the forms

$$^{(0)}\psi_{pn}(r_1\, s_1,\, r_2\, s_2) = \frac{1}{\sqrt{2}}\, [\psi_{pn}(1,\, 2) + \psi_{pn}(2,\, 1)]\chi(1,\, 2,\, T = 0;\, M_T = 0) \tag{49}$$

$$^{(1)}\psi_{pn}(r_1\, s_1,\, r_2\, s_2) = \frac{1}{\sqrt{2}}\, [\psi_{pn}(1,\, 2) - \psi_{pn}(2,\, 1)]\chi(1,\, 2,\, T = 1;\, M_T = 0) \tag{50}$$

$$^{(1)}\psi_{pp}(r_1\, s_1,\, r_2\, s_2) = \psi_{pp}(1,\, 2)\, \chi(1,\, 2,\, T = 1,\, M = +1) \tag{51}$$

$$^{(1)}\psi_{nn}(r_1\, s_1,\, r_2\, s_2) = \psi_{nn}(1,\, 2,)\chi(1,\, 2,\, T = 1,\, M_T = -1) \tag{52}$$

Here, χ (1, 2, T, M_T) is a two-particle isospin wave function of the nucleons, corresponding to a total isospin T and a total of a third component $t_3 = M_T$. Let us take the following example as illustrations:

First of all, consider the energy levels that are the same for each nucleon, since Hamiltonian H is independent of the type of the nucleon (proton or neutron). The following are energy levels (spectra) in general:

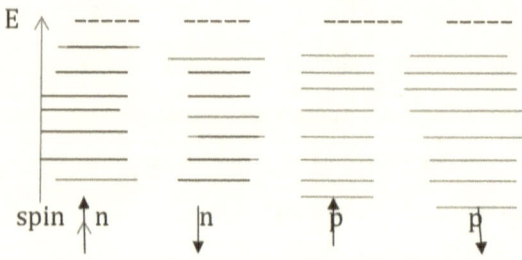

Consider α -particle (2n + 2p) helium nucleus,

T_3 defined as $1/2(N - Z)$ or $1/2(N - Z)$

So $T_3 = 0$, $M_T = 0$

For helium He (6), Z = 2, N = 4; $T_3 = +1$, $M_T = +1$.

The nucleon's distribution takes the form

n⎯⎯n ⎯p⎯p⎯ T_3 is defined as
$1/2$ (N − Z) or $1/2$ (Z − N).→$T_3 = 0 = M_T$.

For helium 6, there are two isotopes: one with z = 2, n = 4; the second with z = 3, n = 3.

The nucleon's distribution can be shown as follows:

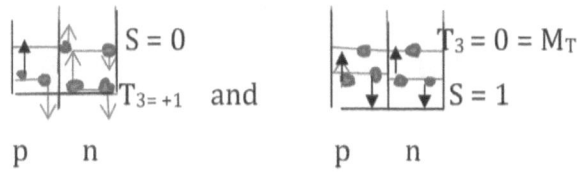

Take also ^{6}Be, where $p = 4$, $n = 2$, $S = 0$, $T_3 = -1$

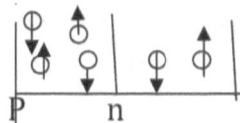

Let us demonstrate these examples on the coordinates of the energy axis as Y-axis and the third component T_3 as X-axis.

^{6}He	^{6} Be	^{6} Li
S=0	S =1 ,S=0 ,T₃=1	
−1	0. 0	+1 T₃

$$(S = 0, T_3 = -1), \quad (S = 1, T_3 = 0), \quad (S = 0, T_3 = +1)$$

As stated before, the generalized Pauli exclusion principle restricts that the wave function has to be antisymmetric where all the coordinates are exchanged (spin and isospin); therefore, ground states of the nuclei are

$$\psi_{sy}(1, 2) = \phi_1(r_1)\phi_2(r_2) + \phi_1(r_2) \phi_2(r_1) \tag{53}$$

$$\psi_{as}(1, 2) = \phi_1(r_1) \phi_2(r_2) - \phi_1(r_2) \phi_2(r_1) \tag{54}$$

sy and as stand for symmetrical and antisymmetric respectively.

From these equations, one can note that if $r_1 = r_2$, it yields that ψ_{sy} becomes large, $\psi_{as} = 0$. This clearly confirms the Pauli exclusion principle; two identical particles cannot occupy the same quantum state (eigenstate). From this, one can conclude that

for the antisymmetric = 0, it is coupled with $T_3 = M_T = 1$, and for S = 1, it is coupled with $T_3 = M_T = 0$. This is a natural rule.

Now, $T_3 = T_3(1) + T_3(2)$, and $T_{max.} = T_{3max} = T_3(1) + T_3(2)$.

Table 2 gives the values of T_3 and T, where $T_3(1)$ and $T_3(2)$ of particle 1 and particle 2 are known.

Table 2

$T_3(1)$	$T_3(2)$	$T_3 T = T_3(max)$
1/2	1/2	1
1/2	-1/2	0
-1/2	1/2	-1

Problem 3:

1. Derive the electrostatic energy for a uniformly distributed charge nucleus (coulomb energy).

2. Consider two mirror nuclei; find the difference of coulomb energy $\Delta E_e = \omega$. Then find ω for light nuclei, where z actually $\sim$ 1/2 A (A = mass number).

I-6—Magnetic Moment and Schmidt Limits

In the view of the nuclear shell theory discovered and developed by Mayer and others in 1947, the magic number phenomenon was well explained successfully. The magnetic momenta for odd-odd, odd-even, and even-even nuclei based on this theory were found by Schmidt. He demonstrated this finding using elegant diagrams named after him, as Schmidt limits lines, which will be clarified later on.

I-6-1—Magnetic Moment of the Deuteron ($^2D_1$)

The deuteron is the simplest known nucleus in the periodic table of elements.It is Z = N = 1, A = 2. The experimental data known today concerning this nucleus can be summarized as follows:

The magnetic moments of the nucleus and its proton and neutron are measured to be

u_d = 0.857 nm (nm = u_{nuc} = eh/4πm$_p$c, m$_p$ = proton mass), for ^{2}D

u_n = −1.91 nm for the neutron

$u_{p=}$ 2.79 nm for the proton

Here, we notice that $u_n + u_p$ = 0.880 nm, which is bigger than u_d (of D) by 0.023 nm, which, quantum mechanically, is appreciable physical quantity. This difference has a physical reasoning that can be clarified briefly next. It is well-known that the u_{op} (operator of the magnetic moment) is given by $u_{op} = u_p\sigma_p + \sigma_n u_n + 1/2L$ (L, σ_i-vectors).

L(v) = Angular momentum of the system

σ_i(v) = Pauli matrix, where $\sigma_p = 2S_p$, $\sigma_n = 2S_n \rightarrow S_p + S_n = 1/2(\sigma_p + \sigma_n)$

From the atomic physics, we know that the magnetic moment operator u_{op} for any atom is given by $u_{op} = gL_v$, where g is so-called Lande g-factor.

Using this method, one can find this magnetic moment operator for a nucleus such as

$$u_{op} = 1/2[((u_p + u_n) + 1/2)(\vec{S} + \vec{L}) + (u_p + u_n) - 1/2)(\vec{S} + \vec{L}) + (u_p - u_n)(\sigma_p - \sigma_n)] \tag{55}$$

Now, let us assume that $\vec{I} = \vec{L} + \vec{S}$ = the total angular momentum of a nucleus, and let it be a good quantum number, and its projection m_I is a good quantum number too. Also, the parity is a (GQN) taking the value ± 1.

$S = 0$ or 1.

$S = 0$ is a singlet state, which is antisymmetric.

$s = 1$ is a triplet state; it is a symmetrical.

So for $I = 1$, we have $^{2s+1}L_I$, which gives the spectra

$^{3}S_1 {}^{1}P_1 {}^{3}P_1 {}^{3}D_2$

Angular momentum for all $S = 1$

$L = 0, 1, 1, 2$

So if one thinks that the $^{3}S_1$ is the dominating state with the state $^{3}D_1$, then quantum mechanically, one can write the wave function of the deuteron system as a combination of S and D states as follows:

$\psi = \alpha\,\psi_s + b\,\psi_d$ (this is the wave function describing the deuteron nucleus.)

Here, $\psi_s = {}^{3}S_1$, $\psi_d = {}^{3}D_1$, then $\psi = \alpha\,{}^{3}S_1 + b\,{}^{3}D_1$

Now if $(u_{op})_z$ is the operator of u along Z direction, then the expected value of the magnetic moment u is given by $<u>_= <\alpha\, \text{III}(u_{op})_z| \alpha\, \text{II}>$, which leads to $u = |a|^2(^3S_1 l (u_{OP})_z|^3S_1) + |b|^{23}D_1 l (u_{OP})_z|^3D_1)$. By normalization, one gets $|a|^2 + |b|^2 = 1$. The calculation found that b = 0.04, which means that $4^0/_0$ of $^3D_1$ is mixed with $^3S_1$. This mixed configuration of the states of deuteron affects the magnitude of its magnetic moment compared to the total magnetic moment of its constituents (p, n). This calculation did not take into consideration the effect of the mesons' exchange between the nucleons (p, n). For exact calculations, one has to consider this effect clearly. Remember, we are talking about the simplest nucleus, it is the deuteron nucleus. For complex nuclei, the effect will be larger. Now if we take into consideration the eigenvalues of the physical operators as given quantum mechanically, the magnetic moment [equation (55)] in general is given by

$$u_D = 1/2[(u_p + u_n + 1/2) + (u_p + u_n - 1/2)] \dfrac{s(s+1) - l(l+1)}{I(I+1)} \qquad (56)$$

where $S^2\psi_s = \hbar\, s(s+1)\psi_s$

and $L^2\,\psi_D = \hbar\, l\,(l+1)\,\psi_d$

As an exercise, show equation (56), find u_d for the states S = 0.879, P = 0.689, P = 0.500, D = 0.310 as pure states and $^3S_1 + {}^3D_1(4^0/_0) = 0.88$ within $2^0/_0$.

Let us extend the discussion to heavy nuclei where the mass number (A = Z + N) is >2. Here, the Lande g-factor, as in atomic physics, play a main role in deriving the magnetic moment for the nuclei.

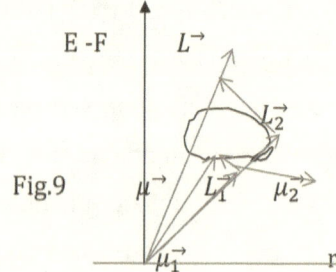

Fig.9

$u_1^{\rightarrow} = (e\hbar/2mc)g_1 L_1^{\rightarrow}$　　　　(57a)

The same figure cab be drawn for the particle 2 to find that

$$u_2^{\rightarrow} = (e\hbar/2mc)\cdot g_2\, L_2^{\rightarrow} \qquad 57b$$

If $g_1 = g_2$, we get the same Lande g-factor, then u_v is along L_v, but this is not the case, i.e. $g_1 \pm g_2$; therefore,

$$u^{\rightarrow} = u_1^{\rightarrow} + u_2^{\rightarrow} = e\hbar/m_p c[g_1\, L_1^{\rightarrow} + g_2\, L_2^{\rightarrow}] \qquad (58)$$

If the system is subjected to a magnetic field, $u^{\rightarrow}$ will process about $L^{\rightarrow}$, as shown in figure 9. Then, $u_{\uparrow\uparrow\ (parallel)}^{\rightarrow} = e\hbar/2\,m_p c\,(g\,L^{\rightarrow})$　　(59)

and $u^{\rightarrow} \cdot L^{\rightarrow} = u_{\uparrow\uparrow}^{\rightarrow} \cdot L^{\rightarrow} = e\hbar/2m_p c\,[g_1\, L_1^{\rightarrow} \cdot L^{\rightarrow} + g_2\, L_2^{\rightarrow} \cdot L^{\rightarrow}]$　　(60)

Using quantum mechanics, eigenvalue equation one gets

$L^2\psi = l(l+1)\{\hbar\}^2\psi \rightarrow l(l+1)(\hbar)^2$ is the eigenvalue of the operator L^2.

Now, since $L^{\rightarrow} = L_1^{\rightarrow} + L_2^{\rightarrow}$, then, $L_1^{\rightarrow} = L^{\rightarrow} - L_2^{\rightarrow}$, $L_2^{\rightarrow} = L^{\rightarrow} - L_1^{\rightarrow} \rightarrow$

$L^2 = \{L_1^{\rightarrow} + L_2^{\rightarrow}\}^2 \rightarrow (L_2^{\rightarrow})^2 = (L_2.)^2 = L^2 + {}^2L_1 - 2L_1 \cdot L_2 \rightarrow$

$L_1 \cdot L = 1/2[L^2 + (L_1)^2 - (L_2)^2],$　　　　　　　　(61)

By the same way,

$L_2 \cdot L = 1/2[L^2 + (L_2)^2 - (L_1)^2]$　　　　　　　　(62)

Here, all L are vectors.

Now substituting into (59), one gets

$$L_1 \cdot L = \hbar^2 / 2 \; mp \; c[l(l+1) + l1(l1+1) - l2(l2+1)] \tag{63}$$

$$L_2 \cdot L = \hbar^2 / 2 \; mpc[l(l+1) + l2(l2+1) - l1(l2+1)] \tag{64}$$

From 59 and 60 one gets; $g = (g_1 L_1 \cdot L + g_2 L_2 \cdot L)/L^2$ (65)

Substitute (63) and (64) into (65); g takes the form

$$g = 1/2 \, (g_1 + g_2) + 1/2(g_1 - g_2)[l_1(l_1+1) - l_2(l_2+1)/l(l+1) \tag{66}$$

Let us take the deuteron as a simple example to apply (66)

$I_p = 1/2$ spin of the proton, $u_p = I_p g = 1/2g \rightarrow g = 2u_p = 2(2.79 \text{ nm}) = 5{-}58$ nm for the proton. In = 1/2 the spin of the neutron, $u_n = I_n g = 1/2 \, g \rightarrow g = 2 \, u_n \, 2(-1.91) = -3.82$ nm.

If we assume no orbital motion for the deuteron, i.e., $l = 0$, s-state, then $s_1 \uparrow s_2 \uparrow \rightarrow I_d = 1$ the spin of the deuteron. Hence, $g = 1/2 \, (g_1 + g_2) = 1/2(1.78) = 0.88$ nm, where $u_d = g \, I_d = 0.88$nm, then it is consistent with $u_d = u_p + u_n = 0.88$nm, which is larger than the experimental result with about 2.6% (the reason was clarified before mixing S and D states.)

I-6-2—Schmidt Limits and More Complicated Nuclei

The previous section was dedicated to the simplest nucleus with a mass number of two nucleons (A = 2 = p + n). In this very brief section, the nuclei of mass number A>2 will be considered; two cases will be taken: one with odd proton and the other with odd neutron.

1. Odd Z (one proton): In this case, there is a single particle outside the shell, so the system is core (closed shell) and one particle outside moving around the core, so the total angular momentum of the nucleus j is either $l^\rightarrow + s^\rightarrow$ or $l^\rightarrow {}_- s^\rightarrow$, i.e., either l and s parallel or antiparallel. This model of a nucleus is called shell model (individual-particle or single-particle model). The orbital angular momentum of the proton is $l^\rightarrow$, which is a vector. Its intrinsic spin ($s^\rightarrow$) is, as stated before, either $_r \uparrow\uparrow$ or $\uparrow\downarrow$, with $l^\rightarrow$; therefore, $J^\rightarrow = l^\rightarrow \pm s^\rightarrow$ is the total angular momentum of the nucleus.

Since the intrinsic spin of the nucleon is (1/2) in unit of $\hbar$, then,

$J^\rightarrow = l^\rightarrow + 1/2$ or $l^\rightarrow - 1/2$. Now comparing with equation (65), where here j stands for L, l for l_1, and s for l_2, the single particle is proton, so $g_1 = 1, g_2 = 5.58$.

Using equation (66), one finds

$$g = 1/2(g_1 + g_2) + 1/2(g_1 - g_2)[l_1(l_1 + 1) - l_2(l_2 + 1) /j(j + 1)] \qquad (67)$$

There are two cases:

i) $J^\rightarrow = l^\rightarrow + s^\rightarrow$ or $l^\rightarrow = J^\rightarrow - s^\rightarrow$

Using equation (67), this gives $g = 1 + \dfrac{2.29}{j}$; therefore, $u = j\,g^\rightarrow$

$$u = [j + 2.29] \qquad (68)$$

ii) For j = l–1/2 or l = j + 1/2, this gives g = (1–2.29/j + 1), since u = jg, then,

$$u = (j - 2.29j/j + 1) \tag{69}$$

If u is drawn versus j, one gets Schmidt limits diagram for the odd-proton nuclei, it is well known to physicist.

2. The odd-neutron case: Here, (n) is neutral; i.e., the charge is zero. Hence, $g_l = 0$, $g_s = -3.82$. As in the case of the proton, there are two possible values to depend on the possible values of j or l, i.e., $l = j + 1/2$ or $l = j - 1/2$.

Following the same steps in the case of the odd proton, you find for the neutron case

$$U = (1.91 \, \frac{j}{j+1}) \tag{70}$$

for $l = j + 1/2$, $g = 1.91/j + 1$

and

$$u = -1.91 \tag{71}$$

for $l = j - 1/2$, $g = -1.91/j$

If u is drawn versus j, one gets Schmidt limits lines for odd-neutron nuclei, which is well known to physicist.

Treatment of nuclei with more than one nucleon outside the nucleus core, each moving in its own orbit and their mutual interaction, will be dealt with in the next chapters. Also, their configuration and the buildup of their wave function will be worked out.

Chapter Two

Representation of Three-Dimensional Rotation Group

(Theory and Application to Nuclear Physics)

II-1—Introduction

In general, the nuclear system is a group of particles (nucleons, p and n). These nucleons are bound into a very small dimension ($\sim$ 1fm = 10^{-13} cm); these particles are in a moving state, each in its own orbit. At the same time, each is spinning about its intrinsic axis. So each has an orbital angular momentum $\vec{l}$ and spin s = 1/2 in unit of $\hbar$. Therefore, their physical properties are fairly well affected by these rotations. Since the nucleus properties are an average of its constituent's properties, it was found that the mathematical rotational group theory is the the well known tool to deal with these physical properties. As we shall see, this group very well describes the physical status of nuclear systems, such as the nucleus structure and property, the nucleons' motion in their orbits and spinning about their intrinsic axes. Due to these physical facts, the wave functions describing these physical entities are affected by these rotational operations. So if ψ is denoting the wave function, that describes the system under the investigations; under rotation, it will be transformed as

$\Psi(\vec{r}) \rightarrow$ under rotation $\rightarrow \psi(\vec{r}^{-})$ from r coordinates system $\rightarrow$ r‾ prime coordinates system.

From quantum mechanics, it is known that

$R(\alpha,\beta,\gamma)$ = rotational operator = $R(\omega)$, where $(\alpha\beta\gamma)$, represented by (ω), are the known Euler angles.

Hence, $\psi(r^{-\rightarrow}) = u(\omega)\psi(r^{\rightarrow})$. Multiply by ψ_i; one gets

$$\psi_k(r^{-\rightarrow}) = \Sigma_i \psi_i(r^{\rightarrow}) D_{ik}(\omega) \tag{1}$$

where $D_{ik}(\omega) = (<\psi_i, u(\omega)\psi_K> = <i|u|k>$ \hspace{1cm} (2)

Let us use angular momentum representation and Dirac notation for the state representation such as

$|jm> \propto |jm^->$, then $D^{(j)}_{mm^-} = <jm|i|j\, m^->$
\hspace{3cm} (3)

where $D^{(j)}_{m\, m^-}$ is called D-matrix; i.e., it is (2J + 1)(2J + I) matrix. Therefore, we have

$$\psi_{jm}(r^{-\rightarrow}) = \Sigma \psi_{jm}^-(r^{\rightarrow}) D^{(j)}_{m^-m} \tag{4}$$

II-2—Properties of D-Matrix

$D^{(j)}_{mm^-}(\alpha,\beta,\gamma)$ = <jm|R(α,β,γ)|jm->,

where $R(\alpha,\beta,\gamma)$ is the rotation operator given by

$$R(\alpha,\beta,\gamma) = e(\exp.-i\gamma J_z)e(\exp.-i\beta J_Y)e(\exp.-i\alpha J_z) \tag{5}$$

Hence,

1. $D^{-1} = D^+$, which implies that

$$D^{-1}(\alpha,\beta,\gamma) = D_{mm^-}(-\gamma,-\beta,-\alpha) = D^*_{mm^-}(\alpha,\beta,\gamma))$$

2. $\hat{J_z}\Psi_{jm} = m\Psi_{jm}$, which implies that

$$D^{j}_{mm^-}(\alpha,\beta,\gamma) = <jmlR(\alpha\beta\gamma)ljm^-> = e(\exp.-im\gamma)\,d^{j}_{mm^-}\,e(\exp.-im^-\alpha)$$
(6)

where $d^{j}_{mm^-}$ = $<jml\,e$ (exponential $(-\,i\beta J_y)ljm^->$, which is called Wigner-Eckart small D-matrix is real, and $d^{j}_{mm^-}(-\beta) = d^{j}_{mm^-}(\beta)$. This gives that

$$d^{j}_{mm^-}(\beta) = (-)^{m-m^-}d^{j}-m,-m^- \text{ (See \textit{Angular Momentum in Quantum}}$$

\textit{Mechanics}, Edmonds.) Using this relation, one can write

$$D^{j}_{mm^-}(\alpha,\beta,\gamma) = D^{j}_{mm^-}(-\alpha,-\beta,-\gamma) = (-)^{m-m^-}D^{j}_{mm^-}(\alpha,\beta\,\gamma)$$
(7)

3. The group characteristics of $D^{j}(\omega)$ are

$$D^{j}(\omega) = D^{j}(\omega_2)\,D^{j}(\omega_1)$$
(8)

To verify this, let us apply $u(\omega_2)u(\omega_1)$ to $\psi_{jm}(r\rightarrow)$; we get

$$u_1\,u_2\,\Psi_{jm^-} = u_2\,\Sigma_{m^-}\,\Psi_{jm}^-(r)\,D^{j}_{mm^-}(\omega_2) = \Sigma u_2\,\Psi_{jm^-}\,D^{j}_{m^-m}(\omega_1).$$

Now, $u_2\Psi_{jm}^- = \Sigma_{m^{--}}\Psi_{jm}^{--}\,D^{j}_{m^-m}(\omega_2)$

Therefore, $u_2 u_1\,\Psi_{jm^-} = \Sigma\,\Psi_{jm}^{--}(D^{j}_{m^-m}(\omega_2)\,D^{j}_{m^-m}(\omega_1) = \Sigma_m\,^{--}\Psi_{jm}^{--}\,D^{j}(\omega)$

$$\rightarrow D^{j}(\omega) = D^{j}(\omega_2)\,D^{j}(\omega_1)$$
(9)

Let us take an example for $j = \ell$, $m = 0$ for a coordinate system (S):

$$D^{\ell}_{m0}(\alpha,\beta,\gamma)_,\,D^{\ell}_{0m}(\alpha,\beta,\gamma),\,D^{\ell}_{00}(\alpha,\beta,\gamma)$$

Now, we rotate to (S^-) system (see figure 10).

Then, $\psi_{jm^-}(r^-) = \sum_m \psi_{jm}(r\rightarrow)_{mm^-}(\alpha,\beta,\gamma)$

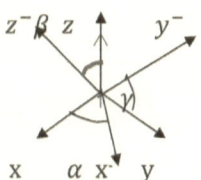

Multiply both sides by $[D^j{}_{mm^-}(\alpha\beta\gamma)]^{-1}$; you get

Figure (10)

$\psi_{jm}(r\rightarrow) = \sum \psi_{jm}{}^-(\rightarrow r^-)\, D^{j^-}{}_{mm}(\alpha,\beta,\gamma)$

for $j = \ell$ = integer, then $\psi_{\ell m}(r^-) = Y_{\ell m}(\Theta^-,\varphi^-)$ in the primed system (S^-) (figure 11)

$$Y_{\ell m^-}(0,0) = \sqrt{\frac{2\ell+1}{4\pi}}\,\delta_{m^-0} = \sqrt{\frac{2\ell+1}{4\pi}} \quad \text{for m = 0}$$

$$\psi_{\ell m}(\beta,\alpha) = \sum_{m^-} Y_{\ell m^-}(0,0)\, D^{\ell^*}_{mm^-}(\alpha,\beta,\gamma) = \sqrt{\frac{2\ell+1}{4\pi}} D^{\ell^*}_{m0}$$

But $D^{\ell^*}_{m0} = (\frac{4\pi}{2l+1})^{1/2} Y^*{}_{\ell\,m}(\beta,\alpha)$, therefore, and $D^{\ell}_{0m}(\alpha,\beta,\gamma) =$

$$\sqrt{\frac{2\ell+1}{4\pi}}Y_{\ell m^-}(\beta,\alpha) \tag{10}$$

This implies that $D^{\ell}_{00}(\alpha,\beta,\gamma) = P_{\ell}(\cos\beta)$ (11)

$P_{\ell}(\cos\phi^-) = (4\pi/(2\ell+1))_{=-}\sum_m^{\ell} Y_{\ell m} Y^*{}_{lm}(\beta\alpha)$ (12)

Legendre polynomial

where $\cos\phi^- = \cos\beta\,\cos\varphi + \sin\beta\,\sin\varphi\,\cos(\alpha-\beta)$

$Y_{lm}(\theta,\varphi)$ is the spherical harmonic function.

II-3—Addition of Angular Momentum

In nuclear shell model, there is a core and shell of levels occupied by nucleons, protons, or neutrons, but some of the levels may not be fully occupied, so the outer level may contain one or more nucleons but less than its occupation capability (occupation number). Usually, the core is fully occupied, so the nucleons outside the core play the main rules in describing the nucleus properties, especially its total angular momentum and its wave function. Therefore, studying the angular momenta of the nucleons and their addition is quite important. For the addition of angular momenta, there are mathematical rules. The addition actually is a vector addition in general; therefore, the mathematical vector analysis is quite important. (A good reference for this is the *Nuclear Shell Theory* by De-Shalit, Lawson, and Billzard.)

We will start with the case of two nucleons of angular momentum j_1 and j_2. As mentioned previously that for one nucleon outside the core, its angular momentum (j) is the angular momentum of the nucleus, either $j = l + s$ or $j = l-s \equiv j = l + 1/2$ or $j = l-1/2$, where, (l) is the orbital angular momentum and (s) is the intrinsic spin in unit of $h/2\pi, = \hbar$, h= Plank constant. Let us start with the coupling of two angular momenta.

II-3-1—Coupling of Two Angular Momenta

Take two nuclear systems their eigenstates are $lj_1\ m_1>$ and $lj_2\ m_2>$respectively, such as S_1 weakly interacts with S_2.

The problem is, how can one describe these systems as a whole?

Let $lj_1\ j_2\ m_1\ m_2> = lj_1\ m_1> lj_2\ m_2>$ be assumed as a representative of the eigenstate of the whole system (1 and 2)which is a complete set (contains all the information about the system). The projections of the angular momenta j_1 and j_2 $(m_1 \, m_2)$ imply the dimensionality of the system eigenstate, which is given by $(2j_1 +1)(2j_2 +1) =$ for a given j_1 and j_2.

As it is well-known that (m_j) takes the values given by $(2j_i + 1)$, this kind of representation is called uncoupled representation. The dimension of the space spanned by $lj_1 \, m_1 > lj_2 \, m_2 >$ is given by (N), as mentioned above.

For strong coupling, tackling the system as a whole, the total angular momentum $\vec{J}$ has to be introduced. It is the vector some of $\vec{J_1}$ and $\vec{J_2}$; i.e., $J_t = J_1 + J_{2,}$ and that J and J satisfy the commutation relation →

$\vec{J} x \vec{J} = i \, \vec{J}$, i.e, $\vec{J}$ also satisfies the commutation relation.

Hence, for a coupled system of angular momentum $\vec{J}$, which usually has a magnetic quantum number $M(M_j = $ projection of $\vec{J}$), it is important to find its eigenstate lJM> or the eigenket, as it is called according to Dirac notation. Using the eigenvalue equation, one gets (J is a vector, but J^2 is scalar)

$J^2 lJM> = j(j + 1) lJM>$, setting $\hbar = 1$

$J_z \, lJM> = M \, lJM>$, J_z *is the component of J along Z-coordinate with the conditions that*

$$lj_1 - j_2 l \le j \ge j_1 + j_2, \, -j \le M \le j \qquad (13)$$

M takes the values given by $\sum 2j + 1 = \{2j_2 + 1/2\}[2(j_1 - j_2) + 1 + 2(j_1 + j_2) + 1]$

$= (2j_1 + 1)(2j_2 + 1) = N$, which is the dimensionality, as stated before.

II-3-2—Strong Coupling

To couple the two systems strongly, the mathematical approach requires strong coupling coefficients, so the two systems, S_1 and S_2 eigenstates, can be combined to be related to the eigenstate of the new united system denoted by |JM >. These coefficients were suggested by Racah and developed extensively by Clebsch and Gordan; therefore, they are called Clebsch-Gordan coefficients, denoted by (C-G-C). To do that, let |JM> be written as

$$= \sum_{m_1 m_2} \left| j_1 \, j_2 > \right| j_2 \, m_2 > < j_1 m_1 j_2 m_2 \left| \text{JM} > \right. \tag{14}$$

Equation (14)is the coupling equation.

Let us write |JM> as

$$|\text{JM}> = \sum_{m_1 m_2} C_{m_1 m_2} 1 \; j_1 m_1 > \text{I} j_2 m_2 > \tag{15}$$

Then if one multiplies both sides of (15) by the bra $<j_1 \; m_1 \; j_2 \; m_2$| one will get

$$< j_1 \; m_1 \; j_2 \; m_2 \; |\text{JM}> = \sum_{m_1 m_2} C_{m_1 m_2} \tag{16}$$

Substituting (16) for $\sum C$ into (15), one gets

|JM> =$\sum$l $j_1 \; m_1$>l$j_2 \; m_2$><$j_1 \; m_1 \; j_2 \; m_2$ l|JM> as equation (14).

This leads to the Clebsch-Gordan coefficients given by

$C_{m1 \, m2}$ =<$j_1 \; m_1 \; j_2 \; m_2$ l|JM>.

Also, they are called Wigner coefficients or vector addition coefficients. They are a coupling matrix of $N \times N$ dimensions, i.e., C = $(\leftarrow_N) \downarrow$ N .; C^{-1} = C^+ = $C^{+\sim}$. It may be written as $C_{JM}^{j_1 m_1 j_2 m_2}$.

II-3-3—Properties of C-G Coefficients

The properties of the C-G coefficients can be summarized in the following:

a. $<JM|j_1 m_1 j_2 m_2> = <j_1 m_1 j_2 m_2 |JM>^* = <j_1 m_1 j_2 m_2 | JM>$ (17)

This implies that C is real.

b.   C = 0 unless $M = m_1 + m_2$, $J_z = J_{1z} + J_{2z}$; the following is the proof of this:

$J_z |JM> = M|JM> = \sum (m_1 + m_2)|j_1 m_1><j_2 m_2><j_1 m_1 j_2 m_2 | JM>$ (sum over m1 m2). By subtraction, we have $\sum [(m1 + m2)-M]|j1m1> |j_2 m_2>< j_1 m_1 j_2 m_2|JM>=0$ (18)

Now, since the eigenstates $|j_1 m_1>$, $|j_2 m_2>$ linearly independent equation (18) is true only if $(m_1 + m_2 - M)< j_1 m_1 j_2 m_2|JM> = 0$, but $< j_1 m_1 j_2 m_2 |JM> \neq 0$; therefore, $[(m_1 + m_2) - M]$ has to be zero, which implies that $m_1 + m_2 = M$, so $C \neq 0$, and equation (18) is satisfied.

The triangle conditions should be satisfied too, i.e., $|j_1 - j_2 \leq j \leq j_1 + j_2$, $\Delta(j_1 j_2 j_3) = 0$, for a diagonal.

c.   C-G coefficients are orthonormal, i.e., that $CC^{-1} = 1 = C C^+$; hence,

$\sum <j_1 m_1 j_2 m_2 |JM><j_1 m^-_{(1)} j_2 m^-_2 |JM> = \delta_{m_1 m_{\bar{1}}} \delta_{m_2 m_{\bar{2}}}$

or $\sum <j_1 m_1 j_2 m_2 |JM><JM|j_1^- m_1 j_2^- m_2> ============= \uparrow$

or $\sum <J M| j_1 m_1 j_2 m_2>^* <j_1 m_1 j_2 m_2 | J^- M^-> = \delta_{JJ^-} \delta_{MM^-}$

or $\sum <j_1 m_1 j_2 m_2 |JM><j_1 m_1 j_2 m_2 | J^- M-> \delta_{JJ^-} \delta_{MM^-}$

d.   Computation of C-G coefficients

Consider a stretched case where $J_{max.} = J_1 + J_2$ and $M_{max} = J_{max.} = J_1 + J_2$

Then, $J_1 J_2$ couple to form J as in the figure

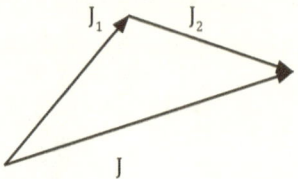

$1 J_{max} J_{max}> = <J_1 J_1 J_2 J_2 1 J_{max} J_{max}> 1 J_1 J_1>1 J_2 J_2>$, then, $j^{\rightarrow}$

$1 <J_1 J_1 J_2 J_2 1 J_1 + J_2, J_1 + J_2>1 = +1$; to show this, multiply by $1 J_{max} J_{max}>$ then,

$< J_{max} J_{max} 1 J_{max} J_{max}> = 1 = [1 < J_1 J_1 J_2 J_2 1 J_1 + J_2 J_1 J_2> 1^2]$ $\times$ $(1 <J_1 J_1 1 <J_2 J_2$

$1 l^2 = 1$, *normalization effect)*; therefore, [ ]=1.

So $1 <J_1 J_1 J_2 J_2 1 J_1 + J_2, J_1 + J_2>1 = 1$; the absolute value, the negative value has no physical meaning, so it is neglected.

Hence, $1 <j_1 m_1 j_2 m_2 1 JM>1_{max} = 1$ (16)

Note: C-G coefficients values are usually tabulated in certain tables such as ORNL 171 by Simond et al.

Problem: Show that $[(j + M)(j - M + 1)]^{1/2} <j_1 m_1 1 j_2 m_2 1 JM - 1> = [j_1 + m_1 + 1][j_1 - m_1)]^{1/2} \times [<j_1 m_1 + 1, j_2 m_2 1 jM> + [(j_2 + m_2 + 1)(j_2 - m_2)]^{1/2} \times <j_1 m_1 j_2 m_2 + 1 1 j M >$

(Hint: use $J_- = J_{1-} + J_{2-}$)

II-3-4—Wigner j-Symbols

The previous example was for the case where the nucleons were only two outside the nucleus. Now how can one deal with nuclei of more than two nucleons outside their cores where building up a wave function becomes quite complicated? For that, Wigner and other physicists developed a clever method to tackle this case; it is the so-called 3j-symbols, 6j-symbols, and 9j-symbols. These tools are very well established. There are tables contained in them, in addition to very well-known books that deal with the detailed demonstrations of the theory of the angular momentum (see references at the end of the book). In this book, these j-symbols will be used where they are needed. So their main mathematical and physical properties will be understood.

A. 3j-symbols

By definition, 3j-symbols proposed by Wigner is given by

$$\begin{bmatrix} J_1 & J_2 & J_3 \\ m_1 & m_2 & -M \end{bmatrix} \equiv \frac{(-1)^{J_1 - J_2 + M}}{(2J+1)^{1/2}} <j_1 j_2 \, m_1 m_2 | JM> \tag{17}$$

This is for the case where $J_1 + J_2 + J_3 = 0$; they are coupled to zero, with corresponding $m_1 \, m_2 \, m_3$ respectively coupled to zero too, i.e.,

$m_1 + m_2 + m_3 = 0$

B. Principal properties of 3j-symbols

1. All are reals, i.e., $[<j_1 \, j_2 \, m_1 \, m_2 \, |JM>]^* = <j_1 \, j_2 \, m_1 \, m_2 \, |JM>$.

2. The required selection rules are

    i) $m_1 + m_2 = M$

    ii) $|j_1 - j_2| \leq j \leq j_1 + j_2$, —(triangle condition)

These two conditions are a must; otherwise, $<j_1 \, j_2 \, m_1 m_2 \, |JM> = 0$. This implies that C-G coefficients vanish.

C. $\begin{bmatrix} j_1 & j_2 & j_3 \\ m_1 & m_2 & m_3 \end{bmatrix}$ is numerical; therefore

1. It is invariant in a circular permutation of three columns;

2. It is invariant under even permutation of raws;

Hence,

$$\begin{bmatrix} j_1 & j_2 & j_3 \\ m_1 & m_2 & m_3 \end{bmatrix} = \begin{bmatrix} j_2 & j_3 & j_1 \\ m_2 & m_3 & m_1 \end{bmatrix}$$

3. For a permutation of two columns, it has to be multiplied by the factor

$$(-1)^{j_1+j_2+j_3}$$

4. If the signs of m_1, m_2, and m_3 change simultaneously, they should be multiplied by the factor $(+1)^{j_1+j_2+j_3}$ therefore, according to these characteristics, the following can be written :

(a) $\begin{pmatrix} j_1 & j_2 & j_3 \\ m_1 & m_2 & m_3 \end{pmatrix} = (-1)^{j1+j2+j3} \begin{pmatrix} j_2 & j_1 & j_3 \\ m_2 & m_1 & m_3 \end{pmatrix}$ (19)

(b) $\begin{bmatrix} j_1 & j_2 & j_3 \\ m_1 & m_2 & m_3 \end{bmatrix} = (1)^{j_1+j_2+j_3} \begin{bmatrix} j_1 & j_2 & j_3 \\ m_1, & -m_2, & -m_3 \end{bmatrix}$ (20)

In (b), changing the signs of m_1, m_2, and m_3 means changing the quantization only.

(c) $\sum (2j + 1) (j_1 j_2 j_3) (j_1 j_2 j_3^-) = \delta_{j_3 j_3^-} \delta_{m_3 m_3^-}$ (21)

$(m_1 m_2 m_3) (m_1 m_2 m_3^-)$ one bracket, sum over $m_1 m_2$

(d) $\sum_{j3} (2j_3 + 1) \begin{pmatrix} j_1 & j_2 & j_3 \\ m_1 & m_2 & m_3 \end{pmatrix} \begin{pmatrix} j_1 & j_2 & j_3 \\ m_1^- & m_2^- & m_3 \end{pmatrix} = \delta_{m1m\bar{1}} \, \delta_{m2m\bar{2}}$ (22)

So the total eigenstate IJM> is given by

$$IJM> = \sum \frac{(-1)^{j1-jz+M}}{(2_{j3}+1)^{1/2}} \begin{pmatrix} j_1 & j_2 & j_3 \\ m_1 & m_2 & m_3 \end{pmatrix} Ij_1 m_1 > Ij_2 m_2 >$$ (23)

Equations 21 and 22 represent the orthonormality relations of 3j-symbols. The consequences of (a)–(d) related to C-G coefficients are the following:

i) $< j_1 j_2 m_1 m_2 | J M> = (-1)^{j_1 + j_2 - J} <j_2 j_1 m_2 m_1 IJM>$ (24)

or $= (-1)^{j_1 - J + m_2} (\sqrt{2J + 1})/2j_1 + 1 <J j_2 M, -m_2 | j_1 m_1>$ (25)

ii) $(-1)^{j_2 - J - m_1} \sqrt{2J + 1}/2j_2 + 1 <j_1, J, -m_1, M | j_2 m_2>$ (26)

iii) $(-1)^{j_1 + j_2 - J} <j_1 j_2, -m_1, -m_2 IJ, -M>$ (27)

The relations between 3j-symbols and the spherical harmonics functions $Y_{lm}(\theta, \varphi)$ are

1) $\int Y_{\ell_1}^{m_1}(\omega) Y_{\ell_2}^{m_2}(\omega) Y_{\ell_3}^{m_3}(\omega) d\omega = [(2l_1 + 1)(2l_2 + 1)(2l_3 + 1)/4\pi]^{1/2}$

$\times \begin{bmatrix} l_1 & l_2 & l_3 \\ 0 & 0 & 0 \end{bmatrix} \begin{bmatrix} l_1 & l_2 & l_3 \\ m_1 & m_2 & m_3 \end{bmatrix}$ (28)

These 3j-symbols are to be rewritten as usual later.

Also, these (ys) are $Y_{11}^{m_1}$, $Y_{12}^{m_2}$, and $Y_{13}^{m_3}$; ω is a solid angle,

where

$$Y_{l_1}^{m_1}(\omega)\, Y_{l_2}^{m_2}(\omega) = \sum_{L=l_1-l_2}^{l_1+l_2} \sum_{M=-L}^{M=L} (2l_1+1)[(2l_2+1)/4\pi\,(2L+1)]^{1/2}$$

$$\times < l_1 l_2\ 0\,0\,|\,L\,0 > < l_1 l_2\ m_1 m_2\,|\,L\,M >\, Y_L^M(\omega) =$$

or

$$= \sum_{L=l_1-l_2}^{l_1+l_2} \sum_{M=-L}^{M=L} (-)^M\ [(2l_1+1)(2l_2+1)(2L+1)/4\pi]^{1/2} \begin{bmatrix} l_1 & l_2 & l_3 \\ 0 & 0 & 0 \end{bmatrix}$$

$$\times \begin{bmatrix} l_1 & l_2 & L \\ m_1 & m_2 & M \end{bmatrix} Y_L^{-M}(\omega) \qquad\qquad (30)$$

Problem given:

a) $<j_1\ m_1\ j_2\ m_2\ l j_3\ m_3> = A<j_2,\ -m_2\ j_3\ m_3\ l j_1\ m_1>$

b) $<j_1\ m_1\ j_2\ m_2\ l j_3\ m_3> = B<j_1,\ -m_1\ j_2,\ -m_2\ l j_3,\ -m_3>$

Find A and B in terms of 3j-symbol:

(Hint: use the CGC properties, then look for 3j-symbols in their special tables such as C-G coefficients by Rodenberg et al.)

To show how 3j-symbols work, let us use a simple example; take a system of one proton and a neutron, where

$|\psi_p> = |1/2,\pm 1/2>$, $j = 1$ or 0

$|\psi_n> = |1/2\pm 1/2>$, $j = 1$ or 0

where 0 is a singlet and 1 is a triplet.

Consider the singlet state where $J = 0$, $m = 0$; hence,

$l00> = \sum l1/2\ m_n > l1/2\ m_p > < 1/2m_n\ 1/2\ m_p\ ljm>$

$= <1/2,\ 1/2,\ 1/2,\ -1/2l00>l1/21/2>_n l1/2,\ -1/2>_p + <1/2,\ -1/2,\ 1/2,$

$1/2l00>l1/2,-1/2>_n l1/2,1/2>_p = 1/\sqrt{2}\ l1/2\ 1/2>_n l1/2,-1/2>_n\ -1/\sqrt{2}$

$$l1\left[/2,\ -1/2>l1/2,\ 1/2>\right] = 1/\sqrt{2}\ \uparrow\downarrow\ -1/\sqrt{2}\ \downarrow\uparrow = 1/2\ \begin{array}{ccc} 1/2 & -1/2 & 0 \\ -1/2 & 0 \end{array}\ \substack{\downarrow\uparrow}.$$

$$\begin{array}{c} 1 \\ = 0 \\ -1 \end{array}$$

For triplet case, J = 1, m}

Take J = 1, m = 1, i.e., the eigenstate is l11>; hence,

$l11> = <1/2,\ 1/2,\ 1/2,\ 1/2l11>l1/2,\ 1/2>_n l1/2,\ 1/2>_p$

$= (-1)^{1/2-1/2+1}\ /\left(\sqrt{3}\dfrac{1/2\ 1/2}{1/2\ 1/2}\Big)_{-1}^{\ 1}\ \uparrow\uparrow\right\}$

$= \sqrt{3}(-1)^{1/2+1/2+1}\left(\dfrac{1/2\ 1/2}{1/2\ 1/2}\Big)_{1}^{1}\right.$

$= -\sqrt{3}\ [(1/2+1/2) + (1/2+1/2+1)/3.2.1]^{1/2} = -\sqrt{3}\ [-1/3]^{-1/2} = -1\uparrow\downarrow$

As an exercise:

A. Find l1, −1>, and l10>

B. For an electron of $l = 1,\ s = 1/2$, p-state

$j = 1/2,\ 3/2,\ j_v = l_v \pm s_v$, find l3/2, 3/2> in terms of $Y_{11}(\theta,\varphi)$ and $\chi_{1/2\ 1/2}$

C. Addition theorem of D-matrices (Clebsch-Gordan Series)

Suppose there are two particles systems such as

$$l\ j_1 m_1 > a \quad\Longleftrightarrow\quad b\ l\ j_2\ m_2 >$$

Let *a* be the origin system, and the system *b* is the rotated system; that is to say, a $\;\rightarrow\; D(\alpha,\beta,\gamma) \longrightarrow$ b, or in ket representation, it can be written as

$1\,j_1\,m_1 > \rightarrow D(\alpha,\beta,\gamma) \longrightarrow 1\,j_2\,m_2 >$, hence $lj_1m_1>lj_2m_2> = lj_1m_1>lj_2m_2>$

$lJ_1M_1>lJ_2M_2 >$; this transformation can be formally represented as

$$1\,J_1M_1>lj_2M_2> = \sum_{m_1} |\; j_1\,m_1> \; D^{j_1}_{m_1M_1} \sum_{m_2} |j_2\,m_2> \; D^{j_2}_{m_2M_2}$$

$$= \sum_{m_1m_2} D^{j_1}_{m_1M_1} D^{j_2}_{m_2M_2}\, 1\,j_1m_1>lj_2\,m_2 > \qquad (31)$$

Take $lJ_1\,M_1> <J_2M_2\,l = \sum_j < j_1M_1j_2\,M_2l\,J\,M>.l\,J\,M>$

But $j1m1> = D^{(-j)}\,lj\,M>$ or $l\,J\,M > =D^{\wedge}j\,l\,j1m1>$; therefore

$$lJ1M1> 1\,J_2M_2 > = \sum_j |j_1\,M_1\,j_2\,M_2 >l\,J\,M>.\sum_m |jm > D^j_{mM}$$

$$= \sum_{j,m} < j_1M_1\,j_2\,M_2\,l\,J\,M >lj\,m > D^j_{mM} =$$

$$\sum_{mm_1m_2} j_1\,m_1\,j_2\,m_2lJ\,M><j_1m_1\,j_2m_2ljm>|\,j_1m_1 >|\,j_2m_2 > D^j_{mM} \qquad (32)$$

$$But\; ljm > = \sum_{M_1m_2} <j_1\,m_1\,j_2\,m_2lj\,m>lj_1\,m_1 >< j_2\,m_2\,l$$

$\underrightarrow{\hspace{4cm}}$

(CGC)

Comparing (31) with (32), the required formula is obtained:

$$D^j_{mM}(\alpha,\beta,\gamma) = D^{j_1}_{m_1M_1}(\alpha,\beta,\gamma)\; D^{j_2}_{m_2M_2}(\alpha,\beta,\gamma) \qquad (33)$$

Coupling of Three Angular Momentum

Consider a system of j, j' , and j", where $J = j + j' + j'' \equiv (j_1 + j_2 + j_3)$, and it is required to find the total angular momentum in the dimension of space of $(2j + 1)(2j' + 1)(2j'' + 1)$ dimensionality; the space is spanned by the vector $|m\, m'\, m''> \equiv |jm> |j'm'> |j''m''>$ (where j, j', j" are given and m, m', m" are variables). The subspace of angular momentum (JM) is usually of more than one dimension [2J and 2M, integral numbers, even or odd, like $2(j + j' + j'')$; $\min | j \pm j' \pm j''| \leq J \leq j + j' + j''$; $-J \leq M \leq J$]. The two coupling systems may be clarified in the following. (Note: the order in which the different vectors are coupled is quite important.) Let us consider these two possible ways of coupling a and b:

(a) – j' + j = g', then g' + j" = J

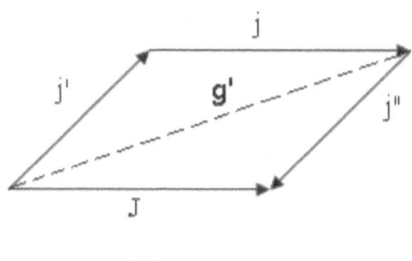

(a)

(b) – j + j" = g"

$$j' + g'' = J$$

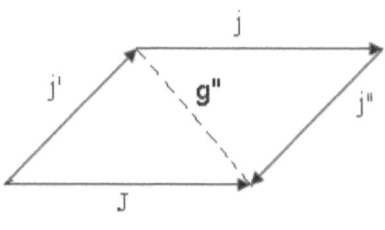

(b)

To go from (a) to (b), one has to use a certain unitary transformation, where this unitary transformation is related to Wigner "6j"-symbols, which are defined as

$$\begin{Bmatrix} j_1 & j_2 & j_3 \\ J_1 & J_2 & J_3 \end{Bmatrix}$$

Therefore,

$$< j',(jj")g",JM \left| (j'j)g',j",J'M' >= \delta_{JJ'}\delta_{mm'}[(2g'+1)(2g"+1)]^{\frac{1}{2}}(-1)^{j+j'+j"} \begin{Bmatrix} j' & j & g' \\ j" & J & g" \end{Bmatrix} \right.$$

(34)

i.e.

$$|(j'j)g',j",TM >= \sum_{g"} | j',(jj")g",TM > \sqrt{(2g'+1)(2g"+1))}](-1)^{j+j'+j"+J} \begin{Bmatrix} j' & j & g' \\ j" & J & g" \end{Bmatrix}$$

$$Notice: (j'j)g' \equiv j' + j = g'$$
$$and (j,j")g" \equiv j + j" = g"$$

(35)

II-4—Some Important Properties of "6j"-Symbols

1. "6j" are real, i.e., "(6j) = "(6j)*"

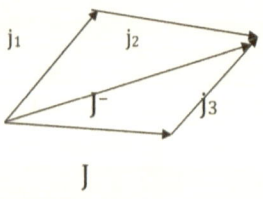

2. Selection rules are to have

$$\begin{Bmatrix} j_1 & j_2 & j_3 \\ J_1 & J_2 & J_3 \end{Bmatrix} \neq 0$$

The three angular momentum represented by each face of the tetrahedron must be such that it is possible for one of them (j's) to be the sum of the other two. In other words, it is necessary that the elements of each of $(j_1 j_2 j_3)$, $(j_1 J_2 J_3)$, $(J_1 j_2 J_3)$, $(J_1 J_2 j_3)$ to

1. satisfy the triangular inequalities.

2. have an integral sum.

(Note: either the six j's are integral, or the three j's of the same face are integral, or two j's corresponding to opposite edges are integral.)

II-4-1—Symmetry Relations

The "6j"-symbols are invariant

1. under a permutation of its columns, i.e.

$$\begin{Bmatrix} j_1 & j_2 & j_3 \\ J_1 & J_2 & J_3 \end{Bmatrix} = \begin{Bmatrix} j_2 & j_1 & j_3 \\ J_2 & J_1 & J_3 \end{Bmatrix}$$

2. 2. under an exchange of two elements of the first line with the corresponding elements in the second line as

$$\begin{Bmatrix} j_1 & j_2 & j_3 \\ J_1 & J_2 & J_3 \end{Bmatrix} = \begin{Bmatrix} J_1 & J_2 & j_3 \\ j_1 & j_2 & J_3 \end{Bmatrix}$$

In other words, to define "6j," it is enough to give the six angular momentum and their relative positions on the associated tetrahedron (as shown above).

II-4-2—Fundamental Relations with CGC

It is important to be acquainted with the relations between "6j"-symbols and Clebsch-Gordan coefficients to be able to tackle complicated nuclei. Therefore, notice the following important relations:

$$
\sum_{\substack{M_1 M_2 M_3 \\ m_1 m_2 m_3}} -(1)^{J_1+J_2+J_3+M_1+M_2+M_3} \begin{pmatrix} J_1 & J_2 & j_3 \\ M_1 & -M_2 & m_3 \end{pmatrix} \begin{pmatrix} J_2 & J_3 & j_1 \\ M_2 & -M_3 & m_1 \end{pmatrix} \begin{pmatrix} J_3 & J_1 & j_2 \\ M_3 & -M_1 & m_2 \end{pmatrix} \begin{pmatrix} j_1 & j_2 & j_3' \\ m_1 & m_2 & m_2 \end{pmatrix}
$$

$$
= \delta_{j_3 j_3'} \delta_{m_3 m_3'} \frac{1}{2j_3+1} \times \begin{Bmatrix} j_1 & j_2 & j_3 \\ J_1 & J_2 & J_3 \end{Bmatrix}
$$

$$----- (36)$$

$$
\sum_{M_1 M_2 M_3} -(1)^{J_1+J_2+J_3+M_1+M_2+M_3} \begin{pmatrix} J_1 & J_2 & j_3 \\ M_1 & -M_2 & m_3 \end{pmatrix} \begin{pmatrix} J_2 & J_3 & j_1 \\ M_2 & -M_3 & m_1 \end{pmatrix} \begin{pmatrix} J_3 & J_1 & j_2 \\ M_3 & -M_1 & m_2 \end{pmatrix} \begin{pmatrix} j_1 & j_2 & j_3' \\ m_1 & m_2 & m_2 \end{pmatrix}
$$

$$
= \delta_{j_3 j_3'} \delta_{m_3 m_3'} \frac{1}{2j_3+1} \times \begin{Bmatrix} j_1 & j_2 & j_3 \\ J_1 & J_2 & J_3 \end{Bmatrix} \dots\dots\dots(37)
$$

$$
\sum_{g m_g} -(1)^{g+m_g} \begin{Bmatrix} j_1 & J_1 & g \\ g_2 & j_2 & f \end{Bmatrix} \begin{pmatrix} j_1 & J_1 & g \\ m_1 & M_1 & -m_g \end{pmatrix} \begin{pmatrix} j_2 & J_2 & g \\ m_2 & M_2 & m_g \end{pmatrix}
$$

$$
= (-1)^{j_2+j_1+f+g} \sum_f (-1)^{f+m_f} \begin{pmatrix} j_1 & j_2 & f \\ m_1 & m_2 & -m_f \end{pmatrix} \begin{pmatrix} J_1 & J_2 & f \\ M_1 & M_2 & m_f \end{pmatrix} \quad ------- (38)
$$

[Note: the relations of "6j" with CGC come about through the "3j" [equation (17)].

6-"9j"-Symbols: Definition and Properties

In this symbolic notational representation, there are four angular momenta to be coupled to give the total angular momentum of the system J, i.e., $J = j_1 + j_2 + j_3 + j_4$.

In the $\displaystyle\prod_{i=1}^{4}(2j_i+1)$ dimensional space spanned by the vectors

$$|m_1 m_2 m_3 m_4> \equiv \prod_{i=1}^{4}|j_i m_i >$$ where j_i is given and m_i is variable.

The Wigner "9j"-symbols are coming out of the following two schemes that lead to two distinct systems of basis vectors for the subspace of the angular momentum:

1. $j_1 + j_2 = j_{12}$; $j_3 + j_4 = j_{34}$; $j_{12} + j_{34} = J$

2. $j_1 + j_3 = j_{13}$; $j_2 + j_4 = j_{24}$; $j_{13} + j_{24} = J$ (two possible ways of getting J)

Therefore, the vectors are

$|(j_1 j_3)J_{13}; (j_2 j_4)J_{24}; JM>$ and $|(j_1 j_2)J_{12}; (j_3 j_4)J_{34}; JM>$.

Hence the "9j" within a constant, are the coefficients of the unitary transformation that takes one basis to another basis. Therefore, by definition, "9j"-symbols are defined by the following expression:

$$\langle (j_1 j_2)J_{12},(j_3 j_4)J_{34} JM\,|\,(j_1 j_3)J_{13};(j_2 j_4)J_{24};\overline{JM}\rangle =$$

$$\delta_{J\bar{J}}\delta_{M\bar{M}}\sqrt{(2J_{12}+1)(2J_{34}+1)(2J_{24}+1)(2J_{13}+1)}\begin{Bmatrix} j_1 & j_2 & J_{12} \\ j_3 & j_4 & J_{34} \\ J_{13} & J_{24} & J \end{Bmatrix} \quad (39)$$

Also, "9j"-symbols may be defined as

$$\langle [(j_1 j_2)J_{12},j_3]J_{123};j_4;JM\,|\,[j_4 j_2]J_{423}j_1;\overline{JM}\rangle = (-1)^{J_{423}+j_1-J_{123}-j_4}$$

$$\delta_{J\bar{J}}\delta_{M\bar{M}}\sqrt{(2J_{12}+1)(2J_{123}+1)(2J_{42}+1)(2J_{423}+1)}\begin{Bmatrix} J_2 & J_{12} & j_1 \\ J_{42} & j_3 & J_{123} \\ j_4 & J_{123} & J \end{Bmatrix} \quad (40)$$

"9j" is related to the 3j as the following:

$$\begin{pmatrix} J_{13} & J_{24} & J \\ M_{13} & M_{24} & M \end{pmatrix} \begin{Bmatrix} j_1 & j_2 & J_{12} \\ j_3 & j_4 & J_{34} \\ J_{13} & J_{24} & J \end{Bmatrix} = \sum_{\substack{m_1 m_2 m_3 m_4 \\ M_{12} M_{34}}} \begin{pmatrix} j_1 & j_2 & J_{12} \\ m_1 & m_2 & M_{12} \end{pmatrix} \begin{pmatrix} j_3 & j_4 & J_{34} \\ m_3 & m_4 & M_{34} \end{pmatrix}$$

$$\begin{pmatrix} j_1 & j_3 & J_{13} \\ m_1 & m_3 & M_{13} \end{pmatrix} \begin{pmatrix} j_2 & j_4 & J_{24} \\ m_2 & m_4 & M_{24} \end{pmatrix} \begin{pmatrix} J_{12} & J_{34} & J \\ M_{12} & M_{34} & M \end{pmatrix} \qquad (41)$$

II-5—Symmetry Relations

_____ The symmetrical relation of _____ are the following:

1. If two lines or two columns are exchanged, then this has to be multiplied by $(-)^R$, where $R = \sum_{i=1}^{g} J_i$

2. It is invariant under a reflection through one of the diagonals:

The Orthonormality Relation

$$\sum_{J_{13} J_{24}} (2J_{13}+1)(2J_{24}+1) \begin{Bmatrix} j_1 & j_2 & J_{12} \\ j_3 & j_4 & J_{34} \\ J_{13} & J_{24} & J \end{Bmatrix} \begin{Bmatrix} j_1 & j_2 & \bar{J}_{12} \\ j_3 & j_4 & \bar{J}_{34} \\ J_{13} & J_{24} & J \end{Bmatrix} = \frac{\delta_{J_{12} \bar{J}_{12}} \delta_{J_{34} \bar{J}_{34}}}{(2J_{12}+1)(2J_{34}+1)}$$

$$= \frac{1}{(2J_{12}+1)(2J_{34}+1)} \delta_{J_{12} \bar{J}_{12}} \delta_{J_{34} \bar{J}_{34}} \qquad (42)$$

"9j" Relation to "6j"

$$\begin{Bmatrix} j_1 & j_2 & j_{12} \\ j_3 & j_4 & j_{34} \\ j_{13} & j_{24} & J \end{Bmatrix} = \sum_g (-)^{2g} (2g+1) \begin{pmatrix} j_1 & j_2 & j_{12} \\ J_{34} & J & g \end{pmatrix} \begin{pmatrix} j_3 & j_4 & j_{34} \\ j_2 & g & J_{24} \end{pmatrix} \begin{pmatrix} J_{13} & J_{24} & J \\ g & j_1 & j_3 \end{pmatrix}$$

$$(43)$$

If one of j is null, then,

$$\begin{Bmatrix} j_1 & j_2 & f \\ j_3 & j_4 & \overline{f} \\ g & \overline{g} & 0 \end{Bmatrix} = \delta_{f\overline{f}}\delta_{g\overline{g}} \frac{(-)^{j_2+j_3+f+g}}{\sqrt{(2f+1)(2g+1)}} \begin{pmatrix} j_1 & j_2 & f \\ j_4 & j_2 & g \end{pmatrix}$$

$$(44)$$

II-6—Irreducible Tensor: Definition and Properties

1. Tensor operator ≡ set of operator that linearly transforms one into another under rotation.

2. Irreducible tensor operator

    The (2k + 1) operator $T_q{}^{(k)}$ (q = −k . . . k) are, by definition, the standard components of irreducible tensor operator of rank (k), $T^{(k)}$, if they transform under rotation according to

$$RT_q^{(k)}R^{-1} = \sum T_{q'}^{(k)} R_{q'q}{}^{(k)}$$

$$(45)$$

3. Vector operator

    A vector operator (V) is an irreducible tensor operator of order k = 1.

    (Note: these physical dynamic variables are so important in theoretical studies of physics in general and in nuclear physics specially, as it will be shown in the next sections.)

    Now let us clarify these beyond merely a definition. Hence, consider a vector (V), if V_x, V_y, and V_z are its components along (oxyz) (see figure 12), so its standard components are

$$V_+ = -\frac{1}{2}\sqrt{2}(V_x + iV_y)$$

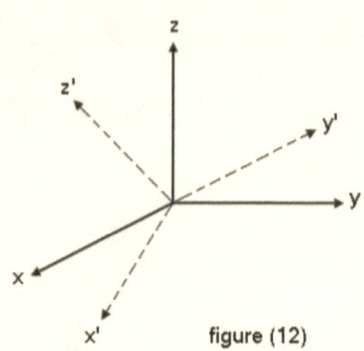

figure (12)

$$V_0 = V_z$$

$$V_- = \frac{1}{2}\sqrt{2}(V_x - iV_y)$$

But for the ordinary vector, the Cartesian coordinates are

$$\vec{r}(x, y, z); P(x, y, z)$$

$$\overset{=}{r}(x, y, z); P(x, y, z)$$

$$\overline{X}_i = \sum_k a_k X_k$$

and

$$\| a_{ik} \| = 1$$

To keep the length of the vector unchanged, then AA⁺ = 1, so

$$\sum_k a_{ik} a_{lk} = \delta_{il} = \begin{cases} 1 & if \quad i = l \\ 0 & if \quad i \neq l \end{cases}$$

The definition of a vector in the Cartesian coordinates is

$$\vec{V} = \vec{V}(v_1, v_2, v_3) \equiv \vec{V}(v_x, v_y, v_z)$$

in (xyz) system.

And

$$\overrightarrow{\overrightarrow{V}}_i = \sum_k a_{ik} V_k$$

in (x′ y′ z′) system; it is a tensor of rank 1.

If Λ is the tensor rank, then $\Lambda = 1$ for a vector in Cartesian coordinates. So a Cartesian tensor of rank Λ is a three-component number. For $\Lambda = 0$, one component (scalar) for $\Lambda = 1$, three components (vector); for $\Lambda = 2$, nine components (tensor), and so on.

4. *Scalar*—it is quantity that does not change direction under transformation. So it is a tensor of rank zero. Or it is an irreducible tensor operator of order zero.

Now after the necessary clarification of these terminologies, which will be of important use in this section and the following sections in the coming chapters, it is important to go through their characteristics and their relations to angular momentum and other physical operators.

II-6-1—Commutation Relations with J (TAM)

$$\left[J_{\pm}, T_q^{(k)}\right] = \sqrt{k(k+1) - q(q \pm 1)} T_{q\pm1}^{(k)} \tag{46}$$

$$\left[J_z, T_q^{(k)}\right] = q T_q^{(k)} \tag{47}$$

II-6-2—Hermitian Conjugate of $T_q^{(k)}$

$$S^{(k)} = T^{(k)^+} \quad f \quad S_q^{(k)} = (-)^q T_{-q}^{(k)^+} \tag{48}$$

According to this, $Y_l^{(m)}$ of a given order (l) forms a Hermitian tensor operator. Also, as it is mentioned and operated in many quantum mechanical books, there is a fundamental property concerning the calculation of an average value of the tensor operator $T_q^{(k)}$, which is by Wigner-Eckart theorem:

$$\langle TJM \mid T_q^{(k)} \mid \overline{TJM} \rangle = \frac{(-)^{2k}}{(2J+1)^{1/2}} \langle TJ \parallel T^{(k)} \parallel \overline{TJ} \rangle \langle \overline{Jk} \overline{M}_q \mid JM \rangle \tag{50}$$

Notice here the CGC and the reduced matrix ($\parallel \ldots \parallel$), which is M independent; therefore, it is reduced by neglecting the dependency on M.

By remembering the relation between CGC and "3j," one finds

$$\langle TJM \mid T_q^{(k)} \mid \overline{TJM} \rangle$$

$$\langle TJM \mid T_q^{(k)} \mid \overline{TJM} \rangle = (-)^{J-M} \langle TJ \parallel T^{(k)} \parallel \overline{TJ} \rangle \begin{pmatrix} J & k & \overline{J} \\ -M & q & \overline{M} \end{pmatrix} \tag{51}$$

where $\langle TJ \parallel T^{(k)} \parallel \overline{TJ} \rangle \equiv \text{RME} \equiv$ reduced matrix elements.

And the conjugation relation (k-integer) is

$$\langle TJ \parallel T^{(k)} \parallel \overline{TJ} \rangle = (-)^{J-\overline{J}} \langle TJ \parallel T^{(k)^+} \parallel \overline{TJ} \rangle$$

$$\tag{52}$$

II-6-4—Special Tensor Operators: Identity Operator

1. $\quad \langle \alpha J \parallel 1 \parallel \overline{\alpha J} \rangle = \delta_{\alpha \overline{\alpha}} \delta_{J \overline{J}} \sqrt{2J+1} \tag{53}$

2. Total angular momentum

$$\langle \alpha J \parallel 1 \parallel \overline{\alpha J} \rangle = \delta_{\alpha \overline{\alpha}} \delta_{J \overline{J}} \sqrt{J(J+1)(2J+1)} \tag{54}$$

II-6-4—Tensor of Rank (Λ = 2)

This tensor has 3^2 = nine components:

$$T_{ij} = \sum_{l,k} a_l \, a_k \, T_k$$

Now the linear combination of $T^{(\Lambda)}$ may be clarified as follows:

Take two vectors: $\vec{U} = (u_1, u_2, u_3) \quad and \quad V = (v_1, v_2, v_3)$

Then,

$T_{ij} = U_i \, V_j$, where $U_i = \sum a_{ik} u_k \quad and \quad V_j = \sum a_{jl} v_l$

$$U_i V_j = \sum a_{ik} a_{il} u_k v_l = \overline{T_{ij}} \Rightarrow T_{ij} = U_i V_j$$

Let S_{ij} be a symmetrical, which is written as

$$S_{ij} = U_i \, V_j + U_j \, V_i \tag{55}$$

and

$$A_{ij} = U_i \, V_j - U_j \, V_i \text{ as antisymmetric} \tag{56}$$

where (by symmetry) $S_{ij} = S_{ji}$ and (by antisymmetric) $A_{ij} = -A_{ji}$ (57)

So for $A_{ij} = 0$, it means two identical particles cannot occupy the same state according to Pauli exclusion principle. These splitting up are independent of the reference system. Hence, in general, a tensor T_{ik} can be written as

$T_{ik} = S_{ik} + A_{ik} \equiv$ (symmetrical tensor + antisymmetric tensor, as nature wants)

Therefore,

$$S_{ki} = S_{ik} = 1/2[T_{ik} + T_{ki}]$$
$$A_{ki} = -A_{ik} = 1/2[T_{ik} - T_{ki}] \tag{58}$$

II-6-5—Reduction of Tensor

a. Take $A_{ki} = -A_{ik}$ (axial vector, antisymmetric)

Since A_{ik} is an irreducible tensor, then,

$$T' = \sum_i \overline{S}_{ii} = \sum_{kl} a_{ik} a_{il} S_l = \sum_{k,l} \delta_{kl} S_{kl} = \sum S_{ll} = T$$

where T is a scalar tensor, i.e., $\Lambda = 0$.

Now define τ_{ik} as $\tau_{ik} = 1/3\delta_{ik}T$; hence,

$S_{ki} = S_{ik} = 1/2[T_{ik} + T_{ki}]$ is a symmetrical tensor with $T = 0$.

But in general, S_{ik} is written as

$S_{ki} = S_{ik} = 1/2[T_{ik} + T_{ki}] - \tau_{ik}$

This implies that a T_{ik} can be written as

$$T_{ik} = S_{ik} + A_{ik} + \tau_{ik} \tag{59}$$

where S_{ik} is the symmetrical part ($T = 0$).

A_{ik} is axial vector (antisymmetric)

τ_{ik} is a scalar $= 1/2\ \delta_{ik}T$.

Take as an example a special tensor of rank $\Lambda = 1$, whose components are $X^\alpha Y^\beta Z^\gamma$, where $\alpha + \beta + \gamma = l$ and $\alpha, \beta, \gamma > 0$, so the tensor rank is the sum of α, β and therefore, the number of independent components N is given by

$N = 1/2(l+1)(l+2)$

Let us construct the tensor $X^\alpha Y^\beta Z^\gamma$; take the spherical harmonics

$$r^l Y_{lm}(\theta,\varphi) = \sum a_{\alpha\beta\gamma} X^\alpha Y^\beta Z^\gamma \tag{60}$$

(As an exercise, show that $(x \pm y) = r\sin\theta^{\pm i\varphi}$)

II-6-6—Rotational Problem

Take

$$r^{\ell} Y_{\ell m}(\theta', \phi') = \sum Y_{\ell m'}(\theta, \phi) D^{\ell}_{mm'}(\omega)$$

This means that $Y_{\ell m}(\theta^-, \varphi^-)$ transform among themselves, so it is a irreducible tensor such that

$$Y^{\ell-2}_{\ell-2,m'}, r^{\ell-4} Y_{\ell-4,m'}, \ldots\ldots Y_0$$

where there are $(2l + 1)$ components that represent the possible values taken by m.

So for $l = 0$, there is only one component (scalar); for $l = 1$, there are three components $(1, 0, -1)$; for $l = 2$, there are five components $(2, 1, 0, -1, -2)$; and so on according to l-values.

Therefore, the dimensionality N is given by $[2l + 1] + [2(l - 2) + 1]$ $[2(l - 4) + 1] + \ldots$

If l is even, then one has $1/2[l/2 + 1][2l + 1 + 1] = 1/2(l + 1)(l + 2)$

And If l is odd, then one has $1/2[(l + 1)/2][2l + 1 + 2] = 1/2(l + 1)(l + 2)$

Here, $X^\alpha Y^\beta Z^\gamma \equiv$ Cartesian basis.

$\equiv$ spherical basis $Y_{\ell m}(\theta, \phi)$

Take as an example the rank $\Lambda = l = 1$, so there are three components for $l = 1$; they are $rY_{11}, \gamma Y_{1,-1}, Y_{10}$.

Now, $P(x, y, z)$, $\vec{r_c}$ are the position of point p in Cartesian coordinates,

and $\vec{r_s}$ $P(T_{11}, T_{1-1}, T_{10})$ are the position of the point P in spherical coordinates.

Let us deal with spherical coordinates of point P. Let us define

$$T_{+1}^{(1)} = \frac{-1}{\sqrt{2}}(x+iy); \quad T_0^{(1)} = \overline{z}; \quad T_{-1}^{(1)} = +(x-iy) \tag{61}$$

Remember that $T_m{}^1$ stands for these in (61), for m values 1, 0, −1, for l = 1. Also, one is familiar with

$$rY_{1,-1}(\theta,\varphi) = \frac{1}{\sqrt{2}}\sqrt{\frac{3}{2\pi}}.r\sin\theta\, e^{\pm i\varphi}$$

$$rY_{1,\pm1}(\theta,\varphi) = \pm\sqrt{\frac{3}{4\pi}}.(r\sin\theta\, e^{\pm i\varphi})$$

$$rY_{10}(\theta,\varphi) = \frac{1}{2}\sqrt{\frac{3}{\pi}}.r\cos\theta\, e^{\pm i\varphi} \quad = \frac{1}{2}\sqrt{\frac{3}{\pi}}.Z$$

and

$$\vec{r}_c = \begin{pmatrix} x \\ y \\ z \end{pmatrix} \quad ; \quad \vec{r}_s = \begin{pmatrix} T_{+1}^{(1)} \\ T_0^{(1)} \\ T_{-1}^{(+1)} \end{pmatrix}$$

To go from Cartesian representation to spherical representation, a transformation has to be done as follows:

$$\vec{r}_s = U.\vec{r}_c \quad where \quad U = \frac{1}{\sqrt{2}}\begin{pmatrix} -1 & i & 0 \\ 0 & 0 & \sqrt{2} \\ 1 & i & 0 \end{pmatrix} \quad Hence;$$

$$\begin{pmatrix} T_{+1}^{(1)} \\ T_0^{(1)} \\ T_{-1}^{(1)} \end{pmatrix} = \frac{1}{\sqrt{2}}\begin{pmatrix} -1 & i & 0 \\ 0 & 0 & \sqrt{2} \\ 1 & i & 0 \end{pmatrix}\begin{pmatrix} x \\ y \\ z \end{pmatrix} \tag{62}$$

II-6-7—Spherical Bases Vectors

Let us define the bases vectors as follows:

$$\left.\begin{array}{l}\vec{\xi}_{+1} = \dfrac{-1}{\sqrt{2}}(\vec{e}_x + i\vec{e}_y) \\[3mm] \vec{\xi}_0 = \vec{e}_z \\[3mm] \vec{\xi}_{-1} = \dfrac{+1}{\sqrt{2}}(\vec{e}_x - i\vec{e}_y)\end{array}\right\} \tag{63}$$

Hence, the spherical vector $\vec{V}$ is written as

$$\vec{V} = \sum_{\mu=0,\pm1} V_\mu \vec{\xi}_{-\mu}(-1)^\mu$$

where V_μ are spherical components, where

$$\vec{\xi}^* = (-1)^\mu \vec{\xi}_{-\mu} \Rightarrow \vec{\xi}_\mu \cdot \vec{\xi}^*{}_{\overline{\mu}} = (-1)^{\overline{\mu}} \vec{\xi}_\mu \cdot \vec{\xi}_{\overline{-\mu}} = \delta_{\mu\overline{\mu}} \tag{64}$$

Therefore,

$$V_\mu = \vec{\xi}_\mu \cdot \vec{V} = \vec{V}\vec{\xi}_\mu$$

Consider a scalar product of two vectors in spherical representation. Let $\vec{A}$ and $\vec{B}$ be such vectors, then,

$$[\vec{A}.\vec{B}] = \sum_\mu (-1)^\mu A_\mu B_{-\mu} \tag{65}$$

Problem: find the spherical components of the second rank tensor, symmetrical:

$$\delta_{ik} = \frac{1}{2}(x_i x_k + x_k x_i) - \frac{1}{3}\tau_k$$

Hint: use $Y_{lm}(\theta, \varphi)$.

II-8—Theory Tackling in a General Way

It is known that irreducible tensor of rank Λ or spherical tensor of rank Λ are, in fact, a set of $(2\lambda + 1)$ components $T_\mu{}^{(\lambda)}$, $\mu = -\lambda, \ldots, 0, \ldots, + \lambda$, λ are stands of where λ is a positive integer, as it should be.

Let $T_\mu{}^{(\lambda)}$ be transformed to $T_{-\mu}^{(\lambda)'}(r')$, i.e.,

$$T_{-\mu}^{(\lambda)'}(r') = \sum T_\mu^{(\lambda)} D_{\mu\mu'}^{(\lambda)}(\omega)$$

(66)

Under rotation of coordinates, one gets

$$T_{-\mu}^{(\lambda)'}(r') = T_{\mu'}^{\lambda'}(R.\vec{r})$$

(67)

II-7—Scalar Product of Two Tensors

Let $T^{(\lambda)}$ and $V^{(\lambda)}$ be two tensors, then,

$$(T^{(\lambda)} \cdot V^{(\lambda)}) = \sum_\mu (-1)^\mu T_\mu^{(\lambda)} V_{-\mu}^{(\lambda)}$$

(68)

To verify this, take

$$(T.V)' = (T'.V') = \sum_{\mu'} (-1)^{\mu'} T_{\mu'}'.V_{\mu'}' \qquad by \quad (66)$$

$$= \sum_{\mu'} (-1)^{\mu'} \sum_\mu T_\mu D_{\mu\mu'}^{(\lambda'')}(\omega).\sum_\upsilon V_{-\upsilon} D_{-\upsilon\mu'}^{(\lambda')}(\omega)$$

It is known that

$$D_{mm'}^{(j)}(-\alpha,-\beta,-\gamma) = (-1)^{m-m'} D_{-m,-m'}(\alpha,\beta,\gamma)$$

Then,

$$(T.V)' = (T'.V') = \sum_{\mu'} (-1)^{\mu'} (-)^{\nu-\mu} D_{\mu\mu'}^{(\lambda)}(\omega) D_{\mu\mu'}^{(\lambda)} . T_{\mu}^{(\lambda)} V_{\nu}^{(\lambda)}$$

$$= \sum (-1)^{\mu'} T_{\mu'}' V_{-\mu'}'$$

Hence,

$$\sum_{\mu'} (-1)^{\mu'} T_{\mu'}' . V_{-\mu'}' = \sum_{\mu} (-1)^{\mu} T_{\mu}^{(\lambda)} V_{-\mu}^{(\lambda)}. \qquad as \quad in\,(68)$$

So it is a scalar product.

Now it is possible to construct a tensor $T_{\mu}^{(\lambda)}$ from two given tensors $T_{\mu_2}^{\lambda_2}$ and $T_{\mu_1}^{\lambda_1}$ as follows:

$$T_{\mu}^{(\lambda)} = \sum \langle \lambda_1 \mu_1 \lambda_2 \mu_2 \mid \lambda \mu \rangle T_{\mu_1}^{\lambda_1} T_{\mu_2}^{\lambda_2} \tag{69}$$

[As an exercise, verify (69).]

Problem: Construct $T^{(2)}$ from two $T^{(1)} = \vec{r}_s$; compute $T_{\pm 2}^{(2)}$, $T_{\pm 1}^{(2)}$, $T_0^{(2)}$. Use $T_{\mu}^{(2)} = A.r^2 Y_{2,\mu}(\theta,\phi)$; compare with

$$\delta_{ik} = \frac{1}{2}(x_i x_k + x_k x_i) - \frac{1}{3} \tau_{ik}$$

II-7-1—Wigner-Eckart Theorem

This theorem is quite important in computing the matrix elements of irreducible tensors in angular momentum representation. Suppose there is an initial state |njm> operator on it by a tensor operator $T_{\mu}^{(\lambda)}$, so a transition occurs to a final state |n´ j´ m´>, then the matrix elements of this type of interaction can be calculated by using this theorem, such as

$$\langle n'j'm' \mid T_\mu^{(\lambda)} \mid njm \rangle = \alpha \begin{pmatrix} j & \lambda & j' \\ m & \mu & m' \end{pmatrix} \langle n'j' \mid T^{(\lambda)} \mid nj \rangle \tag{70}$$

This needs to be verified in detail because it is a powerful tool in quantum theory computations.

Take $T_\mu^{(\lambda)}$ as an irreducible tensor operator operating between the states |njm> and |n′ j′ m′>, and the computing of its matrix elements is required; that is, <n′ j′ m′| $T_\mu^{(\lambda)}$| njm>.

So operate on |njm> by $T_\mu^{(\lambda)}$, i.e., $T_\mu^{(\lambda)}$ |njm>, then rotate it as

$$\left(T_\mu^{(\lambda)} \mid njm \rangle \right) = \sum_{\mu m} T_\mu^{(\lambda)} D_{\mu\mu'}^{(\lambda)} \mid njm \rangle D_{mm'}^{(j)}$$

$$= \sum_{\mu m} D_{\mu\mu'}^{(\lambda)} D_{mm'}^{(j)} T_\mu^{(\lambda)} \mid njm \rangle$$

Using that
$$D_{\substack{\mu\mu' \\ m_1 m_2}}^{(j_1)} D_{\substack{mm' \\ m_2 M_2}}^{(j_2)} = \sum_j \langle j_1 m_1 j_2 m_2 \mid jm \rangle \langle j_1 M_1 j_2 M_2 \rangle$$

Then,

$$\sum_{\mu m} D_{\mu\mu'}^{(\lambda)} D_{mm'}^{(\lambda)} T_\mu^{(\lambda)} \mid njm \rangle = \sum_{\mu m, j^* m^*} \langle \lambda\mu jm \mid j^* m^* \rangle \langle \lambda\mu' jm' \mid j^* m^* \rangle \times D_{m^* m'^*}^{j^*} T_\mu^{(\lambda)} \mid njm \rangle$$

Now introduce

$$\mid nj^* m^* \rangle = \sum_{\mu m, j^* m^*} \langle \lambda\mu jm \mid j^* m^* \rangle T_\mu^{(\lambda)} \mid njm \rangle$$

Also,

$$(T_\mu^{(\lambda)} \mid njm \rangle)^* = \sum_{j^* m^*} \langle \lambda\mu' jm' \mid j^* m^* \rangle . \mid njm \rangle D_{m^* m'^*}^{j^*}$$

Here, we notice that when applying $T_\mu^{(\lambda)}$ to |njm> j→j*, j→m*, then rotating back, one finds

$$T_\mu^{(\lambda)} \,|\, njm\rangle = \sum_{j^*} \langle \lambda\mu\, jm \,|\, j^*m^*\rangle . \,|\, nj^*m^*\rangle$$

Multiply from left by <n′ j′m′|; you get

$$\langle n'j'm' \,|\, T_\mu^{(\lambda)} \,|\, njm\rangle = \sum_{j^*} \langle \lambda\mu\, jm \,|\, j^*m^*\rangle \langle n'j'm' \,|\, nj^*m^*\rangle \qquad (71)$$

Therefore, one notices that

(1) $\langle n'j'm' \,|\, T_\mu^{(\lambda)} \,|\, njm\rangle$ = (CGC) (scalar product).

(2) Scalar product $\langle n'j'm' \,|\, n^*j^*m^*\rangle$ is independent of m′ (scalar).

To show (2)—since<n′ j′ m′| forms a complete set, therefore, it can be expanded as

$$|\, n'j'm'\rangle = \sum_n |\, nJ'm'\rangle a_{nn'}(J'm') \qquad (a)$$

And since |n′ J′ m′> is orthonormal, then by multiplying (a) from left by |n′ J′ m′>, one gets the coefficient a$_{nn'}$ as

$$a_{nn'} = \langle nJ'm' \,|\, n'J'm'\rangle \qquad (b)$$

Note: if m′<j J$_+$ is applied, and if m′ = j′, J$_-$ is applied, so

$$\hat{J}_+ \,|\, n'j'm'\rangle = \sqrt{(j'-m')(j'+m'+1)} \,|\, n'j'm'+1\rangle$$

Then,

$$|\, n'j'm'+1\rangle = \frac{1}{\sqrt{(j'-m')(j'+m'+1)}} \hat{J}_+ \,|\, n'j'm'\rangle$$

From (a)

$$|\, n'j'm'\rangle = \sum a_{nn'}(j'm') \,|\, nj'm'\rangle$$

Therefore,

$$| n'j'm' + 1\rangle = \frac{1}{\sqrt{k}} . \hat{J}_+ \sum | nj'm' > a_{nn'}(j'm')$$

$$= \frac{1}{\sqrt{k}} . \sum J_+ | nj'm' > a_{nn'}(j'm')$$

$$= \frac{1}{\sqrt{k}} . \sum_n \sqrt{(j'-m')(j'+m'+1)} . | n'j'm'+1\rangle a_{nn'}(j'm'+1)$$

$$= \frac{1}{\sqrt{(j'-m')(j'+m'+1)}} . \sqrt{(j'-m')(j'+m'+1)} . \sum_n nj'm'+1\rangle a_{nn'}(j'm'+1)$$

$$= \sum_n nj'm'+1\rangle a_{nn'}(j'm'+1)$$

$$| n'j'm' + 1\rangle = \sum_n a_{nn'}(j'm'+1)nj'm'+1\rangle$$

(c)

Compare (a) with (c); one gets

$$a_{nn'}(j'm') = a_{nn'}(j',m'+1)$$

which indicates that $a_{nn'}$ coefficients are independent of m'. So using the relation between CGC and 3j-symbols, one can write

$$\langle n'j'm' | T_\mu^{(\lambda)} | njm\rangle = (-1)^{j'-m'} \begin{pmatrix} j' & \lambda & j \\ m' & \mu & m \end{pmatrix} \langle n'j' \| T^{(\lambda)} \| nj\rangle$$

(72)

Equation (72) represents Wigner-Eckart theorem, which gives the matrix elements of the operator $T_\mu^{(\lambda)}$. The $\| \dots \|$ notation means the independency on m is removed. So the reduced matrix elements ($\| T^\lambda \|$) are independent of (m).

II-7-2—Some General Remarks on D-Matrices (Summary)

a. Unitarily

1. $D^*(\alpha,\beta,\gamma) = D^{-1} = D(-\gamma,-\beta,-\alpha) \Rightarrow DD^* = 1$

2. $\left[D_{mm'}^{(j)}(\alpha,\beta,\gamma) \right]^* = D_{mm'}^{(j)}(-\gamma,-\beta,-\alpha)$ ..... *conjugation*

3. $\sum_\mu D_{m\mu}^{(j)}(R) D_{m'\mu}^{(j)}(R) = \delta_{mm'}$ *orthogonality*

4. $\sum_\mu D_{\mu m}^{(j)}(R) D_{\mu m'}^{(j)}(R) = \delta_{mm'}$ *orthogonality*

b. Group Property

1. $D(R_2)D(R_1) = D(R_2 R_1)$ *or* $D^{(j)}(\omega) = D^{(j_1)}(\omega_1) D^{(j_2)}(\omega_2)$

2. $\sum_{m'} D_{mm'}^{(j)}(R_2) D_{m'm''}(R_1) = D_{mm''}(R_2 R_1)$

3. $D(R)D(R^{-1}) = D(R)D^{-1}(R) = D(R)D^*(R) = 1$ *Unitarily   again*

4. Orthogonality is

$$\int_0^{2\pi} d\alpha \int_0^{2\pi} d\gamma \int_0^\pi \sin\beta d\beta \left[D_{mm'}^{(j)}(\alpha,\beta,\gamma) \right]^* D_{\mu\mu'}^{(j')}(\alpha,\beta,\gamma) = \frac{8\pi^2}{2j+1}.\delta_{m\mu}\delta_{m'\mu'}\delta_{jj'}$$

II-7-3—Explicit Evaluation of $D_{mm'}^{(j)}(\alpha, \beta, \gamma)$

From quantum mechanics, it is found that

$$D_{mm'}^{(j)}(\alpha,\beta,\gamma) = \exp(-im\,\alpha) d_{mm'}^{(j)}(\beta) \exp(-im\,\gamma)$$

where $d_{mm'}^{(j)}$ is Wigner small D-matrix, which is given by

$$d_{mm'}^{(j)}(\beta) = (-)^{m-m'} \left[\frac{(j+m)(j-m)}{(j+m')(j-m')} \right] (\cos\frac{1}{2}\beta)^{m+m'} (\sin\frac{1}{2}\beta)^{m-m'} \times P_n^{m-m',m+m'}(\cos\beta)$$

where

$$P_n^{(\mu,\upsilon)}(\cos\beta) = \sum_\alpha \binom{\mu+n}{\alpha}\binom{\upsilon+n}{n-\alpha}(-)^{n-\alpha}(\sin^2\frac{1}{2}\beta)^{n-\alpha}(\cos^2\frac{1}{2}\beta)^{j-m}$$

$$= \binom{n+m}{n}\Gamma(-n, n+\mu+\upsilon, \mu+1; (\sin\frac{1}{2}\beta)^2)$$

Hence,

$$P_n^{(\mu,\upsilon)}(\cos\beta) = \binom{n+m}{n}(\cos\frac{1}{2}\beta)\Gamma(-n, -n-\upsilon; \mu+1; -\tan^2\frac{1}{2}\beta)$$

where Γ is a hypergeometric function that is given by

$$\Gamma(a,b,c,x) = 1 + \frac{ab}{c}x + \frac{a(a+1)b(b+1)x^2}{c(c+1)2} \tag{73}$$

There are special tables for Γ function one can use.

Hence, by calculating $d_{mm'}^{(j)}$ values, it is easy to find $D_{mm'}^{(j)}(\alpha,\beta,\gamma)$.

II-7-4—Symmetry Properties of $d_{mm'}^{(j)}$ and $D_{mm'}^{(j)}$

a. For $d_{mm'}^{(j)}$

1. $\quad d_{mm'}^{(j)}(\beta) = d_{-m'-m}^{(j)}(\beta)$

2. $\quad d_{mm'}^{(j)}(\beta) = d_{m'm}^{(j)}(\beta) = (-1)^{m-m'} d_{m'm}^{(j)}(\beta)$

3. $\quad d_{m'm}^{(j)}(\pi-\beta) = (-1)^{j-m'} d_{mm'}^{(j)}(\beta)$

(As an exercise, verify these properties for $d^{(j)}_{mm'}$)

b. For $D^{(j)}_{mm'}$

1. $D^{(j)*}_{mm'} = (-)^{m-m'} D^{(j)}_{-mm'}(R) = (-)^{m-m'} D^{(j)}_{-m-m'}(R) = D^{(j)}_{m'm}(R^{-1})$

When R = R_z is a rotation about z-axis through angle α, then

$D^{(j)}_{mm'}(R_z) = \exp(-im\,\alpha)\delta_{mm'}$

2. When R is a rotation about y-axis through angle π, then

$D^{(j)}_{mm'}(R_y) = (-)^{j-m}\delta_{m,-m'}$

3. If R is a rotation about x-axis through angle π, then

$D^{(j)}_{mm'}(R_x) = (-)^{j}\delta_{m,-m'}$

II-7-5—D-Matrix and Spherical Harmonics Functions $Y_{\ell m}(\theta,\varphi)$

Although such relations have been mentioned before, it is quite useful to summarize that here

1. $D^{l}_{m0}(\alpha,\beta,\gamma) = [\dfrac{4\pi}{2l+1}]^{1/2} Y^{*}_{lm}(\beta\alpha)$

2. $D^{l}_{0m}(\alpha,\beta,\gamma) = (-)^{m}[4\pi/2l+1]^{1/2} Y^{+}_{lm}(\beta\alpha)$

3. $D_{00}(\alpha,\beta,\gamma) = P_l(\cos\beta)$ (Legendre polynomial)

4. $Y_{lm}{}^{-}(r^{-}) \sum_m Y_{lm}(r) D^{l}_{m\,m}{}^{-}(R)$, where $Rr = r^{-}$ or $r = R^{-1}r^{-}$

The detailed properties of these functions are fairly well treated in quantum mechanics books (see I. Schiff, Meresbacker, Saxon, and others).

II-7-6—Type of Matrix Elements

There are two types: either diagonal or non-diagonal matrix elements. These are summarized below:

a. Diagonal matrix elements

i) $j^- = j$, $m^- = m$, $\delta(jj) = 1$, $\delta(mm) = 1$

ii) $\begin{bmatrix} j & \tau & j \\ -m & 0 & m \end{bmatrix} \neq 0$ only if $j^- + \tau + j = 0$, Or $|j^- - j| \leq \tau \leq j^- + j$

If $j^- = j$, then $\tau \leq 2j \rightarrow j \geq \tau/2$, $\tau > 0$

b. Non-diagonal matrix elements

Take the case of the transition of electromagnetic radiations between two nuclear states ljmn> and l j⁻ m⁻ n⁻>, as shown:

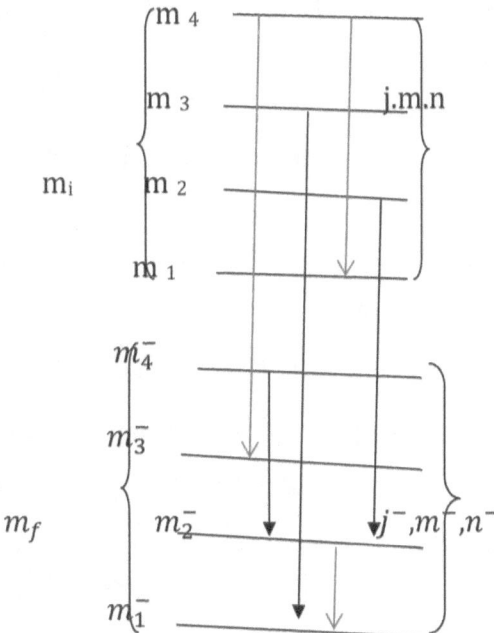

The selection rules of this transition are

$$\begin{bmatrix} j^* & \tau & j \\ m(-) & u & m \end{bmatrix} \neq 0 \quad \text{unless } lj^* - jl \leq \tau \geq j^* + j$$

The intensity of transition $I(m_i \rightarrow m_{-i})$ is proportional to 3j-symbols; can easly be calculated.

II-7-7—Reflection in Space, Parity Operator

The reflection in space is a very important physical phenomenon, where everything in nature is in motion in its domain linearly and rotationally. To describe this property of a physical system, a physical operator called parity denoted by pi (π) is proposed. A rotational motion and transformation were mentioned previously. Here, we are talking about the reflection of the motion of a system in space. This dynamic operator can be defined through describing the state of the moving system by its wave function, which is a function of special coordinates and time. Let the wave function be ψ(x, y, z) in space. Consider it as reflected through the origin of its coordinates system (x, y, z).

So this reflection gives ψ(x, y, z) $\leftrightarrow$ ψ(-x, -y, -z) $\rightarrow$

$\pi_0 \psi$(xyz) = ψ(-x, -y, -z); this happens under the effect of the reflection operator. Parity operator has the following properties:

1. $\pi^{-1} = \pi \rightarrow \pi^2 = 1$, then $\pi = \pm 1$ so $\pi = 1$ or -1

2. π is a linear operator

3. π is a hermitian operator

Let us verify these properties briefly.

Take $(\pi \psi, \phi) = \int_{-\infty}^{\infty} \psi^*(-x, -y, -z) \phi(x, y, z)$ dx dy dz

Applying the reflection, one gets

$- \int_{-\infty}^{\infty} \psi^*(x, y, z) \phi(-x, -y, -z)$ dx dy dz =

$+ \int_{-\infty}^{\infty} \psi^*(x, y, z) \pi\phi(x, y, z)$ dxdydz = (ψ, $\pi\phi$), which means that $(\pi\psi, \phi) = (\psi, \pi\varnothing)$, which implies that $\pi = \pi^*$, and π is a Hermitian operator; consequently, its eigenvalue is real.

Now to show that its eigenvalue is real +1 or −1, take π_0, $\pi_0 \psi = \pi\psi$, π here as its eigenvalue. Operate again with π_0, so $\pi_0 \pi_0 \psi = \pi_0(\pi\psi) = \pi(\pi_0 \psi) = \pi^2\psi$, since $\pi_0 \pi_0 = 1$, then $\psi = \pi^2 \psi \rightarrow$

$\pi^2 = 1 \rightarrow \pi = \pm 1$. So the eigenvalue of the operator π_0 is either $+1$ or -1. These values are either $+1$ or -1; it is even if $\pi = +1$ and odd if $\pi = -1$.

Hence, $\pi_0 \psi = +\psi_+(-x, -y, -z)$ is even. $\Pi = +1$

For $\pi = -1$, ψ is odd, i.e., $\pi\psi(x, y, z) = -\psi_-(-x, -y, -z)$.

Therefore, any eigenstate can be written as a sum of ψ_+, ψ_-, i.e.

$\psi = \psi_+ + \psi_-$

Let us now elaborate at this by considering B_0 as an operator, which is a sum of even and odd operators such as $B_0 = A_0^+ + A_0^-$, where $A_0^- = \pi_0$ $A_0 \pi_0^{-1} = \pm A$, so A is even or odd.

Now let us check the behavior of the oddness and the evenness of the operators.

1. Assume odd $A_0 = \pi_0 A_0 = -A_0$; the matrix elements of A_0 is given by

$< f^\pm| A_0 | i^\pm> = (< f^\pm| \pi_0 A_0 | i^\pm>) = -(< f^\pm| \pi A \pi| i^\pm>)$

Remember that $\pi_0 A = A^-$, $\pi_0 |i> = \pi|i>$

π is Hermitian.

So this $= 0$, $\rightarrow - (< f^\pm | A_0 | i^\pm>) = 0$; this implies the matrix elements of the odd operator vanish.

2. If A^+ is even by the same way $<f^\pm |A| i^\pm> = 0$?

Therefore, in general, $<f\pi_f |A^\pi| i\pi_i> \neq o$ only if π_i π $\pi_{f=} +1$, where i denotes the initial state and f is the final state, so π here is the parity carried by the transition radiation, as shown in the following figure:

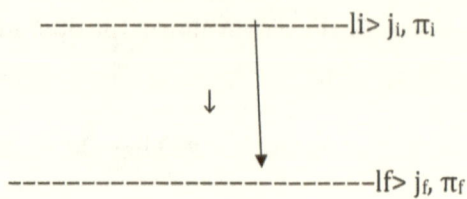

So one can construct the matrix element of the transition from li> to lf> as

$$[< n^- j^- m^- \pi^- l T \mu^{\pi\pi} l n j m \pi >] = 1 + \pi \pi\pi/2(-)^{j-m-} \left(\left\{ \begin{matrix} j & \tau & j \\ -m(-) & \mu & m \end{matrix} \right\} \right)$$

$< n- j^- l T^\tau l n j > \text{---------------}(74)$

Note: If $\pi_i \pi \pi_f = \pi^- \pi_o \pi = 1$, then either $\pi^- = 1, \pi_o = 1, \pi -1$

or $\pi^- = 1, \pi_o = -1, \pi = 1$, then equation (70) goes to zero, which indicates no transition. Therefore, in order for the transition to take place, $\pi_i \pi \pi_f$ must take the value (+1).

Chapter Three

Electromagnetic Properties of Nuclei

III-1—Introduction

As it is clearly known, a nucleus is a buildup of a number of nucleons (2 to more than 250). The nucleon is either a proton or neutron, where the mass of the neutron is bigger than that of the proton, as shown in chapter 1. The proton is a charged particle; its charge is positive equals to (ze), where (z) is the atomic number (number of the protons = the number of the electrons for the nucleus), and (e) is the charge of the electron. Also, orbiting any nucleus electrons, remembering that in any domain of charge, there is an electromagnetic field, based on those properties to that domain. Hence, the nucleus contains positively charged particles (protons) and is orbited by negatively charged particles (electrons), but the neutrons are neutral particles (no charge). Hence, dealing with a nucleus physically, one has to consider all these constituents and the fact that the nucleus is a charged identity. So in studying such system, the following has to be taken care of:

1. The electromagnetic interactions of nuclei

2. The electromagnetic moment of nuclear states

The electromagnetic interaction usually excites the nucleus to higher states of energy, wherein most cases, the nucleus will be in unstable states; to go back to its ground or stable state, the gained energy, due to the interactions, has to be given as electromagnetic radiations. Those are given either through many states or directly to the stable state (see figure 13 below).

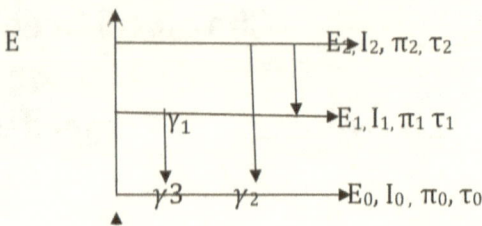

E

E_2, I_2, π_2, τ_2

$E_1, I_1, \pi_1 \tau_1$

γ_1

$\gamma_3 \quad \gamma_2$ E_0, I_0, π_0, τ_0

Figure 13: nuclear states

Each state is defined by many parameters summarized by the energy, angular momentum, parity, and the lifetime. All these are physical dynamic variables treated quantum *mechanically; they play a big role in the building up of the eigenstate of the state that describes the state itself. Each transition between any two states usually carries angular momentum and parity. There are usually selection rules that organize the way of transition, the type of the transition, and whether it is allowed or not.

The nucleus is an isolated system if it is not subjected to an external force; therefore, its Hamiltonian (H) (KE + PE) is invariant under rotation with a well-defined angular momentum. Let its angular momentum be I instead of j, and since (H) is invariant under reflection, then $\pi_0 H = H$. In figure 13, in the transition denoted by (γ), its energy is $E_i = h\nu_i = E_i - E_f$, where h is Planck constant in joule per second; ν_i is the frequency of the radiation (electromagnetic wave), and i and f are the initial and the final states, respectively. So E_i of $\gamma = E_i - E_f$, i = 1, 2, 3—n, f = 2, 3, 4—n + 1.

It is important to realize that the nuclear interactions provide us with very important information about the structures of the nuclei and the properties of the force that combine their constituents (nucleons). This type of interaction (E&M interaction) is demonstrated quite well by the electromagnetism theory, especially Maxwell famous equations, which have united the electric and the magnetic forces in 1865. To proceed in discussing the electromagnetic properties of the nuclei, we start with a brief descriptive explanation of the nucleus, although it was clarified in chapter 1, because it is mostly related to the object of this chapter. The nucleus, as it was discovered in 1911, is a finite extended size; its positive charge is almost uniformly distributed, as shown in chapter 1. Its charge is given by the charge sum of its protons' charge, i.e., q $\equiv$ Ze, where Z is the number of the protons and e is the charge unit (electron charge). This charge provides the nucleus with scalar potential $\emptyset(r)$ = Ze/r, where (q) here is considered as a point charge. The charge density distribution is given by $\rho(\vec{r})$, and the current (electrons in motion in unit time) distribution is given by j($\vec{r_n}$), where r_n is the radius of the nucleus. For the nuclei, there is a basic radius measured to be $r_0 = \cong (1.2) \times 10^{-13}$ cm = 1.2 fm (1 fermi = 10^{-13} cm). The finiteness of the nucleus is defined by this basic radius. But the radius (R) of any nucleus is depending on the mass number denoted by (A), where $R = r_0 A^{1/3}$. Hence, as (A) increases, (R) increases. Also, the radius R is only defined operationally, depending on the measuring experiment used, i.e., the type of the interaction performed. Hence, electromagnetic radius (charge radius) can be measured using electron scattering experiment. The $r_{0=}$ 1.2 fm value was found using such method.

The charge distribution (ρ (r)) is shown in figure 1, chapter 1. The basic radius r_0 was measured for different nuclei according to their weight, for light nuclei was measured to be $\cong$ (1.08–1.2) fm, and (1.2–1.3) fm for the heavy nuclei. It was found that the electron scattering experiment is the most reliable, due to its nature as a negative charge, opposite to the nucleus charge sign.

III-2—Shape of the Nucleus

It was mentioned in chapter 1 that the charge distribution shape of the nucleus takes a spherical shape, or elliptical, prolate, or segar, according to the value of its quadruple moment (Q), as follows:

1. If Q = 0 → spherical shape, i.e., a = b major axis = minor axis

2. If Q > 0 → segar shape (prolate), a > b

3. If Q < 0 → oblate shape a < b, as shown in chapter 1

The current distribution j(r) leads to the magnetic momentum (μ) of the nucleus. Before going in deep details into the electromagnetic properties of the nuclei, a brief discussion might be given about the types of the transition of radiation, emitted or absorbed by excited states of a nucleus, directly or in cascade. This can be shown graphically.

a. Emission to ground state (G_s) or between excited states (figure a)

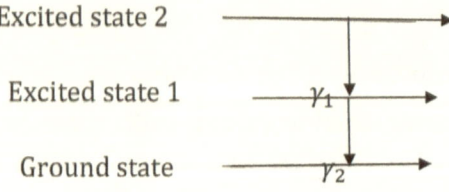

Figure a

b. Absorption: resonance florescence (figure b)

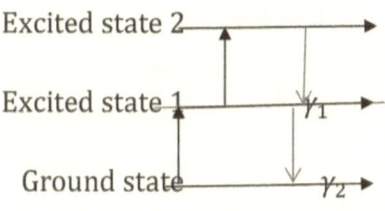

Figure b

c. Recoil, Mossbauer effect

Figure c: recoil nucleus

It is important for light nuclei, because the recoiling energy is quite reasonable quantity that has to be considered in calculations. Or in this case, it is better to get rid of the recoil energy by enclosing the radioactive sample by a crystal cavity frozen by liquid nitrogen, which makes the sample quite heavy, so the recoiling energy can be neglected. This is known as the Mossbauer effect.

d. Radiation in cascade (this is shown in figure d)

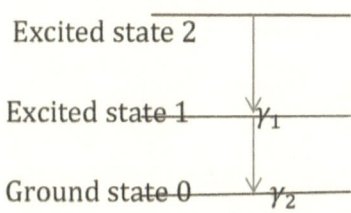

Figure d: cascade transition

If a nucleus is subjected to an interaction, it will gain energy, so it will be excited to some higher energy states,; these are called excited states. In order to go back to its ground states or lower ones, it has to get rid of the excess energy. This energy will be emitted as radiation with a certain energy, angular momentum, and parity, as mentioned in chapter 2; see figure d. (The transition might go through successive excited states or directly to the stable state (G_s). This transition of radiation may be emitted in cascade, as in figure d. This type of transition is so important to be studied by the correlation method to get important information about the nucleus. This will be treated extensively later on. The important thing here is to find the correlation function denoted as $W(\theta) = \sum A\, P_l(\theta)$, where A is a coefficient that contains important information about the nucleus and the nature of radiation's transition; P_l is Legendre polynomial. The schematic diagram of the experiment to measure the correlation function is shown in figure 14 .

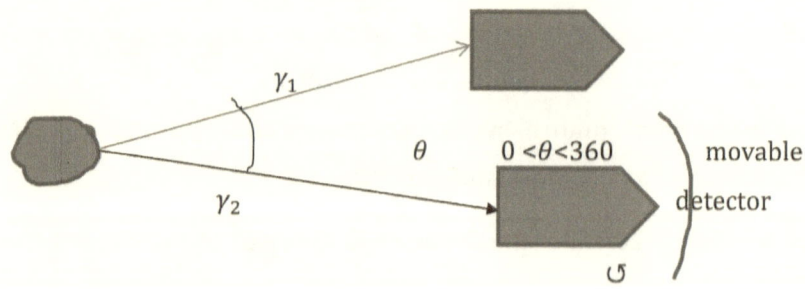

Figure 14: angular correlation setup

e. Coulomb excitation

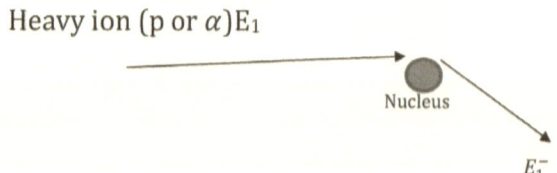

Heavy ions such as a proton or alpha particle might be scattered by a nucleus. The scattering might be nonelastic, a case in which the nucleus gains energy; as a result, the nucleus will be lifted to an excited state. The scattering cross section, which represents the probability that the interaction takes place, is easily calculated, where very important information will be obtained.

f. Electron scattering

This type of scattering is quite important in studying the charge distribution in the nuclei. In this scattering also, the form factor measuring to the nuclei is of great importance, which is related to the charge distribution density ($\rho^{\rightarrow}\left(r^{\rightarrow}\right)$) in the nucleus, and depends on the momentum transfer of the incident electrons. The figure below shows the shape of the scattering.

Beam of electrons

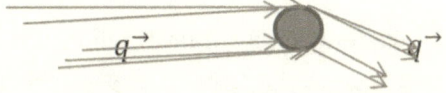

These kinds of interactions usually represent the interaction between the nucleus and the electromagnetic field; remember that each charge is surrounded by such field. To elaborate more about the subject, consider a well-defined state for the nucleus such as the ket l i>, and its Hamiltonian operator is H_0, which implies that H_0 li> = E_i li>, its eigenvalue equation.

Since H_0 is invariant under rotation and reflection, which implies that the state has definite H, I_i, m_i, and parity π_i. Also, the interaction with electric or magnetic field causes splitting to the energy level for the nucleus. See the simple diagram below that demonstrates that schematically:

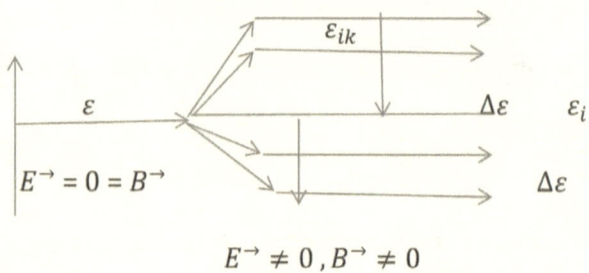

$$E^{\rightarrow} \neq 0, B^{\rightarrow} \neq 0$$

This case is fairly well treated by perturbation method, where H_t = H_o (unperturbed) + H^- (perturbing term) = total Hamiltonian after perturbation due to external force such as electric field ($E^{\rightarrow}$) or magnetic field ($B^{\rightarrow}$).

This problem is of two cases:

1. The case of nondegenerate state or level, where ΔE = <il H^- li>

2. The case of degenerate state or level, which is the state given by li_k>, where ε_I is independent of k, k = 1, 2, 3,— β, which means there is a β-fold degeneracy; li_k> is unperturbed state.

Hence, H_o li_k> = ε_I li_k>. To solve such problem, the matrix elements of the perturbation term (H^-) has to be calculated; that is to say, the H^-_{KK} has to be found, where H^-_{kk} = <i_k l H^- li$_k$>, where

$$H^{\wedge -} = \begin{bmatrix} 1 - - - & K - - - & \beta \\ - - - & H^-_{KK^-} - - - - & - \\ - - - - - & - - - - - - - & - \end{bmatrix}$$

These, in general, are not diagonal, can be diagonalized using the unitary operator U, such as U $H^- U^{-1}$ = ε^- or

$$\varepsilon^- = \begin{bmatrix} \varepsilon_1^- & 0 & 0 \\ 0 & \varepsilon_n^- & 0 \\ 0 & 0 & \varepsilon_\beta^- \end{bmatrix} \Rightarrow \varepsilon_{ik} \rightarrow \varepsilon_i + \varepsilon_k^-$$

Therefore,

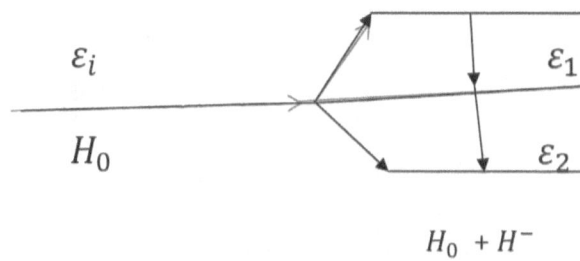

$$H_0 + H^-$$

In the nucleus case, $\beta = 2I + 1$; I is the angular momentum of the nucleus, and $k \rightarrow m$, for static interaction $H_{m^- m}^- = <n^- j^- m^- l \, H^- l \, n \, j \, m>$.

This method dealing with solving the problem due to perturbation is so important. For the convenience of the readers, the summary of the steps dealing with such problems is demonstrated in the following:

1. The Hamiltonian of interaction H^- has to be determined first.

2. Express H^- in terms of irreducible tensors $\sum_u T^u . V^u$ (see chapter 2), i.e., $H^- \rightarrow \sum_u T^u . V^u$.

3. Compute the matrix elements of H^- using Wigner-Eckart theorem (chapter 2), i.e., $<I^- m^- l \, H^- l I \, m> — W\text{-}E \text{ theorem} — \propto (3j)(RME)$.

4. Diagonalize (H^-) (if necessary) to find the perturbation energy ε_{m^-}, i.e., $= U H^- U^{-1} = \varepsilon^-$, then, $\varepsilon_m = \varepsilon_0 + \varepsilon^-$; problem is solved.

III-3—Magnetic Dipole Interaction (MDI)

For each nucleus, there is a magnetic dipole, so if it is subjected to a magnetic field $\vec{B}$, it will interact with this field, introducing a perturbation term added to the original Hamiltonian given by $H^- = -\vec{\mu} \cdot \vec{B} = \mu_z \cdot \vec{B}$ along the Z-axis.

See figure15, where

$\vec{B}$ is the classical field B_0. | *state I m, $B_0 \vec{} m$*

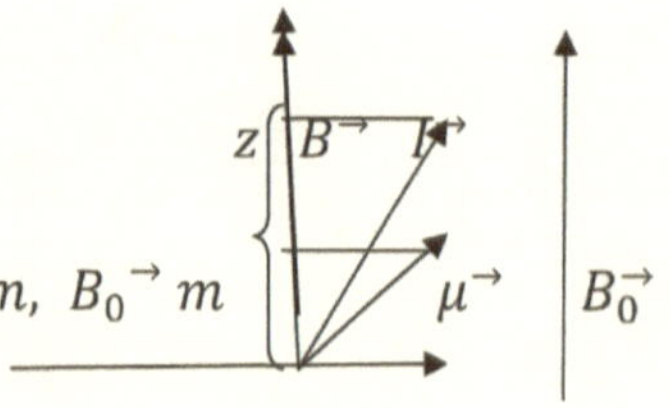

Figure 15

$\mu_z = \mu_0^{\rightarrow}$ *along z − axis,*

$I = \sum i \, Li + Si = \sum i(-iri \times \nabla i + Si)$

If the system is reflected, then $\vec{r} \rightarrow -\vec{r}$, $\nabla_i \rightarrow -\nabla_i$

So $r \times \nabla_i = +1$; S_i is in spin space.

Parity of Z is +1, also $\vec{r_s} = U \vec{r_c}$ —s for spherical; c for Cartesian

U is the transformation factor

$\hat{H}^{\smallsmile} = -\vec{\mu_0} B_0$, where $\vec{\mu_0}$ is a tensor of rank 1; it is a vector.

Now find the matrix elements for perturbation[s] H:

$<I \, m^{'} \, \pi l \, \hat{H} \, | I \, m \, \pi > = -B_0 <I \, m^{'} \, \pi l \, \mu_0^1 | I \, m \, \pi >$

using Wigner-Eckart theorem, the matrix elements given by

$(\hat{H}^{\smallsmile})mm = -B_0(-1)^{l-m^{'}}(\{3j(I, 1 \, I^{'}, -m^{'}, 0, m)\} <I \, ll \, \mu \, ll \, I >$

If m = m`, using the 3j from the 3j table, then the matrix elements given by

$$H_{mm} = -B_0(-1)^{l-m} m/\sqrt{[(2I+1)(I+1)]} \times <RME>$$

RME (*reduce matrix elements*) = $< I\|\, \mu \,\| I >$

For m = I_{MAX}, then, $H_{\|} = -B\, m/\left(\sqrt{2I}+I\right)(I+1) <\|\| \mu \|\| >$ (a)

Now by convention the definition of μ is given by

$$\mu = <I\|\, \mu_o^l \,\| I> = I/\left(\sqrt{2I}+1\right)(I+1) < I\|\, \mu \,\| I >$$ (b)

From (a) and (b), one can find that $H_{m'm}^{\wedge} = <Im|\, H^{\wedge -} \,| Im > =$

$-B_o\, m/I.\, \mu\partial_{m'm}$, for m = m⁻, $H_{mm} = (-B.\, \mu/I).m$; therefore,

$\varepsilon_m = \varepsilon_o + \varepsilon_m' \rightarrow \varepsilon_{m+1} - \varepsilon_m = (-\mu/I)B_0 = \Delta\varepsilon$; it is equidistant splitting drawn.

As it is well-known, the Larmor frequency is given by $\omega_B \hbar$) $= \Delta\varepsilon \rightarrow$

$$\omega_B = -\mu\, B_0/I\hbar = |\mu B_0/I\hbar|$$

It is a unidirectional precession of (I) about the applied magnetic field $\vec{B}$. Notice the graph below. (It is also to be noted that if the magnetic field interacting with the nucleus is due to the electrons surrounding the nucleus, then that will affect the energy spectrum of the nucleus. This is a phenomenon called hyperfine interaction; its result is termed as the hyperfine structure, which is to be treated later.

ω_B $\vec{\mu}$ Larmor frequncy

III-4—Electrostatic Interactions

III-4-1—Electric Multipole Moments

Consider a nucleus with a certain charge distribution, Let the electric field be a classical one given by a fixed charge (e_s),. This is called the surface charge. Also, there is the nucleus charge denoted by (e_n),. $\vec{r_s}$ is the distance between the surface charge and the point into the nucleus. The Hamiltonian (H) of the interaction due to (e_s) and the nucleus charge (e_n) at $\vec{r_n}$ from the nucleus center is given by

$$H' = \sum_{ns} e_s e_n / |\vec{r_s} - \vec{r_n}| \tag{71}$$

It is the Perturbation term

if $r_n < r_s$ such that $r_n/r_s < 1 = x$, then,

$$|\vec{r_s} - \vec{r_n}|^{-1} = 1/r_s [1 - 2x \cos\theta + x^2]^{1/2} = 1/r_s \sum_\lambda (\vec{r_n} / \vec{r_s})^\lambda P_\lambda (cos\theta) \tag{72}$$

Therefore, $H' = \sum_\lambda \sum_{ns} (e_n e_s/r_s)(r_n/r_s)^\lambda P_\lambda (\cos\theta)$ (73)

Using the edition theorem [equation (12), chapter 2], Legendre polynomial $P_\lambda (cos\theta)$ can be expressed in terms of spherical harmonic function $Y_{lm}(\theta, \vartheta)$ such as

$$P_\lambda (Cos\theta) = \frac{4\pi}{2\lambda + 1} \sum_m (-1)^m Y_{lm} (\theta_1, \vartheta_1) Y_l m (\theta_2, \vartheta_2); \text{ hence,}$$

$$H = 4\pi \sum_{\lambda=0}^{\infty} 1/(2\lambda + 1) \sum_m (-1)^m [\sum_n e_n r_n^\lambda Y_{lm} (\theta_n, \vartheta_n)] [\sum_s 1 \, e_s/r_s^{\lambda+1} Y_{lm}(\theta_s, \vartheta_s)]$$

Let $H^{'} = 4\pi\sum_{\lambda=0}^{\infty} M^{\lambda} \cdot V^{\lambda}$ (74)

Where, $M^{\lambda} = \sum_n e_n r_n^{\lambda} Y_{lm}(\theta_n \vartheta_n)$ (75)

$V^{\lambda} = \sum_s e_s / r_n^{\lambda} Y_{lm}(\theta_s, \vartheta_s)$ (76)

By definition, M_{μ}^{λ} is called the operator of static electric 2^{λ} moment. λ is the rank of the tensor M^{λ}. The multipole order of M^{λ}, according to the values of λ, is classified as follows:

$\lambda = 0, M^{(O)}$ is monopole

$\lambda = 1, M^{1}$ is dipole

$\lambda = 2, M^{2}$ is quadruple

$\lambda = 3, M^{3}$ is octuplet

Also, V_{μ}^{λ} is called the field tensor operator; therefore, M and V operator take the forms

$M_{\mu}^{\lambda} = e_n r_n^{\lambda} Y_{\lambda,\mu}(\theta_n, \vartheta_n)$ (75⁻)

$V_{\mu}^{\lambda} = \sum \frac{e_s}{r_s^{\lambda+1}} Y_{\lambda,-\mu}(\theta_s, \varphi_s)$ (76⁻)

Now, $H^{'}$ = H(E$_0$) + H(E$_1$) + H(E$_2$) + —H(E$_n$), where

$\frac{\lfloor H(E,\lambda+1)\rfloor}{\lfloor H(E,\lambda)\rfloor} \cong \frac{r_n}{r_s} = \frac{R_n}{R_{at}} = \frac{10^{-13}}{10^{-8}} = 10^{-5}$ (77)

So the size of the nucleus is one in one hundred thousand the size of the atom, but (99%) of the atomic mass is concentrated in the nucleus. Let us give $\lambda, \mu = 0$ values:

H(E$_0$) = y$_{00}$ = $1/\sqrt{4\pi}$ $\Rightarrow$ No charge distribution involved. It is monopole.

$$M_0^0 = \sum_n e_n / \sqrt{4\pi} = \frac{ze}{\sqrt{4\pi}} \tag{78}$$

and $V_0^0 = \sum_s \frac{e_s}{r_s} Y_{00}(\theta_s, \varphi_s) = \sum_s \frac{e_s}{r_s} \frac{\phi(0)}{\sqrt{4\pi}} = \sum_s \frac{e_s}{r_s}$ (79)

From (74) $H^-(E_0) = z\,e\,\phi(0)$ (80)

Now,

$$M_0^1 = \sum_n e_n \vec{r_n} Y_{10}(\theta, \varphi) = \frac{1}{2}\sqrt{\frac{3}{\pi}} \sum_n e_n \cdot (r_n \cos\theta_n) =$$

$$\sqrt{\frac{3}{4\pi}} \sum_n e_n \vec{z_n} \tag{81}$$

$$M_\pm^1 = \sum_n e_n \vec{r_n} \sqrt{\frac{3}{8}} \sin\theta_n e^{\pm i\phi n} = \sqrt{\frac{3}{8\pi}} \sum_n e_n (x_n \pm iy_n) \tag{82}$$

From basic physics, it is known that the dipole moment d is given by d = $\sum_n e_r \vec{r_n}$. Let the parity operator (π_o) apply on $\vec{r_n}$ so $\pi_o \vec{r_n} = -\vec{r_n} \Rightarrow$ odd parity operator.

As it was shown in chapter 2,

$< \text{Im}\,\pi\,|\,M_\mu^1\,|\,|m'\pi'> \ne 0$ if $\pi_f \pi_o \pi_i = +1$

So in the dipole case, this is −1; hence, in reference to equation (70), this matrix is zero, i.e., no transition is allowed if the parity is a good number, or |Imπ> is an eigenstate to π_o. Therefore, in general,

$\pi_o Y_{lm}(\theta, \vartheta) = (-)^l Y_{lm}(\theta, \vartheta)$, then, $M^\lambda \rightarrow Y_{lm}(\theta, \vartheta) \Rightarrow \pi_o M^\lambda \Rightarrow (-)^\lambda$, if λ is odd $\Rightarrow \pi = -1$, which implies that

$< \text{Im}\,\pi\,|\,M^\lambda\,|\,|m'\pi'> = 0$.

Therefore, any odd static multidipole moment in the nucleus vanishes. But for a non-static case, there is a dipole moment.

For the electric quadruple moment, $H^`$ is given by

$$H^-(E_2) = \frac{4\pi}{2\lambda+1} \sum_{\mu=-\lambda}^{+\lambda} (-1)^\mu M_\mu^2 V_{-\mu}^2 = \frac{4\pi}{5} \sum_{\mu=-\lambda}^{\lambda} (-1)^\lambda M_\mu^2 V_{-\mu}^2$$

where $V_{-\mu}^2$ is a classical field; it is a number. The matrix elements of $H^-(E_2)$ is given as,

$$< I\,m \mid H_\mu^2 \mid I\,m- > = \frac{4\pi}{5} [\sum_{\mu=-\lambda}^{+\lambda} (-1)^\mu < I\,m \mid M_\mu^2 \mid I\,m- >. V(\#).$$

Using W-E theorem, one gets

$$< I\,m \mid M_\mu^2 \mid I\,m- > = \begin{pmatrix} I & 2 & I \\ -m & \mu & m \end{pmatrix} < I \| M^2 \| I >$$

where 3j = 0 unless $-m + m` + \lambda = 0$.

Now the classical field tensor (V^2) has to be taken, which is given by

$$V_{-\mu}^2 = \sum_s \frac{e_s}{r_s} Y_{2,-\mu}(\theta_s \varphi_s) \text{ , } (\lambda=2).$$

Now the charge distribution (e_s) has an axial symmetry where it is possible to choose a certain axis as z-axis to be the symmetry axis. If a rotation happened to be about this axis, then the formalism for describing the rotation of the classical field (V) by the known rotational operator R(z, α) takes the form

$$R(z,\alpha)\, V_{-\mu}^2 = V_{-\mu}^2 \Rightarrow \sum_s \frac{e_s}{r_s} Y_{2,-\mu}^-(\theta_s, \varphi_s^- - \alpha)$$

$$\equiv \sum_s \frac{es}{3} Y_{2.-\mu}(\theta_s, \varphi_s)$$

This implies that $V_{\pm 2}^2 = 0$, so the only term left is

$$V_0^2 = \sum_s \frac{e_s}{r_s^3} Y_{2,0}(\theta_s, \varphi_s) \neq 0 \text{ ; therefore, } V_0^2 = \frac{\sqrt{5}}{16\pi} \sum_s \frac{e_s}{r_s^3}(3\cos^2\theta_s - 1)$$

The potential outside the nucleus is given by

$$\phi(r_s) = \sum_s \frac{e_s}{r_s}, \text{ where } r_s \equiv (x^2 + y^2 + z^2)^{\frac{1}{2}}, \text{ so}$$

$$\frac{\Delta\phi}{\Delta z} = \sum_s e_s z(x^2 + y^2 + z^2)^{\frac{-3}{2}} = \sum_s e_s z. \text{ Therefore,}$$

$$(\frac{d\phi^2}{d^2 z})_0 = \frac{d}{dz}(-\sum_s \frac{es}{r_s^3} z) = \sum_s \frac{es}{r_s^3} (\frac{3z^2}{r_2^2} - 1), \text{ but } \cos = \frac{z}{r_s}$$

Hence, $(\frac{d\phi^2}{d^2 z})_0 = \sum_s \frac{e_s}{r_s^3} (3\cos^2\theta - 1)$, which implies that

$$V_0^2 = \frac{\sqrt{5}}{\sqrt{16\pi}} \sum_s \frac{e_s}{r_s}(3\cos^2\theta - 1) \Rightarrow$$

$$V_0^2 = \frac{\sqrt{5}}{\sqrt{16\pi}} (\phi_{zz}(r_s))_0 \tag{83}$$

Φ_{zz} is a field gradient.

For the axial symmetry, $\mu = 0$, and because $-m + m' + \mu = 0$ so that 3j-symbol $\neq 0$, then m = m', which yields to

< m l H'l m'> =<m l H´lm >, which is diagonal; therefore,
E_n = <Im lH(E$_0$)lIm$^-$>

$$= \frac{4\pi}{5} \sqrt{\frac{5}{16\pi}} (-)^{1-m} \begin{pmatrix} I & 2 & I \\ -m & 0 & m \end{pmatrix} <\text{I ll M}^2\text{lII} > \phi_{zz}$$

Refering to the table of 3j-symbols, E_n takes the form

$$E_n = (\frac{\pi}{5})^{1/2} [\frac{2(3m^2 - I(I+1))}{\sqrt{(2I+3)(2I+2)(2I+1)(2I)(2I-1)}}] \times <\text{I ll M}^2\text{ll I} > \phi_{zz}.$$

With information,

1) $\phi_{zz} = 0$

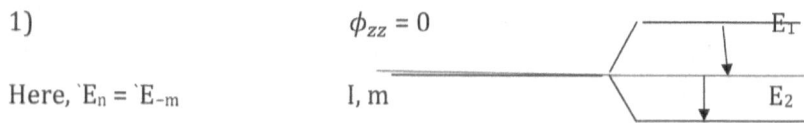

Here, $E_n = E_{-m}$ I, m

2) $E_n^- \propto \phi_{zz}$

3) $E_m \propto <$ Ill M² ll I $>$

4) $E_n = a[3m^2 - I(I + 1)]$

Nuclei with I = 1, 1/2—no quadruple interaction.

For I = 1, m = +1, 0, –1, then,

$E_{\pm 1} = + a$, $E_0 = -2a$. $\phi_{zz} = 0$ a m = ±1

                                          –2a

                                         m = 0

$\phi_{zz} = 0$ 6a   m = ±2

                           –3am = ±1

If I = 2, $E_0 = -6a$, $E_{\pm 1} = -3a$, $E_{\pm 2} = +6a$ –6a

To find a quadruple moment of nucleus, which is usually defined as a measure of the nucleus deviation from the spherical symmetry, which is known to be given by Q = <I ll $M^{(2)}$ ll I>, one takes the following steps:

Take,

$$eQ = <I \; || \; \Sigma_n (3z^2 - r_2^2) || \; I > Y_{2,0}(\theta_n, \varphi_n) = \sqrt{\frac{5}{16\pi}}(3cos_2\theta_n - 1)$$

SO, eQ can be found to be,

$$eQ = \sqrt{\frac{5}{16\pi}} < |\text{II } M_0^2 II | > =$$

$$\sqrt{\frac{5}{16\pi}} (-1)^{\text{I-I}^-} \begin{bmatrix} \text{I} & 0 & \text{I}^- \\ -\text{I} & 0 & \text{I}^- \end{bmatrix} < |IM_0^2 II > \text{,using 3j}$$

symbols the result is

$$[E_m = \frac{1}{4\pi} eQ.\phi_{zz} \frac{(3m^2 - I(I+1))}{I(2I-1)}]$$

(84)

The quadruple frequency ω_E is given by

$$\omega_E = \frac{1}{4} eQ\phi_{zz} \frac{1}{I(I-1)} \Rightarrow \omega_E = \frac{E_m}{3m^2 - I(I+1)}$$

(85)

Hence, $E_m = \omega_E (3m^2 - I(I+1))$

(86)

ΔE_n = n. ω_0 , n = 1, 2, 3, 4,—is the energy difference between adjoined states.

So this means all $V_0^{(2)}$, $V_{\pm 1}^{(2)}$, and $V_{\pm 2}^{(2)}$ are to be taken in consideration.

Let ϕ_{xx}, ϕ_{yy}, and ϕ_{zz} be chosen, then using poison gradient

$$\phi_{xx} + \phi_{yy} + \phi_{zz} = \nabla\phi = 0$$

For a general field, take Φ_{zz}. The asymmetrical parameter η is given

by $\eta = \frac{\phi_{xx} - \phi_{yy}}{\phi_{zz}}$, $| \eta | < 1$. Therefore, $V^{(2)}$ can be

Considered as a function of ϕ_{zz} and η.

For cubical symmetry, $\phi_{zz} = 0 \Rightarrow$ no quadruple splitting.

Problem:

Given a nucleus with $I = 1$, $\mu = \neq 0$, under the effect of a magnetic field $\vec{B}$ (constant), the given information is $Q \neq 0$,

$\cos \beta = 0.1$; choose Z-axis as quantization axis.

$$B = 0, \frac{\Delta\phi^2}{\Delta^2 z} = 0 \rightarrow E_0$$

$$B \neq 0, \frac{\Delta\phi^2}{\Delta^2 z} \neq 0 \rightarrow E_0 + E_0^- + E_1^-$$

Larmor frequency is given as $\omega_I = \omega_E$ (quadruple frequency). With this information, find (E_2) in S' system; rotate S, i.e., $S \rightarrow S'$, $H' = H_s(2) + H_s(\mu)$

$U H U^{-1} = E$ (diagonalization)

Now the question is, What are the matrix elements of the product of two tensors operating in different spaces such as

$$H^-(E_L) = \frac{4\pi}{2L+1} \sum T^{(L)} \cdot V^{(L)}, \text{ where}$$

$T^{(L)}$ is a nuclear tensor operator; $V^{(L)}$ is a classical field.

The answer—this can be treated by using the recoupling of angular momenta, see chapter 2.

III-4-2—Matrix Elements of Scalar Product of Two Commuting Tensor Operators

Take two systems: $|j_1 m_1\rangle$ and $|j_2 m_2\rangle$; let $T_1^{(\lambda)}$ be the operator that operates on the system $|j_1 m_1\rangle$ and the operator $T_2^{(\lambda)}$ that operates on the system $|j_2 m_2\rangle$. By the definition of the scalar product, this product can be written as $\xi^{(0)} = T_{1)}^{(\lambda)} \cdot T_{2)}^{(\lambda)}$.

Here, $\xi^{(0)}$ is a scalar product; its rank is zero. For example, $H(M_1)$ $= -\mu.\vec{B}$, where μ is the magnetic moment of the nucleus belonging to a nucleus space $(g_1. I)$, and $\vec{B}$ is the magnetic field due to the surrounding electrons, so its space is this electrons' space denoted by $(\Psi_{JM})_a$. The space of the nucleus is denoted by $(\psi_{1m})_n$. Now $\xi^{(0)}$ can be written more formally as

$$\zeta^{(0)} = \sum_\mu (-1)^\mu T_{\mu(1)}^{(\lambda)} \cdot T_{\mu(2)}^{(\lambda)}, \; [T_{\mu(1)}^{(\lambda)}, T_{\mu(2)}^{(\lambda)}] = 0, \text{ they}$$

commute.

The commutation is due to the fact that $T_{\mu(1)}^{(\lambda)}$ and $T_{\mu(2)}^{(\lambda)}$ are belonging to a different state. Therefore, the matrix elements of $\xi^{(0)}$, with the using of W-E theorem, are given by

$$\langle j_1^- j_2^- J^- M^- | \zeta^{(0)} | j_1 j_2 JM \rangle =$$

$$(-1)^{J^- - M^-} \begin{pmatrix} j & 0 & j \\ -M^- & 0 & M \end{pmatrix} \langle JM \| \zeta^{(0)} \| J^- M^- \rangle =$$

$$\frac{1}{2J+1} \delta_{JJ^-} \delta_{MM^-} \langle J^- \| \zeta^{(0)} \| J \rangle \tag{87}$$

If $J = J^-$, $M = M^-$, then, $|JM\rangle = \sum_{m1} \langle j1m1\, j2m2 = M - m1 | JM \rangle |j_1 m_1\rangle \times$ $|j_2 M - m_1\rangle = \text{SUM}_{m1} \langle j_1 m_1 J_2 M - m_1 | JM \rangle \langle j_1 m_1 | j_2 M - m_1^- \rangle$

Using equation (85), it can be obtained that $<J\ Ml\ T_{(1)}^{(\lambda)}\ .\ T_{(2)}^{(\lambda)}\ Ljm>$

$$= \sum_{\mu 1m1`m1} (-1)^{\mu} <j_1\ m_1\ j_2\ M - m_1\ l\ j_1\ `m_1\ j_2\ M - `m_1\ Ljm> \times$$

$$[<`j_1\ `m_1\ `j_2 M-`m_1\ T_{\mu(1)}^{(\lambda)}\ .\ T_{\mu(2)}^{(\lambda)}\ lj_1\ m_1\ j_2\ M - m_1 >]^{\cdot}$$

*can be written as

$<`j_1\ `m_1\ T_{\mu(1)}^{(\lambda)}\ l\ j_1\ m_1><`j_2\ M - `m_1\ T_{\mu(2)}^{(\lambda)}\ lj_2\ M - m_1>$, so by substituting this instead of (*), the following is obtained:

$$< jMl\ T_{\mu(1)}^{(\lambda)}\ .\ T_{\mu(2)}^{(\lambda)}\ ljM> = \sum_{\mu m1`m1} [(-1)^{\mu} + j1 - j2`+M + j1`-j2`+M + j1`$$

$$- m1` + j2` -M + m1`] / \sqrt{(2j +1)(2j`+1)} \left\{ \begin{bmatrix} j1 & j2 & J \\ m & M-m & M \end{bmatrix} \right.$$

$$\times \left. \begin{Bmatrix} j1 & j2 & J \\ m1` & M-m1 & -M \end{Bmatrix} \right\} \times$$

$$\begin{bmatrix} j1 & \lambda & j1 \\ -M` & \mu & M1 \end{bmatrix} \begin{bmatrix} j2` & \lambda & j2 \\ -M - m1` & -\mu & M - m1 \end{bmatrix} \times <j_1 llT_1 llj_1 ><j_2 llT_2 ll j_2> \quad (86)$$

If $j, j_1, j_2,$ and μ are given, then the interaction is known. All 3j-symbols can be found from their tables, with the aid of the relations between the 3js and the C-G coefficients, so equation (86) takes the form

$$<(j_1 `j_2)jMl\ T_1^{(\lambda)} . T_2^{(\lambda)}\ l(j_1\ j_2)jm> \equiv (-)^{`j_1 +`j_2 +j} \begin{bmatrix} j1 & j2 & j \\ j2` & j1` & \lambda \end{bmatrix} \times$$

$$<`j_1 llT_1 llj_1 ><`j_2 llT_2 ll j_2 > \quad (87)$$

III-4-3—Application to Hyperfine Structure of Free Atoms

As it is well-known, any nucleus is surrounded by moving electrons; therefore, an internal magnetic field $B^{\rightarrow}$ is created, since the nucleus has its own nuclear magnetic momentum denoted by (μ_1).

An interaction between the magnetic moment and the magnetic field will take place. (I) indicates the nucleus spin. This interaction was found experimentally to affect the structure of the atomic spectra. This interaction is represented by the equation

$H'(M_1) = -\mu_J^{\rightarrow} \cdot B_J^{\rightarrow} = \sum_{\mu} (-1)^{\mu} T_{\mu}^{(1)}(I) T_{-\mu}^{(1)}(J)$. These T tensors represent, respectively, $\mu_1^{\rightarrow}$ and $B_J^{\rightarrow}$. Both of them are tensors of rank 1; i.e., they are vectors. As a consequence to this reaction, a coupling between the angular momentum of the nucleus ($I^{\rightarrow}$) and that of the atom ($J^{\rightarrow}$) will take place. It is represented by the vector sum $F^{\rightarrow} = I^{\rightarrow} + J^{\rightarrow}$ = the angular momentum of the combined system (see the diagram, which is to be completed). Such problem is time independent (stationary). So by applying equation (87), the matrix elements of $H'(M_1)$ can be found to be $F^{\rightarrow}$ $F^{\rightarrow}$

$< FIJ \mid H'(M_1) \mid FIJ > \equiv J^{\rightarrow} I^{\rightarrow}$

$(-1)^{I+J+F} \begin{bmatrix} \acute{E} & J & F \\ \acute{E} & J & 1 \end{bmatrix} < I \parallel \mu \parallel I > < J \parallel B \parallel J > \text{------(A)}$

If the electric quadruple interact tion is considered, then,

$$H^-(E_2) = \frac{4\pi}{5}(\mathbf{M}^{(2)} \cdot \mathbf{V}^{(2)}), \lambda = 2 \text{ ; hence,}$$

$E_F^{'}(E_2) = \text{<FIJ 1 H}^{'}(E_2)\text{IFIJ>} =$

$$= (-1)^{I+J+F} \begin{pmatrix} E^- & J & F \\ E^- & J & 2 \end{pmatrix} \text{<I || M}^{(2)}\text{|| I> x < }J\text{||V}^{(2)}\text{||}J > \qquad (88)$$

where $<\text{Ill } \mu \text{ lll} >$ is the reduced matrix elements of the quadruple

moment. $<J\text{ll V}^{(2)}\text{ ll}J >$ is $\frac{\Delta\phi^2}{\Delta^2 z} = (\phi_{zz})J$. Using the values of the 6j-symbols from their special tables, the energy splitting will be determined. [As an exercise verify (A) and (B).]

Chapter Four

Emission of Electromagnetic Radiations

(Electromagnetic Transitions)

IV-1—Introduction

In chapter 3, we have considered the static case, where the charge and the current distribution in the source of electromagnetic radiation are described by $\vec{\rho}(\vec{r}), \vec{J}(\vec{r})$, i.e., $\vec{\rho}(\vec{r})$ and $\vec{J}(\vec{r})$ are independent of time, where this case is special one. Hence, let us consider now the more general dynamic case, where the charge and the current distributions are functions of time. Thus, the charge and the current distributions are to be denoted by $\vec{\rho}(\vec{r}_n, t)$ and $\vec{J}(\vec{r}_n, t)$, respectively. Let us demonstrate these concepts by figure 1:

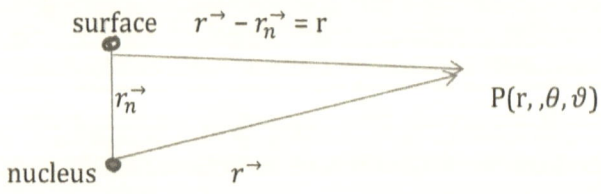

surface $\quad \vec{r} - \vec{r_n} = r$

$\vec{r_n}$

$P(r, ,\theta, \vartheta)$

nucleus $\quad\quad \vec{r}$

Figure 1

In this general dynamic case, in addition to $\vec{\rho}(\vec{r}_n, t)$ and $\vec{J}(\vec{r}_n, t)$, there is a magnetization distribution too, denoted by $M(\vec{r}n, t)$. Hence, there are

116

$\vec{\rho}(\vec{r}_n, t)$ = charge density

$\vec{J}(\vec{r}_n, t)$ = current density

$M(\vec{r}_n, t)$ = magnetization density

In this generalized dynamic problem, it is required to study the

fields produced by $\vec{\rho}(\vec{r}_n, t)$ and $\vec{J}(\vec{r}_n, t)$ at the point P(r, θ, φ), figure 1. The time dependence of ρ, J, and M will be assumed through the factor

exp.(−iωt). These fields are electric $\vec{E}(\rho)$, magnetic $\vec{H}(\rho), \Phi(\rho)$, and

$\vec{A}(\rho)$, which are to be well defined next.

These dynamic variables are physically given by

$$\left. \begin{array}{l} \vec{\rho}(\vec{r}, t) = \rho(\vec{r}_n, 0)e^{-i\omega t} + \rho^*(\vec{r}_n, 0)e^{i\omega t} \\ \vec{J}(\vec{r}, t) = J(\vec{r}_n, 0)e^{-i\omega t} + J^*(\vec{r}_n, 0)e^{i\omega t} \\ \vec{M}(\vec{r}, t) = M(\vec{r}_n, 0)e^{-i\omega t} + M^*(\vec{r}_n, 0)e^{i\omega t} \end{array} \right]$$

(1)

where ω = 2πf; f is the frequency, where ΔE is the energy difference.

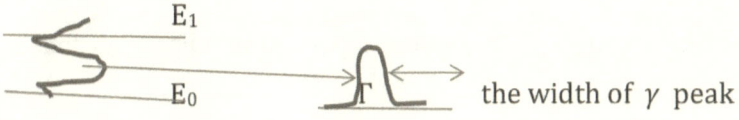

the width of γ peak

$\Delta E = \hbar\omega = E_1 - E_0$

According to this data, one can note the following:

1) Point charge $\rho \sim e^{i\omega t}$

Current $j \sim e^{i\omega t}$

Magnetization $M \sim e^{i\omega t}$

2) Fields produced by ρ, j, and M at the point $P(r, \theta, \phi)$ are $\vec{H}$ (magnetic field) and $\vec{E}$ (electric field).

3) $\vec{E}$ and $\vec{H}$ relation is quite well demonstrated by the familiar Maxwell equations.

4) $\vec{A}$, Φ, the vector potential and the scalar potential, satisfy an inhomogeneous equation (which will be shown later).

5) The solution of the wave equation is known to be a retarded potential (refer to *Classical Electrodynamics* by Jackson), which is given by

$$\Phi(\vec{r}) = \int_{nucleus} \rho(\vec{r}_n) \frac{e^{i\kappa|\vec{r} - \vec{r}_n|}}{|\vec{r} - \vec{r}_n|}$$

$$, \quad \vec{r}_n < \vec{r} \tag{2}$$

Since $\vec{r}_n < \vec{r}$, then $\frac{1}{|\vec{r} - \vec{r}|}$ can be expanded, as it has been done in

chapter 3; therefore, $\frac{1}{|\vec{r} - \vec{r}|} \propto \sum Y(\theta, \varphi) Y(\theta^- \varphi^-)$ should be here!

$$\tag{3}$$

Hence,

6) $\vec{A}_{L,M}$ is an irreducible tensor of rank L.

7) $\vec{A}(\vec{r}) = \sum \vec{B}^{(L)} \cdot \vec{A}^{(L)}$

8) $\vec{H} = \dfrac{-ie\hbar}{m\ell} \vec{A} \cdot \vec{\Delta}$ is Hamiltonian due to the source $\vec{B}^{(L)}$.

8 is well-known in quantum mechanics for a particle in a magnetic field

$(\vec{B})$.

To understand these quite well, one has to study the Maxwell equations. Therefore, we will briefly review these important equations. They are

1) $\text{div}(\vec{H} + 4\pi M) = 0$

2) $\text{div}\vec{E} = 4\pi\rho$

3) $\Delta \times \vec{H} = -\dfrac{1}{c}\dfrac{\partial \vec{E}}{\partial t} + \dfrac{4\pi}{c}\vec{j}$; c is light velocity

4) $\Delta \times \vec{E} = -\dfrac{1}{c}\dfrac{\partial}{\partial t}[\vec{H} + 4\pi \vec{M}]$

where the real (observed) $\vec{E}$ and $\vec{H}$ strengths are described by

$\vec{E}$ and $\vec{H}$. The source of the field is described by $\vec{\rho}$, $\vec{j}$, and $\vec{M}$.

For monochromatic radiation of frequency $f = \dfrac{\omega}{2\pi}$, complex time-independent amplitudes are introduced. Therefore, each of the real fields and the quantities of the source may be expressed as follows:

$$\left[\begin{array}{l} \underline{\vec{E}}(\vec{r},t) = \vec{E}(\vec{r},t) + \vec{E}^{*}(\vec{r},t) \\ \text{where, } \vec{E}(\vec{r},t) = \vec{E}_{o}(\vec{r})e^{-i\omega t} \; ; \vec{E}^{*}(\vec{r},t) = \vec{E}_{o}^{*}(\vec{r})e^{i\omega t} \end{array}\right\} \tag{4}$$

Equation (1) is for ρ, j, and M, but $\vec{H}$ is given by

$$\left[\begin{array}{l} \underline{\vec{H}}(\vec{r},t) = \vec{H}(\vec{r},t) + \vec{H}^{*}(\vec{r},t) \\ \text{where, } \vec{H}(\vec{r},t) = \vec{H}_{o}(\vec{r})e^{-i\omega t} \; ; \vec{H}^{*}(\vec{r},t) = \vec{H}_{o}^{*}(\vec{r})e^{i\omega t} \end{array}\right\} \tag{5}$$

The special case of a static field is the limit $\omega \to 0$. By introducing the wave number, $k = \dfrac{\omega}{c} = \dfrac{2\pi f}{c} = \dfrac{\lambda}{2\pi} = \lambda$, Maxwell equations become

$$\left[\begin{array}{l} \text{curl } \vec{H}_{o} = \dfrac{4\pi}{c}\vec{J}_{o} - i\kappa\vec{E} \\ \text{curl } \vec{E}_{o} = i\kappa(\vec{H} + 4\pi\vec{m}_{o}) \\ \text{div } \vec{E}_{o} = 4\pi\rho_{o} \\ \text{div } (\vec{H}_{o} + 4\pi\vec{m}_{o}) = 0 \end{array}\right] \tag{6}$$

where J_{o}, ρ_{o}, m_{o} are the amplitudes, and the continuity equation is

$$\left[\text{div}J_{o} - i\omega\rho_{o}\right] = 0 \tag{7}$$

In the usual way, the vector potential $\vec{A}$ can be introduced by

$$\text{Curl } \vec{A} = \vec{H} + 4\pi\vec{m} \tag{8}$$

As it was considered in (4) and (5), $\vec{A}$ can be written as

$\vec{A} = \vec{A}_o(\vec{r})e^{-i\omega t} + A_o^*(\vec{r})e^{i\omega t}$, then substituting for $\vec{A}$ into (8), we get

$$\text{Curl } \vec{A}_o = \vec{H}_o + 4\pi\vec{m}_o \tag{9}$$

From (6) and (9), $\vec{E}$ becomes

$$\vec{E} = -\frac{1}{c}\frac{\partial\vec{A}}{\partial t} - \text{grad}\phi \tag{10}$$

where ϕ is a real scalar (scalar potential), where

$$\phi = \phi_o(\vec{r})e^{-i\omega t} + \phi^*(\vec{r})e^{i\omega t}$$

The time-independent form of (10) is

$$\vec{E}_o = i\kappa\vec{A}_o - \text{grad}\phi_o \tag{11}$$

The aim in introducing these potentials is to formulate the theory in terms of them alone. In order to have decoupled wave equations for the potentials, the Lorentz condition(Lorentz gauge) has to be imposed, which is

$$\text{div}\vec{A} + \frac{1}{c}\frac{\partial\phi}{\partial t} = 0 \tag{12}$$

It is important to note that the field strengths are derivable from equations (8–11). Equation (12) gives $\text{div}\vec{A}_o - i\kappa\phi_o = 0$ $\qquad$ (13)

As it is known from the electrodynamics and the classical field theory, the vector potential $\vec{A}$ and the scalar field ϕ build up, together, a four vector denoted by $F^{(4)}$, such as $F^{(4)}(\vec{A}, i\phi)$, i.e., $\vec{A}, i\phi$ are components building up $(F^{(4)})$, where

$$\vec{\Pi}^{(4)} \cdot \vec{F}^{(4)} = 0 \tag{14}$$

$F^{(4)}$ is also called a four potential. It is necessary to remember from the theory of relativity that $X_4 = ict$ and that $\sum_\mu \dfrac{\partial}{\partial x_\mu} F_\mu = 0$, $\mu = 1, 2, 3, 4$, which is equivalent to (14). Using the well-known mathematical identity that $\nabla \times \nabla = \nabla - \nabla^2$, i.e., curl curl = grad div- $\nabla^2 \Rightarrow \nabla(\nabla \cdot f) - \nabla^2 f = \nabla \times (\nabla \times f)$, the well-known wave equations can be obtained. These equations are

$$\left[\begin{array}{l} \nabla^2 \underline{\vec{A}} - \dfrac{1}{c^2} \dfrac{\partial^2 \vec{A}}{\partial t^2} = \dfrac{-4\pi}{c} (\vec{J} + c \text{ curl } \vec{m}) \\[4mm] \nabla^2 \varphi - \dfrac{1}{c^2} \dfrac{\partial^2 \varphi}{\partial t^2} = -4\pi\rho \end{array} \right] \tag{15}$$

As it is well-known, the time-independent forms of (15) are

$$\left[\begin{array}{l} \nabla^2 \vec{A}_\circ + \kappa \vec{A}_\circ = \dfrac{-4\pi}{c} (J_\circ + c \, \nabla \times \vec{m}) \\[4mm] \nabla^2 \varphi - \dfrac{1}{c^2} \dfrac{\partial^2 \varphi}{\partial t^2} = -4\pi\rho \end{array} \right] \tag{16}$$

$$[\phi = \phi_0, m = M_{0,} \varrho = \rho_0] \tag{16}$$

The total current (J), usually, constitutes two components: (1) the convection current $\vec{j}$ and (2) the Amperian current $(c \, \nabla \times \vec{m})$. From now on, the index in the amplitudes will be dropped. Also, for practical interest, the field sources are to be considered within a bounded region and established within a finite interval of time in the past. With these two conditions imposed on the field sources, the retarded potential solutions are

$$\left[\begin{array}{l} \vec{A}(\vec{r}) = \dfrac{1}{c} \int\limits_{nuc.} \vec{J}\, \dfrac{e^{i\kappa|\vec{r}-\vec{r}'|}}{|\vec{r}-\vec{r}'|}\, d\overline{V} \\[4mm] \text{and} \\[2mm] \phi(\vec{r}) = \int\limits_{nuc.} \rho(\vec{r}')\, \dfrac{e^{i\kappa|\vec{r}-\vec{r}'|}}{|\vec{r}-\vec{r}'|}\, d\overline{V} \end{array} \right] \tag{17}$$

Equations (17) are solutions for the case, considering that J and ρ are restricted to a finite region, where they are of zero values outside this bounded region, i.e., J = ρ = 0.

nucleus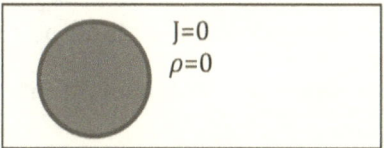

The solutions for the outside case (J = ρ = 0) represent solutions of free-space Maxwell equations. Therefore, equations (16) become

$$\left[\begin{array}{l} \left(\nabla^2 + \kappa^2\right)\vec{A} = 0 \\[2mm] \left(\nabla^2 + \kappa^2\right)\vec{\phi} = 0 \end{array} \right] \tag{18}$$

which are, in fact, Helmholtz equations.

Now let us introduce the source fourth vector $\vec{S}^{(4)}$, where

$$\vec{S}^{(4)} = \left\{ \frac{i}{c}J, = \frac{i}{c}\big(j + c(\nabla \times \vec{m})\big), \rho \right\} \equiv \left\{ \frac{i}{c}J, \rho \right\}$$

and $\Pi^2 \vec{F}^{(4)} = -4\pi S^{(4)}$ \hfill (19)

[Equation (19) is the basic equation of the electromagnetic field.]

The energy density (ε) is given by

$$\varepsilon = \rho\varphi - \frac{1}{c}\vec{A}\cdot\vec{j} = \vec{S}^{(4)}\cdot\vec{F}^{(4)}.$$

So the problem now is to solve equations (15), in which inside the nucleus, the solutions are given by equations (17). It is necessary to notice that inside the field source, one can use guage condition i.e.,

$$\vec{\nabla}\cdot\vec{A}(\text{div}\vec{A}) = \vec{\nabla}\cdot\vec{A}(\nabla\cdot\vec{A}) \rightarrow 0, \phi \rightarrow 0.$$

IV-2—Hamiltonian of the Interaction H

It is known that the eigenvalue equation of a system, described by an eigenstate Ψ_n, with its Hamiltonian of interaction being H, is given by

$\hat{H}\Psi_n = \varepsilon_n\Psi_n$, or its Schrödinger equation is $H^{\wedge}\psi = i\hbar\frac{\partial}{\partial t}\psi$. If the system is subjected to a magnetic field, then, $\vec{P}$ (momentum)

$\rightarrow -i\hbar\vec{\nabla} - \frac{e}{c}\vec{A}$, so

$$\hat{H} = \frac{1}{2m}\vec{P} + e\Phi + V_n, [V_n\text{=nucleus potential}]$$

$$= \frac{1}{2m}\left(-i\hbar\vec{\nabla} - \frac{e}{c}\vec{A}\right)^2 + e\Phi + V_n, \text{ hence,}$$

$$\hat{H} = \frac{-\hbar^2}{2m}\vec{\nabla}^2 + \frac{ie\hbar}{mc}\vec{A}\cdot\vec{\nabla} + \frac{ieh}{2mc}\left(\vec{\nabla}\cdot\vec{A} + \vec{A}\cdot\vec{\nabla}\right) + \frac{e^2}{2mc^2}\vec{A}^2 + e\Phi + V_n$$

Now if V_n is a strong interaction between nucleons,

then $\hat{H} = H_o + H_{int.}$, where, $H_0 = -\dfrac{\hbar^2}{2m}\nabla + V_n$ and

$$H_{int.} = \frac{1}{2m}\left[\frac{i\hbar e}{c}(\vec{\nabla}\cdot\vec{A} + \vec{A}\cdot\vec{\nabla}) + \frac{e^2}{c^2}\vec{A}^2 + e\Phi\right]$$

By using the Lorentz gauge, $\Phi = 0$, $\vec{\nabla}\cdot\vec{A} = 0$, and $\dfrac{e^2}{2mc^2}A^2$ are so small they can be neglected. Hence, $\hat{H}$ becomes

$$H^\wedge = H_0^\wedge + H_{int} \text{ and } [H_{int} = \frac{ie\hbar}{mc}A^{\rightarrow}\cdot\nabla^{\rightarrow}] \tag{20}$$

So $H_{Int} = H_I$ is the perturbation term, due to interaction, which causes the transition from state (i) to state (f).

Hence, to find the transition probability due to this interaction, one has to use the fermi golden rule no. 2, i.e.,

$$dN = \frac{2\pi}{\hbar}\left|\langle f|H_I|i\rangle\right|^2 .n(E)dE = \frac{2\pi}{\hbar}\left|(H_I)_{fi}\right|^2 .n(E)dE.$$

If the spin effect is considered, then $\vec{A}$ has to be found, which is required to introduce the spherical harmonic.

IV-2-1—Vector Field Properties

The vector fields are quite important in studying the electromagnetic fields theory. This vector has three components at each point of space. These three functions, as components of $\vec{A}$, transform under the D^L (Wigner D-matrix of transformation). This property indicates that these functions are irreducible tensors of rank L. Therefore, the total set of these tensors in a vector field would be of number 3 (2L + 1)

functions. These functions can be regrouped into three new sets of
(2J + 1) functions, where J = L + 1, L, and L − 1, assuming L > 0 and
that each of these sets constitutes the components of an irreducible
tensor of rank J. Hence, the angular momentum for the vector field is
the operator $\hat{J} = L + s$, where $L = -i\vec{r} \times \vec{\nabla}$ represents the orbital
angular momentum operator, and s is a set of 3 × 3 matrices for the
intrinsic spin (as it is known). These mentioned functions above are
eigenfunctions of the angular momentum; besides, there are three
components associated with a spin or function that is an eigenfunction
of the spin S, where the eigenvalue of s² is s(s + 1) in ħ unit = 2 or S =
1, i.e.,

$s^2\chi_{1/2,-1/2} = s(s+1)\chi_{1/2,-1/2}$ is the eigenvalue equation of the spin
operator.

In the following discussion, the aim is to obtain these results. The
vector fields $\vec{A}(\vec{r})$ can be written in a Cartesian representation as

$$\vec{A}(\vec{r}) = \begin{pmatrix} A_x(\vec{r}) \\ A_y(\vec{r}) \\ A_z(\vec{r}) \end{pmatrix}$$

(21)

Let us impose a rotation on $\vec{A}(\vec{r})$ as an infinitesimal rotation
through an angle (dα) about the z-axis such as

$$A^{-\rightarrow} = R(d\alpha)A^{\rightarrow}(x,y,z) \quad \text{or} \quad \begin{pmatrix} A_x^- \\ A_y^- \\ A_z^- \end{pmatrix} = R(d\alpha)\begin{pmatrix} A_x \\ A_y \\ A_z \end{pmatrix}$$

(22)

By expanding to first order in $(d\alpha)$, we obtain

$$\vec{A}'_x(x,y,z) = A_x(x+d\alpha y, y - d\alpha x, z) - d\alpha A_y(x+\alpha y, y - \alpha x, z)$$

$$= \vec{A}_x(x,y,z)\left[1+d\alpha(y\frac{\partial}{\partial x} - x\frac{\partial}{\partial y})\right] - d\alpha\vec{A}_y(x,y,z) \tag{a}$$

Also,

$$\vec{A}'_y(x,y,z) = \left[1+d\alpha(y\frac{\partial}{\partial x} - x\frac{\partial}{\partial y})\right]\vec{A}_y(x,y,z) + d\alpha\vec{A}_x(x,y,z) \tag{b}$$

$$\vec{A}'_z(x,y,z) = \left[1+d\alpha(y\frac{\partial}{\partial x} - x\frac{\partial}{\partial y})\right]\vec{A}_z(x,y,z) \tag{c}$$

In matrix notation, where $I = \begin{pmatrix} 1 & 0 & 0 \\ 0 & 1 & 0 \\ 0 & 0 & 1 \end{pmatrix}$, one can write

$$\begin{pmatrix} A'_x \\ A'_y \\ A'_z \end{pmatrix} = \left\{ \left[1+d\alpha(y\frac{\partial}{\partial x} - x\frac{\partial}{\partial y})\right]1 + d\alpha\begin{pmatrix} 0 & -1 & 0 \\ 1 & 0 & 0 \\ 0 & 0 & 0 \end{pmatrix}\right\}\begin{pmatrix} A_x \\ A_y \\ A_z \end{pmatrix}$$

Compare this with (22); you find that

$$j_3 = -(x\frac{\partial}{\partial y} - y\frac{\partial}{\partial x})1 + \begin{pmatrix} 0 & -1 & 0 \\ 1 & 0 & 0 \\ 0 & 0 & 0 \end{pmatrix} \tag{23}$$

where $L_z = -1(x\dfrac{\partial}{\partial y} - y\dfrac{\partial y}{\partial x})$ is the z-component of L and

$\begin{pmatrix} 0 & -1 & 0 \\ 1 & 0 & 0 \\ 0 & 0 & 0 \end{pmatrix}$ is s_z. If the infinitesimal rotation is considered about x and y axes, the components s_x and s_y of the intrinsic spin s are determined by a cyclic permutation. Hence, the spin operators for a vector field in Cartesian representation are

$$s_x = \begin{pmatrix} 0 & 0 & 0 \\ 0 & 0 & -1 \\ 0 & 1 & 0 \end{pmatrix}; \quad s_y = \begin{pmatrix} 0 & 0 & 1 \\ 0 & 0 & 0 \\ -1 & 0 & 0 \end{pmatrix}; \quad s_z = \begin{pmatrix} 0 & -1 & 0 \\ 1 & 0 & 0 \\ 0 & 0 & 0 \end{pmatrix} \quad (24)$$

$\vec{A}(\vec{r})$, as in equation (22), are tensors and transformed under D-matrices (D^L), where each one has (2L + 1) components. So as stated before, this vector field has 3(2L + 1) components. These do not make an irreducible tensor; therefore, it is a requirement to rearrange 3(2L + 1) tensor components into a set of three components—that is,

$$\left. \begin{array}{lll} 1)J = L+, 2L+2+1 = 2L+3 \\ 2)J = L; & 2L+1 \\ 3)J = L-1; & 2L-1 \end{array} \right] \Rightarrow 6L+3, L > 0$$

Hence, for $L = 0 \Rightarrow 3$ components,

$L = 1 \Rightarrow 9$ components

$L = 2 \Rightarrow 15$ components

The total angular momentum of a vector field is $\vec{J} = \vec{L} + \vec{S}$, where $\vec{L} = -i\vec{r} \times \vec{\nabla}$ and $\vec{S}$ is the intrinsic spin.

From equation (24), it is easy to verify the usual commutation rules, for angular momentum operators are satisfied by the spin components $S_x S_y - S_y S_x = 1 S_z$, etc.

Vector $\vec{A}$: Analysis of $\vec{A}(\vec{r})$ under Rotation

It is known that if $\psi(r) \rightarrow \psi(r')$ are under transformation without change of direction and quantity, it is a scalar field. But $\vec{A}(\vec{r})$ is a vector field at point P(x, y, z), figure 17.

Therefore, it is needed

1. to define P(x, y, z);

Figure 17: S system is

rotated to S^- system

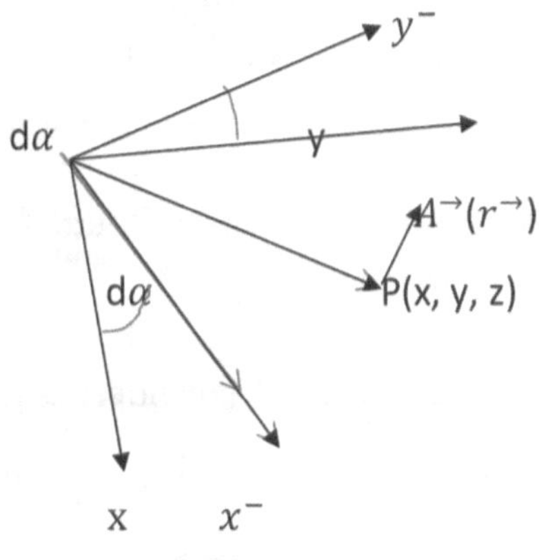

Figure 17

2. to define the bases of coordinates $\vec{e}_x, \vec{e}_y, \vec{e}_z$.

Let (x, y, z) system be drawn and point out the point P(x, y, z) and the vector field $\vec{A}(\vec{r})$ at this point, which is then to be rotated with $d\alpha$ about x-axis, as shown in figure 17.

Here, $R(d\alpha) = \begin{pmatrix} 1 & -d\alpha & 0 \\ d\alpha & 1 & 0 \\ 0 & 1 & 1 \end{pmatrix}$

$\vec{A}(\vec{r}') = \vec{A}(R\vec{r})$; let the rotation for the whole system S, i.e.,

$A^\rightarrow(r^\rightarrow) = R^{-1}A^\rightarrow(Rr^\rightarrow)$ where

$$R^{-1} = \begin{pmatrix} 1 & -d\alpha & 0 \\ d\alpha & 1 & 0 \\ 0 & 0 & 1 \end{pmatrix}$$

So the components of $\vec{A}(\vec{r})$ $\vec{A}_x, \vec{A}_y, \vec{A}_z$ are those in (1,2,3), page 10, and by expansion, neglecting $(d\alpha)^2$, it was found that J_3 is given by (23), then R = 1-idαJ_3. Now if the rotation is about the x-axis, then,

$\vec{A}' = U(d\alpha,0,0)\vec{A}$, where U (dα, 0, 0) = {1-idαJ_z} and $J_z = L_z + S_z$ (as mentioned before).

This can be generalized, considering $\bar{n}$ as any direction. So $U(\bar{n}, d\varepsilon)$ can be written as $[U(\bar{n}, d\varepsilon) = 1 - id\varepsilon\bar{n}\vec{J}]$ (25)

The eigenfunctions of the operators L^2 and L_z, as it is well-known, are the spherical harmonics functions. The intrinsic spin operators given in equation (24) affect a recording of the Cartesian components of the vector field $\vec{A}$; hence, it would be seen that the eigenfunctionions of S^2 and S_z are connected with the unit vectors $\vec{e}_x, \vec{e}_y$ and $\vec{e}_z$ along the Cartesian axis. Therefore, from equation (24), the eigenvalue equation $S_z A = \mu A$ becomes

$$S_z \begin{pmatrix} A_x \\ A_y \\ A_z \end{pmatrix} = \begin{pmatrix} -iA_y \\ iA_x \\ 0 \end{pmatrix} = \mu \begin{pmatrix} A_x \\ A_y \\ A_z \end{pmatrix}, \quad \text{where } \mu = \pm 1, 0.$$

So the solutions are given by

$$\begin{cases} \vec{A} = \pm A_x(\vec{e}_x + i\vec{e}_y) & \text{for } \mu = 1 \\ \vec{A} = \pm A_x(\vec{e}_x - i\vec{e}_y) & \text{for } \mu = -1 \\ \vec{A} = A_z\vec{e}_z & \text{for } \mu = 0 \end{cases} \quad \text{with incoherent signs}$$

These eigenfunctions can be normalized if one chooses $A_\alpha = \dfrac{1}{\sqrt{2}}$

, so $\dfrac{1}{\sqrt{2}}$ is a normalization factor, where the phases are arbitrary and are chosen so that the intrinsic spin eigenfunctions are just represented by the spherical bases vectors, which are

$$\left[\begin{array}{l} \vec{\xi}_x = -\dfrac{1}{\sqrt{2}}(\vec{e}_x + i\vec{e}_y) = \vec{\xi}_1 \\ \vec{\xi}_\circ = \vec{e}_z \qquad\qquad\quad = \vec{\xi}_0 \\ \vec{\xi}_y = \dfrac{1}{\sqrt{2}}(\vec{e}_x - i\vec{e}_y) = \vec{\xi}_{-1} \end{array} \right] \tag{26}$$

This, again, shows that the eigenfunctions of S^2 and S_z are connected with the unit vectors $\vec{e}_x, \vec{e}_y$ and $\vec{e}_z$ along the Cartesian axes. This is in complete analogy with what is known that the eigenfunctions for a unit orbital angular momentum are Y_1^m, where $m = \pm 1, 0$, which are proportional to

1) $\dfrac{-1}{\sqrt{2}}(x+iy$ for m = +1

2) Z   for m = 0

3) $\dfrac{1}{\sqrt{2}}\left(x-iy\right)$ for m = -1

It might be found to be more convenient to transform S matrices [equation (24)] to a representation where S_z is diagonal, where S (S_x, S_y, S_z), the intrinsic spin, is $\vec{S}$, which is represented as

$$\bar{S} = USU^{-1}; \quad U = \begin{pmatrix} -1 & -1 & 0 \\ 0 & 0 & 2 \\ 1 & -1 & 0 \end{pmatrix} \cdot \frac{1}{\sqrt{2}}$$

This will give the following components of S:

$$\bar{S}_x = \frac{1}{\sqrt{2}}\begin{pmatrix} 0 & 1 & 0 \\ 1 & 0 & 1 \\ 0 & 1 & 0 \end{pmatrix}; \quad \bar{S}_y = \frac{1}{\sqrt{2}}\begin{pmatrix} 0 & -i & 0 \\ i & 0 & -i \\ 0 & i & 0 \end{pmatrix}; \quad \bar{S}_z = \begin{pmatrix} 1 & 0 & 0 \\ 0 & 0 & 0 \\ 0 & 0 & -1 \end{pmatrix} \quad (27)$$

With this kind of representation, the spherical bases components are

$$\begin{bmatrix} S_1 = -\frac{1}{\sqrt{2}}\left(\bar{S}_x + i\bar{S}_y\right) = \begin{pmatrix} 0 & -1 & 0 \\ 0 & 0 & -1 \\ 0 & 0 & 0 \end{pmatrix} \\[2em] S_0 = \bar{S}_z = \begin{pmatrix} 1 & 0 & 0 \\ 0 & 0 & 0 \\ 0 & 0 & -1 \end{pmatrix} \\[2em] S_{-1} = \frac{1}{\sqrt{2}}\left(\bar{S}_x - i\bar{S}_y\right) = \begin{pmatrix} 0 & 0 & 0 \\ 1 & 0 & 0 \\ 0 & 1 & 0 \end{pmatrix} \end{bmatrix} \quad (28)$$

Hence, in any representation,

$$S^2 = S_x^2 + S_y^2 + S_z^2 = \begin{pmatrix} 1 & 0 & 0 \\ 0 & 1 & 1 \\ 0 & 0 & 1 \end{pmatrix} \cdot 2 \tag{29}$$

Which is diagonal, i.e., $2 \begin{pmatrix} 1 & & 0 \\ & 1 & \\ 0 & & 1 \end{pmatrix}$

So $S^2 = 2 \begin{pmatrix} 1 & & 0 \\ & 1 & \\ 0 & & 1 \end{pmatrix}$, therefore,

$$S^2 \begin{pmatrix} A_x \\ A_y \\ A_z \end{pmatrix} = S(S+1) \begin{pmatrix} A_x \\ A_y \\ A_z \end{pmatrix} = 2 \begin{pmatrix} A_x \\ A_y \\ A_z \end{pmatrix},$$

hence,

$S(S+1) = 2 \Rightarrow S^2 + S - 2 = 0 \Rightarrow (S+2)(S-1) = 0,$ which implies that S = 1, or S = –2 (physically nonsense). Hence, S = 1, which means that (1) is the eigenvalue of S, which, again, implies that the vector field (A) is an eigenfunction of the spin S. So the intrinsic spin of a vector field is unity. Since the electromagnetic field is classified as a vector field; this conclusion is applied to it too, which concludes that the spin (intrinsic) of the photon is unity too. Therefore,

a. The eigenfunctions of L^2 and L_z are $Y_1^{(m)}(\theta, \varphi)$.

b. The eigenfunctions of S^2 and S_z are $\vec{\xi}_{\pm 1}$ and $\vec{\xi}_0$.

The question is, what are the eigenfunctions of J, which is the total angular momentum of the vector field, where $\vec{J} = \vec{L} + \vec{S}$? Referring to the theory of coupling two types of angular momentum (such as L.S and jj coupling), one might define the following spherical harmonic function,

$$\left[Y_J^M = \sum_m \langle 1, M - m, 1, m | JM \rangle Y_1^{M-m}(\theta, \varphi) \vec{\xi}_m \right] \tag{30}$$

These functions are called, after Blatt and Weisskopf, vector spherical harmonics (VSH), because their roles are similar to Y_1^m with L^2 and L_z. If these VSH are constructed by this method, their properties can be summarized as follows:

$$J^2 \vec{Y}_{jl}^M(\theta, \phi) = j(j+1) \vec{Y}_{jl}^M(\theta, \phi), J_z Y_{jl}^M = M Y_{jl}^M \tag{31}$$

and

$$\left[\begin{matrix} L^2 Y_{jl}^M(\theta, \phi) = l(l+1) Y_{jl}^M(\theta, \phi) \\ S^2 Y_{jl}^M(\theta, \phi) = 2 Y_{jl}^M(\theta, \phi) \end{matrix} \right] \tag{31}$$

where for each value of $J^2 = j(j + 1)$, with $j \geq 1$, there will be three kinds of Y_{jl}^m corresponding to the values of l, i.e., l = j + 1, j, j−1.

Also, one can write the eigenvalue equations for $\vec{S}^2$ and S_z as

$$\left[\begin{matrix} \vec{S}^2 \vec{\xi}_\mu = S(S+1) \vec{\xi}_\mu \\ S_z \vec{\xi}_\mu = \mu \vec{\xi}_\mu \end{matrix} \right] \tag{32}$$

For $J = \vec{L} + \vec{S}$ also, Y_{Jl}^{m} can be written in ket form as

$$|JlM\rangle = \sum_{m} \langle lm - mim | JM \rangle\, Y_{1,M-m}\vec{\xi}_{m}$$, where this VSH represents

a set of vectors. This implies $\vec{Y}_{Jl}^{M}\vec{\xi}_{m} = \langle l, M - m, l, m | JM \rangle\, \vec{Y}_{1,M-m}$ is an irreducible tensor of rank l, where, as stated before,

$$l = \begin{bmatrix} J+1 \\ J \\ J-1 \end{bmatrix} \quad \text{three components of L}$$

IV-2-2—Parity of the Vector Spherical Harmonics (VSH)

The parity, as a physical dynamic variable, has been well dealt with in chapter 2. So the vector spherical harmonic $\vec{Y}_{Jl}^{m}$ has parity denoted by $(-1)^{l}$. Therefore, for a given J, there are L = J, J±l; thus, for J = l, the parity is $(-1)^{J}$, and for the other two functions with J = l±1, their parity is $(-1)^{l+1}$. If the parity operator is denoted by π (as before), then, $\pi\, Y_{jl}^{M} = (-1)^{l}\, Y_{jl}^{M}$.

Scalar spherical harmonic $Y_{l}^{(m)}(\theta, \varphi)$ form a complete set of scalar functions; also, $\vec{\xi}_{M}$ forms a complete set of bases vectors in three dimensions. Thus, the vector spherical harmonics (VSH) consequently form a complete set in a three-dimensional vector space. Therefore, one can write any vector field $\vec{A}(\vec{r})$ as a linear combination of the vector spherical harmonics (VSH), with coefficients depending only on the radial coordinates. Thus, $\vec{A}(\vec{r})$ is given by

$$\vec{A}(\vec{r}) = \sum_{J=0}^{\infty} \sum_{M=-J}^{J} C_l \vec{A}_{J,M}(\vec{r}) \tag{33}$$

Next, let L be used instead of J. Let $\vec{A}_{J,M}(\vec{r})$ be written as

$$\vec{A}_{L,M}(\vec{r}) = f_{L,M}(\vec{r})\vec{Y}_{LL}^{M}(\theta,\phi) + g_{L,M}(\vec{r})\vec{Y}_{L,L+1}^{M}(\theta,\phi) + h_{L,M}(\vec{r})\vec{Y}_{L,L-1}^{M}(\theta,\phi)$$
$$= \vec{A}_{LM}^{+}(\vec{r}) + \vec{A}_{LM}^{-}(\vec{r}) \tag{34}$$

where (+) means positive parity, even;

(–) means negative parity, odd.

Some other properties of the VSH are to be considered:

1. The projection of $\vec{Y}_{jl}$ on $\vec{\xi}_m$:

$$\vec{Y}_{jl} \cdot \vec{\xi}_m = \langle L, M-m, l, m | jM \rangle Y_l^{M-m}$$

2. Scalar product of $\vec{Y}_j^m$ with any vector $\vec{a}$ is

$$Y_{jl}^{M} \cdot \vec{a} = \sum_{m} \langle L, M-m, l, m | jM \rangle Y_l^{M-m} a_m$$

where a_m is a spherical component of $\vec{a}$, and $\left(\vec{Y}_j^M \cdot \vec{a}\right)$ is a tensor of rank l.

The parity is $(-1)^l \pi_a$, where π_a is the parity of $\vec{a}$. [As an exercise, show

that $Y_{LL}^{\rightarrow M} = \dfrac{\vec{L} Y_L^M}{\sqrt{L(L+1)}}$, this is called a gradient formula.

Note: $\vec{L} = -\nabla \times \vec{r}$

IV-3—The Multipole Fields (MF)

From the classical electrodynamics course (see Jackson), it is clearly known that a 2^L-pole multiple field is a solution of the free-space Maxwell equations (18), which is called also as Helmholtz equations. These solutions are the eigenfunctions to J^2, the total angular momentum of the vector field, with an eigenvalue of J^2, $j(j + 1) \equiv L(L + 1)$, if j is replaced by L. Also, it is an eigenfunction for J_z, one component of the total angular momentum along Z-axis with an eigenvalue M; also, it is an eigenfunction to the parity operator π_0. Therefore, a potential in 2^L-pole multipole field has to be an irreducible tensor of rank L, with a definite parity, where the parity is usually determined by the magnetic field $\left(\vec{H}\right)$ parity. The eigenvalue of the parity operator is $\pi = \pm 1$.

The identity of the field, electric or magnetic, is determined according to the parity such as

$\pi = (-1)^L$; it is 2^L-pole electric field $\vec{E}$.

$\pi = - (-1)^L$; it is 2^L-pole magnetic field $\vec{H}$.

In order to construct the multipole fields, a simple case of scalar Helmholtz equations (18) might be considered, i.e., $\nabla^2 \vec{A} + \kappa^2 \vec{A} = 0$ and $\left(\nabla^2 + \kappa^2\right)\phi = 0$.

The solutions with the required angular momentum and parity properties are given as $Z_{LM} = f_L\left(\kappa r\right) Y_L^M\left(\theta, \phi\right)$, where $f_L(r)$ is the radial function that satisfies the differential equation,

$$\frac{d^2 f_L}{dx^2} + \frac{2}{x}\frac{df_L}{dx} + [1 - \frac{L(L+1)}{x^2}]f_L = 0, \text{ where } x = kr \qquad (35)$$

Equation (35) has two solutions that are linearly independent: the spherical Bessel function $J_L(x)$, which is a regular solution at the origin,

$$J_L(x) = \left(\frac{\pi}{2x}\right)^{1/2} J_{L+1/2}(x)$$

(with $L \geq 0$) and also . The other solution that is irregular at the origin is the well-known spherical Neumann

$$n_L(x) = (-1)^{L+1} \left(\frac{\pi}{2x}\right)^{1/2} J_{-L-1/2}(x)$$

function, which is given as . Thus, $J_L(x)$ and $n_L(x)$ are linearly independent functions. Let these solutions now be taken for two spherical cases, where $x \to 0$ and $x \to \infty$. From any mathematical functions book, one finds

for $x \to 0$,

$$\left. \begin{array}{l} J_L(x) \underset{x \to 0}{\approx} \dfrac{x^L}{1 \cdot 3 \cdot 5 \cdots (2L+1)} = \dfrac{x^L}{(2L+1)!!} \\[4mm] n_L(x) \underset{x \to 0}{\approx} \dfrac{1 \cdot 3 \cdot 5 \cdots (2L-1)}{L+1} \end{array} \right]$$

(36)

$$\left. \begin{array}{l} J_L(x) \underset{x \to \infty}{\Rightarrow} \dfrac{1}{x} \sin\left(x - \dfrac{L\pi}{2}\right) \\[4mm] n_L(x) \underset{x \to \infty}{\Rightarrow} -\dfrac{1}{x} \cos\left(x - \dfrac{L\pi}{2}\right) \end{array} \right]$$

(37)

(37) represent standing waves, asymptotic solutions.

An important function that combines $J_L(x)$ and $n_L(x)$ is called spherical Hankel function of the first kind, which is defined by

$h_L^{(1)}(x) = J_L(x) + i\, n_L(x)$, which at $x \to \infty$ (asymptotic), it represents

the outgoing spherical waves, i.e., $h_L^{(1)} \underset{x \to \infty}{\longrightarrow} \dfrac{(-i)^{L+1}}{x} e^{ix}$ (38)

Hence, it is easy to distinguish between the standing wave solutions given by $U_{LM} = J_L(\kappa r) Y_L^M(\theta, \phi)$ (39)

and outgoing wave solutions, $V_{LM} = h_L^{(1)}(\kappa r) Y_{LM}(\theta, \phi)$ (40)

Both U_{LM} and V_{LM} have a parity of $\pi = (-1)^L$.

Let us now look for a solution to the Helmholtz equations (18). Two conditions are to be imposed:

a. The vector potential $\left(\vec{A}\right)$ must be an eigenfunction of J, J_z, and the parity operator π_o.

b. The vector potential for the magnetic multipole field with angular momentum L is of parity $(-1)^L$. Then

$$\vec{A}_{LM}(m) = N(L,m) f_L(\kappa r) \vec{Y}_{LL}^M$$ (41)

$\vec{A}_{LM}(m)$ must have all the required properties, where $N(L, m)$ is an arbitrary normalization constant.

The radial function $f_L(\kappa r)$ form is dictated by the physical boundary conditions. For the electric multipole of the same total angular momentum, one might choose

$$\vec{A}_{LM}(e) = N(L+1,e) f_{L+1}(\kappa r) \vec{Y}_{L,L+1}^M + N(L-1,e) f_{L-1}(\kappa r) \vec{Y}_{L,L-1}$$ (42)

This has the right parity and angular momentum.

Going back to the electric $\left(\vec{E}\right)$ and the magnetic $\left(\vec{H}\right)$ fields, where the following important relations are established,

$$\left.\begin{array}{l}\vec{E}_{LM}\begin{pmatrix} e \\ m \end{pmatrix} = -i\kappa \vec{A}_{LM}\begin{pmatrix} e \\ m \end{pmatrix} \\ \vec{H}_{LM}\begin{pmatrix} e \\ m \end{pmatrix} = \vec{\nabla} \times \vec{A}_{LM}\begin{pmatrix} e \\ m \end{pmatrix}\end{array}\right]$$

(43)

These are multipole fields; (e) denotes electric, and (m) denotes magnetic.

IV-3-1—Dual Fields (DF)

Consider the Maxwell equations in free space:

$$\vec{\nabla} \times \vec{H} = \frac{1}{c}\frac{\partial \vec{E}}{\partial t}; \quad \vec{\nabla} \times \vec{E} = -\frac{1}{c}\frac{\partial \vec{H}}{\partial t}$$

$$\vec{\nabla} \times \vec{H} = 0; \quad \vec{\nabla} \times \vec{E} = 0$$

Now if $\underline{\vec{E}}$ and $\underline{\vec{H}}$ are solutions, then $\underline{\vec{E}}'$ and $\underline{\vec{H}}'$ are solutions too if

$\underline{\vec{E}}' = \pm\underline{\vec{H}}'$ and $\vec{H}' = \pm\underline{\vec{E}}'$, where either the upper or the lower signs are used. Thus, the fields $\vec{E}'$ and $\vec{H}'$ are said to be dual fields to $\vec{E}$ and $\vec{H}$, respectively. To have a dual field with a solenoid gauge (Lorentz condition), $\phi = 0$ and $\vec{\nabla} \cdot \vec{A} = 0$, then, we have

$$N(L, m) = +1; \quad N(L+1, e) = +\sqrt{\frac{L}{2L+1}}$$

and
$$N(L-1, e) = +\sqrt{\frac{L+1}{2L+1}}$$

These yields to

$$\vec{E}_{LM}(m) = -\vec{H}_{LM}(e) = -i\kappa f_L \vec{Y}_{LL}^M \tag{44}$$

$$\vec{H}_{LM}(m) = \vec{E}_{LM}(e) = i\kappa[-\sqrt{\frac{L}{2L+1}}f_{L+1}\vec{Y}_{L,L+1}^M + \sqrt{\frac{L+1}{2L+1}}f_{L-1}\vec{Y}_{L,L-1}^M] \tag{45}$$

Equations (44) and (45) are the multipole fields that are the solutions of Maxwell equations in free space.

$\vec{E}_{LM}(m)$ is the eigenfunction of J^2. In other words, this magnetic field conserves a total angular momentum with eigenvalue L(L + 1). Also, for J_z of M eigenvalue and parity (π), $\Pi \to \begin{pmatrix} e \\ m \end{pmatrix}$.

Chapter Five

Emission of Multipole Radiations (EMR)

(Electromagnetic Radiations)

V-1—Selection Rules (SR)

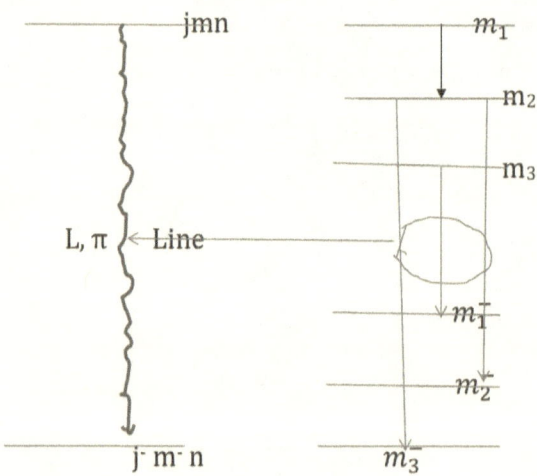

Figure1: schematic diagram to the radiations' transitions between states

We all know that in atomic systems, there is mostly electric dipole radiation, and seldom is there a magnetic dipole or electric quadrupole transitions. In nuclear physics, the problem is quite different where there are higher multiples, a case that is very common. As it is shown in the schematic diagram above (figure 1), all the components together give a line. These components can be split if they are subjected to an electric field (nuclear Zeeman effect). The transition from the state (njm) to the state

$(\overline{n\,jm})$ is given by the fermi golden rule, the second, i.e., the transition probability is given as

$$T_{mjm\bar{j}}(L,\pi) = \frac{2\pi}{\hbar}\left|H_{m\,\overline{j}mj}(L,\pi)\right|^2 \rho(E)$$ -------------------(1)

where the matrix elements of H are given by

$\left\langle n\overline{jm}\middle|H(L,\pi)\middle|\overline{njm}\right\rangle = H_{m\,\overline{j}mj}(L,\pi)$ and L, π are the multipolity

order $\left|\bar{j}-j\right| \le L \le \bar{j}+j$ and π *is* the parity of the line of transitions.

Also, $T_{ij^-}(L,\pi) = \sum_{mm^-}T_{mm^-}$ (see the above diagram)-----(2)

The state vector given as $\psi_L^M(N)$ represents the emitted or absorbed

particle, where ψ_j^m is the initial status of the system and $\psi_j^{\overline{m}}(\overline{n})$ is the final status of the system.

In studying a perturbed system, due to an external force, the perturbation theory is the most useful to tackle such problem. Therefore, the first-order perturbation theory is mostly used. The perturbation term of the interaction is usually denoted by $\overline{H}$ (Hamiltonian term due to interaction). The $\overline{H}(L,\pi)$, the Hamiltonian operator, is a scalar.

Now by replacing $\left\langle njm\middle|\text{and}\middle|\overline{njm}\right\rangle$ by $Y_j^m(n)$ and $Y_j^{\overline{m}}(\overline{n})$, respectively, the matrix element of $\overline{H}$ is

$$\left\langle \psi_L^M(N)\psi_j^{\overline{m}}(\overline{n}), \overline{H}\psi_j^m(n)\right\rangle = \left\langle \psi_L^M(N)\psi_j^{\overline{m}}(\overline{n})\,\overline{H}\psi_j^m(n)\right\rangle$$

$$= \left\langle LMN\overline{jmn}\middle|\overline{H}\middle|jmn\right\rangle$$

The product $\psi_L^M \psi_j^{\overline{m}}$ represents separated parts of the system. It might be represented as a sum of terms, each corresponding to definite eigenvalues of the operators J^2 and J_2 of the total angular momentum. Therefore,

$$\psi_L^M \psi_j^{\overline{m}} = \sum_J \langle LN\overline{jm} | jm \rangle \varphi_j^m \ ; m = M + \overline{m}$$

where $\phi_j^m = \sum_m <LMj^- m^- | jm> \psi_L^M \psi_j^{m^-}$

Now because $\overline{H}$ is a scalar operator, it is a diagonal in j and m. Therefore, the matrix element of $\overline{H}$ is given by

$$\langle LM\overline{jm} | \overline{H} | jm \rangle = \left(\varphi_j^m, \overline{H}\psi_j^m \right) \langle LM\overline{jm} | jm \rangle \tag{3}$$

As it was shown in chapter 2, Wigner-Eckart theorem, the first factor in (48), is a scalar, so it is independent of m (m represents direction!). Thus, (48) can be written as

$$\langle LMN\overline{jmn} | \overline{H} | jmn \rangle = \langle LM\overline{jn} | jm \rangle \times f\left(LN\overline{jjnn} \right) \tag{4}$$

With the aid of the Wigner-Eckart theorem, in chapter 2, the reduced matrix element (RME) of the scalar operator $\overline{H}$ is defined as

$$\langle LMN\overline{jmn} | \overline{H} | jmn \rangle = (-1)^{j-\overline{m}} \begin{pmatrix} \overline{j} & L & j \\ -m & M & m \end{pmatrix} \times \langle LN\overline{jn} \| \overline{H} \| jn \rangle \tag{5}$$

Let us consider a nuclear case where the transition takes place from a state (*i*) to a lower state (*f*) with a multipole order being L and parity π. Again with the aid of W-E theorem, as in (5), the matrix element $\overline{H}_{\overline{mm}}$ is

$$\langle LMN\overline{fmn} | \overline{H}(L, \pi) | Imn \rangle = \overline{H}_{\overline{mm}} = (-1)^{I-m} \begin{pmatrix} f & L & I \\ -m & M & m \end{pmatrix} \langle f \| \overline{H}(L, \pi) \| I \rangle \tag{6}$$

Now combine equation (5) with equation (6); the following selection rules arrive at

1) $\begin{pmatrix} f & L & I \\ -m & M & m \end{pmatrix} = 0$ unless $\Delta(f \quad L \quad I) = 0$, triangle condition

or, i.e., $|f - I| \le L \le f + I$ (7)

2) Due to (44) and (45) equations (chapter 4), multipole radiations with L = 0 will not exist. Thus, irradiative transitions accompanied by an emission of photon between states of angular momentum I and a state of angular momentum $\bar{I} = I = 0$ are absolutely not allowed. In another way of stating this, there are no monopole transitions by a single photon:

3) The transition usually starts from an upper substate (m) and ends at a certain lower substate ($\bar{m}$), as shown in the schematic diagram below.

The Δm selection rule is $I_i = I$

$|\Delta m| \le L$

$\Delta m = 0, \pm 1, \pm 2, \ldots \pm L$

$I_f = I^-$

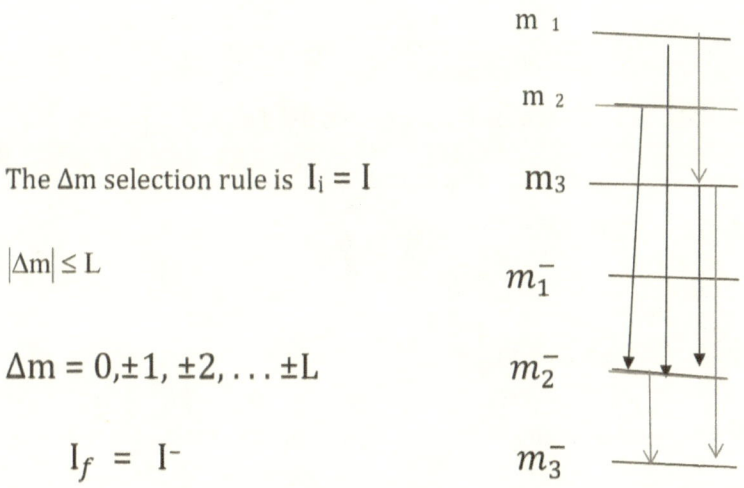

The various (Δm) transitions do not occur with some intensity, because

$$\frac{T\left(m_1 \rightarrow \overline{m_1}\right)}{T\left(m_2 \rightarrow \overline{m_2}\right)} = \begin{pmatrix} \overline{I} & L & I \\ -\overline{m_1} & M & m_1 \end{pmatrix}^2 \begin{pmatrix} \overline{I} & L & I \\ -\overline{m_2} & M & m_2 \end{pmatrix}^{-2}$$

Transition ratio that also explains why some Δm transitions are forbidden.

Parity Selection Rules

According to Wigner-Eckart theorem (chapter 2), the condition for a transition to occur is $\pi_f \pi_{op} \pi_i = +1$, where π_f is the parity of the final state, π_{op} is the parity of the interaction operator causing transition, and π_i is the parity of the initial state [chapter 2, equation (70)]. The operator under consideration is the interaction Hamiltonian $\overline{H}$, where, $H^- = \dfrac{ie\hbar}{mc} A_L^-(r^-)\cdot \nabla^-$,(as shown in chapter 3). For a certain multipole radiation (L, π) $H^-(L,\pi)$ is given as

$$H^-(L,\pi) = \frac{ie\hbar}{mc} A_L^-(r^-)\cdot\nabla^-, \quad \vec{\nabla} = i\frac{\partial}{\partial x} + j\frac{\partial}{\partial y} + k\frac{\partial}{\partial z}$$

therefore, ∇^- is an operator with odd parity. Thus, the parity of $\overline{H}$ is opposite to the parity of the operator $A_L^-(\pi)$. From equation (43), chapter 4, the magnetic field vector is

$$\overline{H}_{LM}\begin{pmatrix} e \\ m \end{pmatrix} = \vec{\nabla}\times\vec{A}_{LM}\begin{pmatrix} e \\ m \end{pmatrix},$$

which also shows that the parity of $\overline{H}_M$ is opposite to the parity of $\vec{A}_{LM}$. Therefore, the parity of $\overline{H}$ is the same as the parity of the magnetic field $\vec{H}(\vec{r})$. This fact clarifies the reason of the choosing of the magnetic radiation as electric or magnetic character of the multipole. This can be summarized as

$\pi_{op} = (-1)^L$ for electric multipole radiation;

$\pi_{op} = (-1)^{L+1}$ for magnetic multipole radiation.

Also, $\pi_f \pi_{op} \pi_i = +1 = (-1)^{2L}$, which leads to

$\pi_i \pi_f = (-1)^L$ for electric multipole radiation;

$\pi_i \pi_f = (-1)^{L+1}$ for magnetic multipole radiation.

Note: The conclusion, in view of this, is that it is not possible to have a mixture of electric or magnetic multipoles of the same order (L). But it is possible for a different order (L).

The probability of multipole emissions is decreasing rapidly with increasing L. Tables A and B summarize the results obtained from these selection rules, where the minimum order $l_{min} = L$ of the multipole radiation in a transition from state (I_a, π_a) to state (I_b, π_b).

Table A: $I_a \neq I_b$

	Electric	**Magnetic**
Parity favored $\pi_a \pi_b = (-1)^{la-lb}$	$L = \left\| I_a - I_b \right\|$	$L = \left\| I_a - I_b \right\| + 1$ except for I_a or $I_b = 0$
Parity unfavored $\pi_a \pi_b = (-1)^{la-lb+1}$	$L = \left\| I_a - I_b \right\| + 1$ except for I_a or $I_b = 0$	$L = \left\| I_a - I_b \right\|$

Table B: $I_a = I_b \neq 0$

	Electric	Magnetic
$\pi_a = \pi_b$	$L = 2$ except for $I_a = I_b = 1/2$	$L = 1$
$\pi_a = -\pi_b$	$L = 1$	$L = 2$ except for $I_a = I_b = 1/2$

Note: the transitions where the electric radiation of the lowest multipole order allowed by the spins is allowed by the parities too are sometimes called a *parity-favored transition*.

V-1-1—Some Remarks to Be Noted

1. If I_a or $I_b = 0$, then only pure E or pure M radiation can occur (2^L pole in either case). The former will occur if $\Delta\pi = (-1)^{L+1}$, because $L_{max} = L_{min} = I_a$ or I_b, so there is one choice of L value only.

2. If $I_a = I_b = 1/2$, then, $L_{min} = 0$ (which is not allowed); $L_{max} = 1$. Hence, only dipole radiation can be emitted by obeying the rules:

   Electric dipole (ED)—if $\Delta\pi$ is *yes*, change *yes*;

   Magnetic dipole (MD)—if $\Delta\pi$ is *no*, change *no*.

3. If L_1 is the same value for the emission of electric and magnetic radiations, the electric radiation emission is much more probable than that of magnetic radiation. Therefore, the radiation in the case of parity favored is practically pure electric emission (multipole) of order $L = \left| I_a - I_b \right|$.

There are two reasons for the weakness of the magnetic radiation: first, it is a magnetic, and second, it requires a higher multiple order (L). Hence, these two effects work against each other in *parity-unfavored* transitions. This phenomenon leads to radiations mixing, i.e., electric and magnetic mixing [more usually E_2 and M_1, i.e., $(E_2 + M_1)$]; note they are different: L, 1, 2. This kind of mixing occurs in certain transitions. It is observed that in most cases, the magnetic radiation is expected to be predominant in parity-unfavored transitions.

V-2—The Sources of Multipole Radiation, the Dynamic Multipole Moments

Cha rge Distribu $= \bigcirc +$. $\downarrow\uparrow + \uparrow\downarrow\uparrow\downarrow$ = core + nucleons' distribution outside the nucleus. Nucleus Core

This proposed status of a nucleus represents the source of radiations being emitted according to the situation of the nucleus and according to the selection rules mentioned before. As we have seen in the previous sections, the multipole field is studied through the following equations:-

$$\vec{A}(r) = \sum_{L,\pi,M} C_L(\pi)\vec{A}_{L,M}(\pi,\vec{r})$$

where, $\vec{A}_{L,M} = -h^{(1)}(\kappa r)Y_L^M(\theta,\phi)$,magnetic.

$$\vec{A}_{LM}(e) = +\frac{1}{\sqrt{2L+1}}\left[-\sqrt{L}h_{L+1}^{(1)}(\kappa r)\vec{Y}_{L,L+1}^M(\theta,\phi) + \sqrt{L+1}h_{L-1}^{(1)}(\kappa r)\vec{Y}_{L,L-1}^M(\theta,\phi)\right] \quad (8)$$

(8) is an irreducible tensor of rank L. It is irregular at the origin.

By the same way, the standing wave vector potential can be written as

$$\vec{B}_{LM}(m,\vec{r}) = -J_L(\kappa r)Y_{LL}^M(\theta',\varphi')$$

$$\vec{B}_{LM}(e,\vec{r}) = +\frac{1}{\sqrt{2L+1}}\left[-\sqrt{L}J_{L+1}^{(1)}(\kappa r)\vec{Y}_{L,L+1}^M(\theta',\varphi') + \sqrt{L+1}J_{L-1}^{(1)}(\kappa r)\vec{Y}_{L,L-1}^M(\theta',\varphi')\right]$$

(9)

The retarded potential solution (as mentioned in chapter 3) is given by

$$\vec{A}(\vec{r}) = \frac{1}{c}\int \vec{J}(\vec{r})\frac{e^{i\kappa|\vec{r}-\vec{r}'|}}{|\vec{r}-\vec{r}'|}dV'$$, where the current is (as it is known) (convection) + c cur m^- (Amperian).

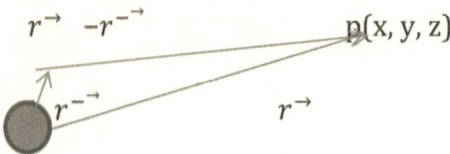

Now following the same procedures as those in static interactions (chapter 3), the expression $\dfrac{e^{i\kappa|\vec{r}-\vec{r}'|}}{|\vec{r}-\vec{r}'|}$ can be expanded in terms of spherical tensors. This kind of expansion (as it is known mathematically) is quite possible. So let

$$G(\vec{r}',\vec{r}) = \frac{e^{i\kappa|\vec{r}-\vec{r}'|}}{|\vec{r}-\vec{r}'|}$$

And by expansion, G becomes

$$G(r',r) = 4\pi i\kappa \sum_{L=0}^{\infty}\sum_{M=-L}^{L}(-1)^M J_L(\kappa r')Y_L^{-M}(\theta',\phi) \times h_L^{(1)}(\kappa r)Y_L^{+M}(\theta,\phi)$$

(10)

For $\vec{r}' < \vec{r}$, G takes the form

$$G(\vec{r}',\vec{r}) = 4\pi i\kappa \sum_{L,M}(-1)U_L^{-M}(r',\theta',\phi')V_L^{+M}(r,\theta,\phi)$$

(11)

where $U_L^{-M} = J_L(\kappa r')Y_L^{-M}(\theta',\phi')$ and $V_L^M = h_L^{(1)}(\kappa r)Y_L^M(\theta,\phi)$ (12)

Now let us introduce a dyadic operator that looks like an identity operator in the matrix case denoted by

$$\Pi = \sum_m (-1)^m \vec{\xi}_m \vec{\xi}_{m'}$$
$$=I$$

(14)

such as, for any vector $\vec{J}$, $\Pi.\vec{J} = \vec{J}$

(15)

Hence,

$$\Pi G(\vec{r}',\vec{r}) = 4\pi i\kappa \sum_{L,M,m}(-1)^M J_L(\kappa r)Y_L^{-M}(\theta',\phi') \times (-1)^m \vec{\xi}_m h_L^{(1)}(\kappa r) \times Y_L^{+M}(\theta,\phi)\vec{\xi}_{-m}$$

$$= \sum_m (-1)^M \vec{B}_{L-M}(\vec{r}') \cdot \vec{A}_{LM}(\vec{r})$$

(16)

Now, vector spherical harmonic $\vec{Y}_{Ll}^M$ has been defined [chapter 3, equation (30)] as

$$Y_{Jl}^M = \sum_m \langle 1, M-m, I, m | JM \rangle Y_l^{M-m}(\theta,\phi)\vec{\xi}_m$$

Hence, the outgoing solution takes the form

$$A_L^M(m) = -\sum_m \langle L, M+m, I, -m | LM \rangle V_L^{M+m}\vec{\xi}_{-m}$$

(17)

and

$$A_L^M(e) = +\frac{1}{\sqrt{2L+1}}[-\sqrt{L}\sum_m \langle L+1, M+m, I, -m | LM \rangle V_{L+1}^{M+m}\vec{\xi}_{-m} + (L+1)$$

$$\sum_m \langle L-1, M+m, I, -m | LM \rangle V_{L-1}^{M+m}\vec{\xi}_{-m}]$$

(18)

Up to this point, a solenoid (Lorentz) gauge $\vec{\nabla} \cdot \vec{A} = 0$ was used, where it is not possible to use it if there are sources around. This implies that the vector potentials ($\vec{A}$) used here do not form a complete set, $\vec{A}' = \vec{A} + \vec{\nabla} \cdot \vec{S}$, where S (x, y, z) is scalar function; $\vec{A}$ gives the same field. The second term gives a field of zero strength. This field is denoted as longitudinal field.

Now, with the help of the gradient formula $\left[\vec{Y}_{LL}^M = \dfrac{\vec{L}Y_L^M}{\sqrt{2L+1}} \right]$, it can be shown that

$$\vec{A}_{LM}(e) = \frac{1}{\kappa} \vec{\nabla} J_L(\kappa R) Y_L^M$$

$$= \frac{1}{\sqrt{2L+1}} \left(\sqrt{L+1} \cdot J_{L+1} \vec{Y}_{L,L+1}^M + \sqrt{L} J_{L-1} \vec{Y}_{L,L-1}^M \right) \tag{19}$$

Therefore, the fields (17, 18, and 19) form a complete set. Hence, we add to (17) and (18) the following:

$$\vec{A}_{LM}(e) = + \frac{1}{\sqrt{2L+1}} [\sqrt{L+1} \sum_m \langle L+1, M+m, I, -m | LM \rangle V_{L+1}^{M+m} \vec{\xi}_{-m}] +$$

$$[\sqrt{L} \sum_m \langle L-1, M+m, I, -m | LM \rangle V_{L-1}^{M+m} \vec{\xi}_{-m}] \tag{20}$$

Now, following the same procedure, similar relations will follow for $\vec{B}_{LM}$, where one can write for the vector($\vec{B}_{LM}(\pi)$) that

$$=IG= \quad \Pi G = 4\pi i \kappa \sum_{L,M} (-1)^M \vec{A}_{LM}(\pi) \cdot \vec{B}_{L-M}(\pi)$$

$$= \sum_{L\mu} \sum_{mm'\mu} \langle L+\mu, M+m, I, -m | LM \rangle \langle L+\mu, M+m', I, -m' | LM \rangle V_{L+\mu}^{M+m} U_{L+\mu}^{-(m+m')} \times (-1)^m \vec{\xi}_{-m} \vec{\xi}_{m'}$$

where $\mu = \pm 1, 0$.

We notice here the sum is order L and μ, keeping $L + \mu = L'$ fixed. Using the orthogonality of CGC will bring in a factor $\delta_{mm'}$, then the sum will be over

$(m + M = M')$, then IG takes the form

$$\Pi G = \sum_{L'M'\pi'}(-1)^{M'} V_{L'}^{M'} U_{L'}^{'M'} \quad \text{or} \quad \Pi G = 4\pi i\kappa \sum_{L,M,\pi} (-1)^M \vec{A}_{LM}(\pi)\vec{B}_{LM}(\pi)$$

(21)

(as it was shown on page 133)

$$\vec{A} = \frac{1}{c} 4\pi i\kappa \sum_{L,M,\pi} (-1)^M \vec{A}_{LM}(\pi,\vec{r}) \int \vec{J}(\vec{r})\vec{B}_{L,-M}(\pi,\vec{r}')d\overline{V}$$

(22)

where $\left[\sum_{L,M,\pi}(-1)^M \vec{A}_{LM}(\pi,\vec{r})\right]$ is evaluated at the field point, and $\int \vec{J}(\vec{r})\vec{B}_{L,-M}(\pi,\vec{r}')dV$ is evaluated at the source point.

For the dual fields, as shown before, we have

$$\vec{E}_{LM}(m) = -i\kappa\vec{A}_{LM}(m) = -\vec{H}_{LM}(e) = -i\kappa f_L \vec{Y}_{LL}^M - \vec{H}_{LM}(\begin{smallmatrix}e\\m\end{smallmatrix})$$

$$= -\vec{\nabla} \times \vec{A}_{LM}(\begin{smallmatrix}e\\m\end{smallmatrix}); \vec{H}_{LM}(m) = E_{LM}(e)$$

(23)

This clearly leads to

$$\left.\begin{array}{l}\vec{A}_{LM}(e) = \dfrac{1}{i\kappa}\vec{\nabla} \times \vec{A}_{LM}(m) \\[2mm] \vec{A}_{LM}(m) = -\dfrac{1}{i\kappa}\vec{\nabla} \times \vec{A}_{LM}(e) \\[2mm] \vec{A}_{LM}(m) = h_L^{(1)}(\kappa r)Y_{LL}^M\end{array}\right\}$$

(24)

Similar equations can be written for $\vec{B}$.

If the current J and the charge distribution of the source are characterized by multipole moments that are valid outside the source, then $\vec{H}(\vec{r})$ might be written as

$$\vec{H}(\vec{r}) = \frac{1}{\kappa} \sum_{L,M} a_{LM}(e) H_{LM}(e) + a_{LM}(m)\vec{H}(m)$$

$$= \sum_{L,M} \left\{ a_{LM}(e) i \vec{A}_{LM}(e) + La_{L,M}(m)\vec{\nabla} \times \vec{A}_{LM}(m) \right. \tag{A}$$

where $\vec{H}(\vec{r}) = \vec{\nabla} \times \vec{A}(\vec{r})$

So equation (A) can be written as

$$= \sum_{LM} \left(\frac{4\pi i\kappa}{c} \right) \left\{ \left(\vec{\nabla} \times \vec{A}_{LM}(e) \right) \int \vec{J}(\vec{r}') B_{LM}^*(e) d\vec{V} + \left(\vec{\nabla} A_{LM}(m) \right) \int \vec{J}(\vec{r}) B_{LM}^*(m) d\vec{V} \right\} \tag{A$'$}$$

By comparing terms in (A, A') equations, one obtains

$$\left. \begin{array}{l} a_{LM}(e) = \dfrac{-4\pi\kappa^2}{c} \int \vec{J}(\vec{r}')\vec{\nabla} \times \vec{B}_{LM}^*(m) d\vec{V} \\[3mm] a_{LM}(m) = \dfrac{4\pi i\kappa^2}{c} \int \vec{B}_{LM}^*(m)\vec{J}(\vec{r}') d\vec{V} \end{array} \right\} \tag{25}$$

From a mathematical theorem, it is known that the general relationship between any two vectors $\vec{R}$ and $\vec{S}$ is given by

$$\vec{\nabla} \cdot \left(\vec{R} \times \vec{S} \right) = \vec{S} \cdot \left(\vec{\nabla} \times \vec{R} \right) - \vec{R} \cdot \left(\vec{\nabla} \times \vec{S} \right) \tag{26}$$

Also, it is known that $\int \vec{\nabla} \cdot \left(\vec{J} \times \vec{B} \right) = 0$ $\tag{27}$

Using (26) and (27), one gets

$$a_{LM}(e) = -\frac{4\pi\kappa}{c} \int \vec{B}_{LM}(m) \cdot \left(\vec{\nabla} \times \vec{J}(r) \right) \cdot d\vec{V} \tag{28}$$

As shown before,

$$J_L(\kappa r) \to \frac{(\kappa r)^2}{(2L+1)!!} \quad , \kappa = \frac{2\pi}{\lambda} \quad ;\lambda(\overset{\circ}{A}) = \frac{12.4}{E(KeV)}$$

as $\kappa r \to 0$

So for energy E = 1.24 MeV $\Longrightarrow \lambda = 10^{-10}$ cm.

Take a nucleus of mass number A = 125, then its radius is R = 1.2 ×

$10^{-13}A^{1/3} = 6 \times 10^{-13}$ cm $\Longrightarrow \frac{\lambda}{R} = 10^{-10}\Big/ 10^{-13} \times 6 = \frac{10^3}{6} \cong 1.67 \times 10^2$

So λ = 167R, i.e., $\lambda >> R$.

Now $\kappa R = R \cdot \frac{2\pi}{\lambda} = \frac{6.28}{10^{-10}} \cdot 6 \times 10^{-13} \cong 37 \times 10^{-3}$

Therefore, $\kappa R \cong 0.04 << 1$, which implies that $\kappa R << 1$. For such value of (κR), an approximation can be possibly used, which might be shown soon.

$B_{LM}^*(m) = J_L(\kappa r)\vec{Y}_{LL}^{M^*}$, where $Y_{LL}^M \equiv$ gradient formula given by

$\vec{Y}_{LL}^M = \frac{-LY_L^M}{\sqrt{L(L+1)}}$.

Now by applying this to equation (24) for $a_{LM}(e)$, then

$$\int J_L(\kappa r)\vec{Y}_{LL}^{M^*}\left(\vec{\nabla} \times \vec{J}(r)\right)d\overline{V} = -\frac{1}{\sqrt{L(L+1)}} \int J_L(\kappa r)\vec{L}Y_{LM'}^{M^*}\left(\vec{\nabla} \times \vec{J}(r)\right)d\overline{V}$$

$$= -\frac{1}{\sqrt{L(L+1)}} \int J_L(\kappa r) \cdot (\vec{L}\left(\vec{\nabla} \times \vec{J}(r)\right)Y_L^{M^*}d\overline{V}$$

$$= -\frac{1}{\sqrt{L(L+1)}} \int (\vec{L}\left(\vec{\nabla} \times \vec{J}\right))Y_L^{M^*}J_L(\kappa r)d\overline{V} \text{ ,because } \vec{L} = i\vec{\nabla} \times \vec{r}' = -i\vec{r}' \times \vec{\nabla}$$

is Hermitian.

Remember the vector identity in vector analysis (CM), which states

that $-i(\vec{r}' \times \vec{\nabla}) \cdot (\vec{\nabla} \times J) = -i[(\vec{r}' \cdot \vec{\nabla} + 1)(\vec{\nabla} \cdot \vec{J}) - \nabla^2(\vec{r}' \cdot \vec{J})]$ from

continuity equation $\vec{\nabla} \cdot \vec{J} + \dfrac{\partial \rho}{\partial t} = 0, \quad \vec{\nabla} \cdot \vec{J} = +i\omega\rho_{\circ}$. Therefore, the

integral mentioned above can be rewritten as

$$\frac{1}{\sqrt{L(L+1)}} \int Y_L^{M^*} J_L(\kappa r') \left[\left(2 + r' \frac{\partial}{\partial r'} \right) i\omega\rho_{\circ} + \kappa^2 (\vec{r}' \cdot J) \right] d\overline{V}$$

If one integrates $\dfrac{\partial \rho_{\circ}}{\partial r'}$ by parts, the integral becomes

$$= -\frac{1}{\sqrt{L(L+1)}} \int Y_L^{M^*} \left\{ \left(r^2 \frac{dJ_L(\kappa r)}{dr'} + J_L(\kappa r) \right) i\omega\rho_{\circ} - \kappa^2 J_L(\kappa r)(\vec{r}' \cdot J) \right\} d\overline{V}$$

The term $\kappa^2 J_L(\kappa r)(\vec{r}' \cdot \vec{J})$ is of the order of $(\kappa r')^2$, and since $\kappa r \ll 1$,

then, $r^2 \dfrac{dJ_L(\kappa r)}{dr'} + J_L(\kappa r) \rightarrow \dfrac{L+1}{(2L+1)!!}(\kappa r')^L$. Therefore, $a_{LM}(e)$

takes the form

$$a_{LM}(e) = -\frac{4\pi\kappa}{c} \left(-\frac{1}{\sqrt{L(L+1)}} \cdot \frac{(L+1)}{(2L+1)!!} \right) i\omega\kappa^2 \times \int r'^L Y_L^{M^*}(\theta', \varphi') \rho(\vec{r}') d\overline{V}$$

$$(29)$$

where $\int r'^L Y_L^{M^*}(\theta', \varphi') \rho(\vec{r}') d\overline{V} \equiv Q_{LM}$, which is called the

dynamic electric multipole moments, external dipole. Hence,

$$[Q_{LM} = \int r'^L Y_L^{M^*}(\theta', \varphi') \rho(\vec{r}') d\overline{V}]$$

$$(30)$$

Following the same procedure, one can find the amplitude of magnetic

multipole radiations. From equation (25), the second one, we have

$$a_{LM}(m) = \frac{4\pi i \kappa^2}{c} \int \vec{J}(\vec{r}') B^*_{LM}(m) d\overline{V}$$

Now $\vec{B}^*_{LM}(m) = -J_L(\kappa r) Y^M_{LL}(\theta', \varphi')$

Therefore, $a_{LM}(m) = -\frac{4\pi i \kappa^2}{c} \int \vec{J}(\vec{r}') J_L(\kappa r) Y^{M^*}_{LL}(\theta', \varphi') d\overline{V}$

The integrand $J_L(\kappa r) \vec{J}(\vec{r}') Y^{M^*}_{LL}(\theta', \varphi')$

$$= -J_L(\kappa r) \frac{1}{\sqrt{L(L+1)}} \left(\vec{L} Y^{M^*}_L\right) \cdot \vec{J}(\vec{r}')$$

where $\vec{L} = -i\vec{r} \times \vec{\nabla}$

$$= +J_L(\kappa r) \frac{1}{\sqrt{L(L+1)}} \left(\vec{r} \times \vec{\nabla} Y^{M^*}_L\right) \cdot \vec{J}(\vec{r}') = -J_L \frac{1}{\sqrt{L(L+1)}} \vec{\nabla} Y^{M^*}_L \left(r \times \vec{J}(r')\right)$$

This is because $\nabla^{\rightarrow}$ is a Hermitian operator. Therefore,

$$= \frac{1}{\sqrt{L(L+1)}} \int J_L(\kappa r') Y^{M^*}_L(\theta', \phi') \vec{\nabla} \cdot \left(\vec{r} \times \vec{J}\right) d\overline{V}$$

$$\int J_L(\kappa r) \vec{J}(\vec{r}') \vec{Y}^{M^*}_L d\overline{V} = -\frac{1}{\sqrt{L(L+1)}} \int J_L(\kappa r) \vec{\nabla} Y^{M^*}_L \left(\vec{r} \times \vec{J}(r')\right) d\overline{V}$$

Hence, $a_{LM}(m) = \frac{4\pi}{c(2L+1)!!} \cdot \frac{\kappa^2 + L}{\sqrt{L(L+1)}} \int r'^L Y^{M^*}_L(\theta', \varphi') \vec{\nabla} \cdot \left(\vec{r}' \times \vec{J}\right) d\overline{V}$

$\Rightarrow$

$$\boxed{a_{LM}(m) = \frac{4\pi}{(2L+1)!!} \cdot \frac{1}{\sqrt{L(L+1)}} \cdot \frac{\kappa^2 + L}{c} \int r'^L Y^{M^*}_L(\theta', \varphi') \vec{\nabla} \cdot \left(\vec{r}' \times \vec{J}\right) d\overline{V}}$$

(30)

By definition, the magnetic multipole moments are given by

$$M_{LM} = \frac{1}{\sqrt{L(L+1)}} \int r'^{L} Y_{L}^{M^*}(\theta', \varphi') \vec{\nabla} \cdot (\vec{r}' \times \vec{J}) d\overline{V}$$

Hence, (31)

$$a_{LM}\binom{e}{m} = \mp \frac{4\pi}{(2L+1)!!} \sqrt{\frac{L+1}{L}} \kappa^{L+2} \binom{Q_{LM}}{M_{LM}}$$

(32)

V-3—Angular Distribution of the Multipole Radiations

The magnetic field (equation A) is given by

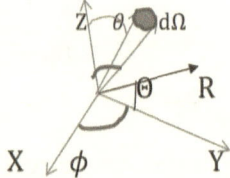

$$\vec{H}(\vec{r}) = \frac{1}{\kappa} \sum_{L,M} (a_{LM}(e) \cdot \vec{H}_{LM}(e) + \sum a_{LM}(m) \vec{H}_{LM}(m)$$

It is known from the electrodynamics that the energy flow of radiation

is described by the pointing vector $\left[\vec{F} = \frac{c}{4\pi} [\vec{E} \times \vec{H}] \right]$; hence,

$E_f = \frac{c}{4\pi} [\vec{E}(\vec{r},t) \times \vec{H}(\vec{r},t)]$, $\vec{E}(\vec{r},t)$ is given as $\vec{E}(\vec{r},t) = \vec{E}(r)e^{i\omega t} + E^*(r)e^{-i\omega t}$

$$\langle \vec{F} \rangle_t = \frac{c}{4\pi} (\vec{E} \times \vec{H}) \Rightarrow |\langle \vec{F} \rangle| = \frac{c}{2\pi} |\vec{E}|^2 = \frac{c}{2\pi} |\vec{H}|^2$$

(Refer to *Classical Electrodynamics* by Jackson.) Therefore, the energy emitted into a solid angle $d\Omega$ is by (see the figure above)

$$dE_\Omega = \left| \langle \vec{F} \rangle \right| \cdot R^2 d\Omega = U(\Omega) d\Omega.$$

$U(\Omega) \equiv U(\theta, \phi)$ is the angular distribution function.

Now take a pure multipole radiation with amplitude a_{LM}. Consider the case of the electric multipole radiation, where

$$U_{LM}(e, \Omega) = \frac{c}{2\pi} |H(e)|^2 R^2 = \frac{c}{2\pi} \cdot \frac{1}{\kappa^2} |A_{L,M}(e)|^2 |a_{LM}(e)|^2 R^2 \Rightarrow$$

$$U_{LM}(e, \Omega) = \frac{c}{2\pi} R^2 h_L^{(1)}(\kappa r) \vec{Y}_{LL}^M(\theta, \varphi) h_L^{(1)^*}(\kappa r) Y_{LL}^{M^*}(\theta, \varphi) |a_{LM}(e)|^2$$

Now,

(1) define $\left[Z_{LM}(\theta, \varphi) = Y_L^M Y_L^{M^*} \right] = \left| \vec{Y}_{LL}^M(\theta, \phi) \right|^2$.

(2) Take the asymptotic limit for $\left[h_L^{(1)}(\kappa r) \to \dfrac{e^{i\kappa r}}{\kappa r} \right]$.

So we have

$$U_{LM}(e, \Omega) = \frac{c}{2\pi} R^2 \frac{e^{-i\kappa r} e^{i\kappa r}}{\kappa^2 R^2} Z_{LM}(\theta, \phi) |a_{LM}(e)|^2 \tag{34}$$

Problems:

1. Prove that the same equation (34) holds for the magnetic multipole radiation.

2. Show that $\left[Z_{LM} = \vec{Y}_L^M \vec{Y}_L^{M^*} = \dfrac{1}{2} \left[1 - \dfrac{M(M+1)}{L(L+1)} \right] |Y_L^{M+1}|^2 \right.$

$$+\frac{1}{2}\left[1-\frac{M(M-1)}{L(L+1)}\right]\left|Y_L^{M-1}\right|^2+\frac{\kappa^2}{L(L+1)}\left|Y_L^M\right|^2\right]$$

(34)

3. Compute Z for L = 2, M = 0, ±1, ±2 and present the result in a polar diagram. There are three diagrams: 0, ±1, ±2.

Example: Take dipole radiation (L = 1) $\Rightarrow\left[Z_1^1\left(\theta,\phi\right)=\frac{3}{8\pi}\left(\sin^2\theta\right)\right]\rightarrow$

z_1^0, at z = 0, $z_1^0\Rightarrow\Theta=0,\pi$

Note: Z_{LM} is the angular distribution function, as it is seen; it is geometrically dependent, not nucleus-model dependent. But $\left|a_{LM}\left(e\right)\right|^2$ depends on the used nuclear model.

So at Z = 0, the angle is θ or π, sin 0 = 0, sin π = 0. This indicates that the dipole does not radiate along its axis if radiation is mixed, such as mixing E_2 with M_1, i.e., (E_2 + M_1), then the radiation pattern contains interference terms. Remember that multiple radiations of some multiple orders (L) do not mix.

V-3-1—Emission (Transition) Probability of Multipole Radiation:

Weisskopf Formula (as an Estimation)

The total energy emitted per second is given by

$$W_{LM} = \int dW_{LM} = \int U_{LM}(\theta,\phi)d\Omega = \frac{c}{2\pi\kappa^2}\int \vec{Y}_{LL}^{M^*}(\theta,\phi)\vec{Y}_{LL}^{M}(\theta,\phi)$$

$$\times \left|a_{LM}\left(\begin{smallmatrix}e\\m\end{smallmatrix}\right)\right|^2 \sin\theta d\theta d\phi = \frac{c}{2\pi\kappa^2}\left|a_{LM}\left(\begin{smallmatrix}e\\m\end{smallmatrix}\right)\right|^2 \int \vec{Y}_{LL}^{M^*}(\theta,\phi)\vec{Y}_{LL}^{M}(\theta,\phi)\sin\theta d\theta d\phi ,$$

where, $Z_{LM}(\theta,\phi) = \vec{Y}_{LL}^{M^*}(\theta,\phi)\vec{Y}_{LL}^{M}(\theta,\phi)$.

We notice that $Z_{LM}(\theta,\phi)$ is the only function that depends on the angle. Therefore, $\int Z_{LM}(\theta,\phi)d\Omega = \int \vec{Y}_{LL}^{M}(\theta,\phi)\vec{Y}_{LL}^{M^*}(\theta,\phi)d\Omega$.

Due to the orthogonality $\int (\ ) = 1$; hence, $\left[W_{LM}\left(\begin{smallmatrix}e\\m\end{smallmatrix}\right) = \frac{c}{2\pi\kappa^2}\left|a_{LM}\left(\begin{smallmatrix}e\\m\end{smallmatrix}\right)\right|^2\right]$

$$(36)$$

The role of quantum mechanics in calculating the transition probability is important and a basic one. Because the system we are dealing with is a microscopic system, the quantum mechanics is the right tool to tackle this system. Therefore, to go for that, one has to remember

1) energy is radiated, as quanta of $\hbar\omega$ or $h\upsilon$;

2) $\vec{\rho}$ and $\vec{J}$ (density and current) should be expressed with the concept of wave function ψ or $|nJM>$, and also, the magnetization (m), i.e., $\rho^\rightarrow$, $J^\rightarrow$, m (density, current, and magnetization), has to

be described formally by quantum mechanics. Hence, the emission probability (quanta per second) is given by

$$T_{L,M}\left({e \atop m}\right) = T_{LM}\left({e \atop m}\right) = \frac{W_{LM}\left({e \atop m}\right)}{\hbar\omega} = \frac{1}{2\pi\hbar\kappa^3}\left|a_{LM}\left({e \atop m}\right)\right|^2$$

Or with the help (aid) of equation (33), $T_{L,M}\left({e \atop m}\right)$ can take the form

$$\left[T_{LM}\left({e \atop m}\right) = \frac{8\pi(L+1)}{L\left[(2L+1)!!\right]}\frac{\kappa^{2L+1}}{\hbar}\left|\frac{Q_{LM}}{M_{LM}}\right|^2\right] \tag{37}$$

To clarify the use of quantum mechanics in describing the radiating system, we would briefly follow these important remarks:

Let the radiating system be represented as an excited state ψ_i to get rid of the excitation energy going back to its ground state. The emission of this energy as gamma radiation has to take place. The problem is of calculating the probability of the transition of γ-radiation, as mentioned and described in chapter 4.

Therefore, $\rho^{\rightarrow}\left(r^{\rightarrow}\right) = e\psi_f^*\psi_i$

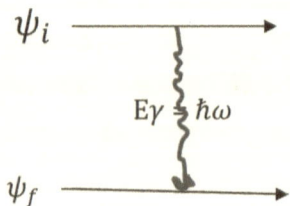

The current density J is given as

$$J = \frac{i\hbar e}{2M}[\psi_f^*\nabla\psi_i - \psi_i\nabla\psi_f^*] \tag{38}$$

Hence,

$$Q_{LM}(i \rightarrow f) = e \int r^L Y^*_{LM}(\theta, \varphi) \psi^*_f(r') \psi_i(r) dV$$

$$= e \langle \psi_f | r^L Y^*_{LM} | \psi_i \rangle$$

and

$$M_{LM}(i \rightarrow f) = \frac{1}{L+1} \cdot \frac{e}{2Mc} \int r^L Y^*_{LM}(\theta, \varphi) \vec{\nabla} \cdot (\vec{r}' \times J'_{if}) dV$$

$(r^{\rightarrow -} \times J^*_{if})$ is a magnetic moment due to current.

It is important to note that the two terms in equation (38), for J_{if} gives the same contribution (i.e., they are equal) and $\vec{L} = -i\vec{r} \times \vec{\nabla}$; therefore, after integration by parts, one gets

$$M_{LM}(i \rightarrow f) =$$

$$\frac{1}{L+1} \cdot \frac{e}{Mc} \int r^{\rightarrow} Y^*_{LM} \nabla \cdot (\psi_f, \psi_i) d\, \mathsf{v} \qquad (39)$$

In this equation where $(\vec{r} \times J_{if})$ was first used, the current J_{if} is J_o (convection) + c $(\vec{\nabla} \times \vec{m})$ (Amperian). We know the nuclear system has an intrinsic spin, so M'_{LM} is due to $\vec{S}$, which is given by

$$M'_{LM} = \frac{e\hbar}{2Mc} \int r^L Y^*_{LM} \vec{\nabla} \cdot \left[\langle f | \sigma | i \rangle \right] dV \qquad (40)$$

By applying a certain approximation to get certain estimation such as

R ⬜ λ̶ ⇒ κR ⬜ 1, then,    $R^{\rightarrow} / c^{\rightarrow}$ nucleus

$$\left|\vec{C}\right| = \vec{R}\left(\frac{d\vec{C}}{d\vec{R}}\right) - \vec{R}\cdot\vec{\nabla}\cdot\vec{C} \Rightarrow \vec{\nabla}\cdot\vec{C} = \frac{\left|\vec{C}\right|}{R}$$

This might be considered as

$$\mathrm{div} \to \frac{1}{R}, \quad \vec{L} = \sqrt{l(l+1)}$$

Also (a part of approximation),l i> is an eigenstate of J, but it is not necessary an eigenstate of $\vec{L}$, but roughly, it is considered so. Also, $\vec{L} \approx \sqrt{l(l+1)} \cong L+1$. Then we can have

$$\frac{\left|M_{LM} + M'_{LM}\right|^2}{\left|Q_{LM}\right|^2} \cong 10\left(\frac{\hbar}{McR}\right)^2 \sim 10^{-2} = 10\left(\frac{\hbar}{McR}\right)^2 \sim 10^{-2}$$

The remarkable note here is that Q_{LM} or M_{LM} are integrals containing nuclear wave functions. These are not quite well-known, which implies that accurate calculations are almost not possible. A crude estimate is the only possible calculation within the order of

magnitude of T_{LM}. Hence, $\quad T_{LM}(m) \cong 10\left(\frac{\hbar}{McR}\right)^2 T_{LM}(e)$;

$$\hbar = m = c = 1, \quad R \approx 3.4\times10^{-13} \text{ cm}$$

(for mass no.) $\to A = 100$.

Hence,

$$T_{LM}(m) \cong \frac{1}{S} A^{\frac{-2}{3}} T_{LM}(e) \cong 10^{-2} \quad \text{,(Weisskopf approximation)}$$

$$Q_{LM} \to p \cong \frac{eR}{R^3} = \frac{e}{R^2}$$

$$M_{LM} \rightarrow m \cong \frac{e\hbar}{Mc} \cdot \frac{1}{R^3} \Rightarrow$$

$$\frac{M_{LM}}{Q_{LM}} \cong \frac{\hbar}{McR} \Rightarrow M_{LM} \cong \frac{\hbar}{McR} \cdot Q_{LM}$$ (This for the convection current.)

Also, $M_{LM}^- \approx g. l\sigma^{\rightarrow}l.\mu_n \cong 3\, M_{LM}$,it is due to spin.

Where g is the Lande g-factor, it is known for a nucleus and can be easily calculated (see chapter 2). These estimations were done by Weisskopf in 1951; it is very much crude. He based his approximations on the basis of the independent-particle model, which refers the properties of a nucleus to the single particle (nucleon) outside the nucleus core. Also, it was assumed that the proton is responsible for the transition. These concepts can be tackled clearly by brief details of the single-particle model. So we have

a nuleus

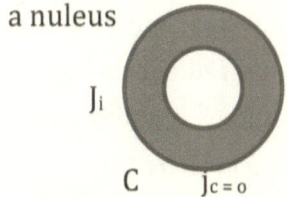

J_i

C $\quad j_{c=0}$

Where,

C is denoting a closed core, where j_c = 0,no current in the core .

J_i is the total angular momentum of the odd nucleon moving around the core, which induces current $J_n^{\rightarrow}$ with the eigenstate l I, M, π >. Thus, for electric transitions, it is assumed that a single proton moves independently in its orbit around the core with angular momentum L and with a total angular momentum $\vec{J} = \vec{L} + \vec{S}$; here, $\vec{L}$ is parallel to $\vec{S}$, then, $\Psi_i = U_i(r')Y_L^M(\theta,\phi)\chi_{1/2}^{1/2}$, where U_i is the radial, Y_L^M is

the angular, and $\chi_{1/2}^{1/2}$ is of the spin. If it is assumed that the final state of

the proton is an S state where $\Psi_f = U_f(r')\sqrt{\dfrac{1}{4\pi}}\chi_{1/2}^{1/2}$, remembering

that $Y_0^0(\theta,\phi) = \dfrac{1}{\sqrt{4\pi}}$,

then, $|i\rangle \equiv U_i(r)Y_{LM}(\theta,\varphi)\chi_{1/2,1/2}$

$|f\rangle \equiv U_i(r)Y_{00}(\theta,\varphi)\chi_{1/2,1/2}$

$\equiv U_f(r)\cdot\sqrt{\dfrac{1}{4\pi}}\chi_{1/2,1/2}$

So to calculate Q_{LM} and/or M_{LM} occurring in T_{LM}, one might assume that the wave functions (radial one) in the final and the initial states are considered as simple rectangles, whose radial extension R is identified with the radius of the nucleus. In other ways, these wave functions are treated as constants throughout the nucleus. Thus,

$$U_i(r) = U_f(r) = \sqrt{\dfrac{3}{R^3}} \qquad \text{for } r < R$$

$$= 0 \qquad \text{for } r > R$$

Hence,

$$Q_{LM} = e\int r^L Y_L^{M*}(\theta,\varphi)U_f(r)\dfrac{1}{\sqrt{4\pi}}Y_L^M(\theta,\varphi)U_i(r)dV$$

$$= \dfrac{e}{\sqrt{4\pi}}\int_0^R r^L U_i(r)U_f(r)r^2 dr$$

$$= \frac{e}{\sqrt{4\pi}} \cdot \frac{3}{R^3} \int_0^R r^{L+2} dr = \frac{3e}{\sqrt{4\pi}} \cdot \frac{1}{R^3} \frac{R^{L+3}}{L+3}$$

So,

$$\left[Q_{LM} = \frac{e}{\sqrt{4\pi}} \cdot \frac{3R^L}{L+3} \right];$$

therefore, from equation (37) and noting that

$$\kappa = \frac{\omega}{c} = \frac{2\pi\upsilon}{c} = 2\pi\lambda^{-1} = \lambda \qquad = \text{wave number,}$$

$$T_L(e) \cong \frac{2(L+1)}{L[(2L+1)!!]^2} \cdot \left(\frac{3}{L+3}\right)^2 \frac{e^2}{\hbar c} \left(\frac{\omega R}{c}\right)^{2L} \cdot \omega, \qquad \text{then,}$$

$$T_L(m) \cong 10 \left(\frac{\hbar}{McR}\right)^2 T_L(e) \cong 10 \left(\frac{\hbar}{McR}\right)^2 T_L(e) \tag{41}$$

(This is Weisskopf formula.)

Note here that U(r) is oscillating and the Weisskopf estimate mentioned above is somehow too large. Now, for I, there are (2I + 1)(M-value), so the probability of transition from M-states is $\frac{1}{2I+1}$, and from all M, S states is $\frac{1}{2I+1}$ .T$(2I+1)$ =T; therefore, T is independent of M. so the lifetime $\tau = \frac{1}{\lambda}$, where the total λ is given by

$$\lambda_t = T_{total} = T_\gamma \left(L, \binom{e}{m}\right) + \sum_x T_{ix} \left(L, \binom{e}{m}\right) .$$

Take as an example a nucleus of mass number A = 100 and $E_\alpha = \hbar\omega$ = 0.5 MeV. The results are shown in the following table:

α-Energy	A	τ_α Second	Type of Multipole Radiation
0.5 MeV	100	10^{-14}	E_1
		10^{-11}	M_1
		10^{-8}	E_2
		10^{-6}	M_2
		10^{-3}	E_3
		10^{-1}	M_3
		10^3	E_4
		10^5	M_4

There is a good agreement, in general, with the experiment, except in rare earth [E_2 = 0.5 MeV; $\tau < 10^{-12}$ sec (actinide region)]. It is found that (as an example)

$$\left[T_{exp\,f}\left(E_2\right) \approx 10^3 T_{Weisskoph}\left(E_2\right) \right] \tag{42}$$

One can conclude that to have a large transition, there must be a large number of particles participating in the transition coherently; therefore, the collective model (deformed nuclei) phenomenon plays the role here. This model was proposed by A. Bohr and Mottelson to solve or explain the observed characters of some nuclei (in the heavy region), which cannot be explained by the individual-particle model in shell theory. In the collective model, nucleons move collectively, contrary to the individual or independent particle model. As shown previously, there are parity-favored and parity-unfavored transitions. For parity-favored transition, $\Delta I = 0$, $\Delta\pi$ = yes, $\pi_{op} = (-)^L = -$, $\pi_i\,\pi_f = (-)$ $(+) = -$, so $\pi_i\,\pi_{op}\pi_f = (-)(-) = +1$, a condition that transition occurs. Such case is represented by the following diagram:

$\pi_i = -\!\!\!-\!\!\!-\!\!\!-\!\!\!-\!\!\!-2^-$

$\pi_f = +\!\!\!-\!\!\!-\gamma\!\!\!-\!\!\!-2^+$

$I_i = I_f = 2, E\gamma, L_\gamma, \pi_\gamma$

The unfavored-parity transition is often mixed multipole radiations, such as $E_2 + M_1$.

V-3-2—Internal Conversion

It is known physically that γ-radiation sometimes has not enough energy to go to a next nuclear state. So it gives its energy to the nearest electron in K, L, M subshells of an atom, which aid the electron to eject out of its orbit. This phenomenon is called an internal convention (IC), referred to in the following diagram: $E_x = \Delta E - B_x = \hbar\omega - B_x$, which is the energy of the ejected electron from x-state, where $K, L,$ M are atomic subshells and B_x is the binding energy of the electron in x-state.

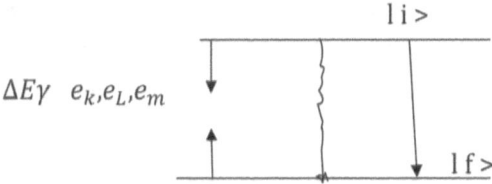

Definition:

By definition, the internal conversion coefficient, I_{cc}, for electric 2^L is given as $\alpha_x(L) = \dfrac{T(L, e_x)}{T(L, \gamma)}$, and for a magnetic

2^L, it is given as $\beta_x(L) = \dfrac{T(L, e_x)}{T(L, m)}$ for pure multipole. So

the total T is given as $T_{total} = T(L, e) + T(L, e)\sum_\alpha \alpha_x(L) =>$

$$T_{total} = T(L, e)[1 + \sum_\alpha \alpha_x(L)] \tag{43}$$

the lifetime $\tau = \dfrac{1}{T_{total}} = \dfrac{1}{T(L, e)[1 + \sum_\alpha \alpha_x(L)]} = \dfrac{\tau_\gamma}{1 + \sum_\alpha \alpha_x(L)}$,

the lifetime, where $\tau_\gamma = \dfrac{1}{T(L, e)}$ for electric multipole radiation.

Hence $\tau = \dfrac{\tau_\gamma}{1 + \sum_\alpha \alpha_x(L)}$ and $\tau = \dfrac{\tau_\gamma}{1 + \sum_x \beta_x(L)}$ $\qquad$ (44)

for 2^L electric and 2^L magnetic, respectively.

V-3-2-1—The General Idea of Computing I_{cc}

Let the following simple diagram give a crude description of an atom:

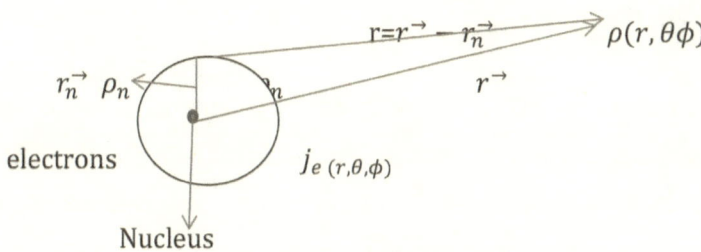

electrons

Nucleus

The ρ_e and j_e are given by electron wave, K-subshell state. The experimental data confirms that T_{total} is given by

$$T_t = T_\gamma + T_{\bar{e}} = \left(\frac{1}{\tau}\right)$$

(45)

Now, using Fermi golden rule no.2 , it can be written as

$$T\left(L, \bar{e}_x\right) = \frac{2\pi}{\hbar}\left|\langle f|\overline{H}|i\rangle\right|^2 \frac{dN}{dE}$$

Consider the initial state $|I_i, m_i, J_m>$ and the final state $| I_f, m_f, K, J>$; this can be demonstrated as

$$|I_i m_i, J_m\rangle I_i, m_i,$$

$$\bar{e}_x$$

$$|I_f m_f, K, J\rangle I_f, m_f,$$

where J_m refers to $\bar{e}$. So the matrix elements of the interaction H^- are given by $\left\langle I_f m_f, \vec{\kappa} \cdot \vec{\sigma} | H | I_i m_i, J_m \right\rangle$, $|J_m\rangle$ is a relativistic wave function for the electron, and $\vec{\kappa}$ is treated by using relativistic quantum mechanics; the nucleus is assumed as a point. Therefore, one can separate between the nuclear interaction as $H^=$ and that of electron interaction as H^-, such that the matrix element can be written as $\langle \vec{\kappa} \cdot \vec{\sigma} | \overline{H} | J_m \rangle \langle I_f m_f | \overline{\overline{H}} | I_i m_i \rangle$.

The matrix $< k^{\rightarrow}.\sigma^{\rightarrow} \, / \, H^- \, / \, j_m >$ can be calculated. It was computed by M. E. Rose. It is a very complex calculation. This confirmed that the nucleus is not a point charge but has a finite size has to be taken into consideration for accurate calculation.

For the purpose of this book, at this level, the theory of internal conversion will be simplified, assuming that

a) nonrelativistic wave function $\psi_c \ll 1$;

b) the initial state of K-electron is S-state, therefore,

$$|i\rangle = \left(\frac{1}{\pi a^3}\right)^{1/2} e^{\frac{-r}{a}} \, , \qquad a = \frac{a_o}{Z} = \left(\frac{\hbar^2}{m_e c^2}\right) \cdot \frac{1}{Z}$$

Z is atomic number, and (a_o) is Bohr radius.

Now $|f\rangle$ is plane wave with a wave number

$$\vec{\kappa}; \hbar\vec{\kappa} = \vec{P}, \quad E(\bar{e}_\kappa) = \frac{p^2}{2m_e} = \frac{\hbar^2 \kappa^2}{2m_e} \rangle \, \beta_\kappa \Rightarrow \quad \text{plane wave. Hence,}$$

$$|\vec{\kappa}\rangle = \frac{1}{v^{1/2}} e^{+i\vec{\kappa}\cdot\vec{r}} \quad \text{is the final state } |f\rangle, \text{ where } v^{1/2} = \text{normalization factor.}$$

c) $\dfrac{v}{c} \langle\langle 1$ (Nonrelativistic implies that radiation effects are to be neglected.)

Now, for static interaction, $\overline{H} \cong \dfrac{e_{el} \cdot e_n}{|\vec{r} - \vec{r}_n|}$; hence,

$$T\left(L, \overline{e}_x\right) = \frac{2\pi}{\hbar} \left|\langle \psi_f | \overline{H} | \psi_i \rangle\right|^2 \frac{dN}{dE} \quad \text{(golden rule, the second).}$$

$\dfrac{dN}{dE}$ = number of states of the ejected electron per unit energy range

$\dfrac{dN}{dE}$. For an electron direction within a solid angle element, $d\Omega$ is given

as $\dfrac{dN}{dE}$ = V(volume)$\dfrac{m\hbar\kappa}{\left(2\pi\hbar\right)^3} d\Omega = V \dfrac{m\hbar\kappa}{\left(2\pi\hbar\right)^3} d\Omega = \dfrac{Vm\kappa}{8\pi^3 \hbar^2} d\Omega$;

therefore,

$$T\left(L, \overline{e}_x\right) = \frac{Vm\kappa}{4\pi^2 \hbar^3} \int |H_{fi}|^2 d\Omega \qquad (46)$$

Now, consider the electrostatic interaction between the protons of the nucleus and the electrons that are given by

$$\overline{H} = \sum_{i=1}^{Z} \frac{e^2}{|\vec{R} - \vec{r}_i|} = \sum_{i=1}^{Z} \frac{e^2}{|R - r_i|}.$$

Let $|f\rangle$ and $|i\rangle$ be the nuclear wave functions of nuclear states f and i, and υ_f, υ_i are the wave functions of the electrons *after ejection* and in *K-shell*, respectively. Therefore, we have the following information:

(1) $\upsilon_i = \left[\pi a^3\right]^{-1/2} e^{-R/a}$ —in K-shell

(2) $\upsilon_f = \left[\dfrac{1}{V^{1/2}}\right] e^{-i\vec{\kappa}\cdot\vec{R}}$ —(plane wave) after ejection

(3) $|\psi_i\rangle = [\pi a^3]^{-1/2} e^{-R/a} |i\rangle$ —initial state of the system

(4) $|\psi_f\rangle = [V]^{-1/2} e^{-i\vec{\kappa}\cdot\vec{R}} |f\rangle$ —final state

So the matrix elements of $\overline{H}_{fi}$ can be written as

$\overline{H}_{fi} = [\pi a^3 V]^{-1/2} \sum\limits_{i=1}^{Z} \int d^3R \int d\tau \langle \psi_f | \dfrac{e^2}{|R-r_i|} | \psi_i \rangle$, using (3) and (4), $\overline{H}_{fi}$ is

written as
$$\overline{H}_{fi} = (\pi a^3 V)^{-1/2} \sum\limits_{i=1}^{Z} \int d^3R \int d\tau e^{-i\vec{\kappa}\cdot\vec{r}} \langle f | \dfrac{e^2}{|R-r_i|} | i \rangle e^{-R/a} \tag{47}$$

$\int d^3R$ extends over the coordinates of electrons.

$\int d\tau$ extends over the coordinates in nuclear wave functions.

Here, (R) are electron coordinates, and (r_i) are nucleon (proton) coordinates.

Now as it has been done in previous chapters, $\dfrac{1}{|R-r_i|}$ can be expanded

as $\left(\dfrac{r_i}{R} \langle\langle 1 \right)$.

$$\dfrac{1}{|R-r_i|} = \sum\limits_{l=0}^{\infty} \sum\limits_{-l}^{+l} \dfrac{4\pi}{2l+1} \cdot \dfrac{r_i^l}{R^{l+1}} Y_{LM}(\theta_e, \phi_e) Y_{LM}^*(\theta_n, \phi_n) \tag{48}$$

(Refer to the addition theorem of SHF.)

By substituting (48) into (47), one obtains

$$\overline{H}_{fi} = \dfrac{e}{(\pi a^3 V)^{1/2}} \sum\limits_{l=0}^{\infty} \sum\limits_{m=-l}^{l} \dfrac{4\pi}{2l+1} Q_{lm}(f,i) J_{LM} \tag{49}$$

where $Q_{lm}(fi) = e \sum_{\kappa=1}^{Z} \int r_\kappa^l Y_{lm}^* (\theta_\kappa, \phi_\kappa) \langle f|i \rangle$

and $J_{lm} = \int e^{-R/a} e^{-i\vec{\kappa}\cdot\vec{r}} R^{-(l+1)} Y_{lm}(\theta_e, \phi_e) d^3R$

If it is assumed that the wavelength of the plane wave is $\lambda \langle\langle a$ of K-shell $\Rightarrow \kappa a \rangle\rangle 1$, then $e^{-R/a} \cong 1$. Hence,

$J_{lm} = \int e^{-i\vec{\kappa}\cdot\vec{r}} R^{-(l+1)} Y_{lm}(\theta_e, \phi_e) d^3R$.

But $e^{-i\vec{\kappa}\cdot\vec{r}}$ can be expanded (as done before) as

$e^{-i\vec{\kappa}\cdot\vec{r}} = 4\pi \sum_{l=0}^{\infty} \sum_{m=-1}^{1} i^l j_l(\kappa r) Y_{lm}^* (\theta, \phi) Y_{lm}(\theta', \phi')$.

Substitute for $e^{-i\vec{\kappa}\cdot\vec{r}}$ in (J_{lm}) above; J_{lm} becomes, after integration,

$$\left[J_{lm} = 4\pi i^{-1} \frac{\kappa^{l-2}}{(2l+1)!!} Y_{lm}(\theta, \phi) \right] \tag{50}$$

If (50) is substituted into (49), the following approximated expression for $\overline{H}_{fi}$ is obtained:

$$\overline{H}_{fi} = \frac{16\pi^2 e}{[\pi a^3 V]^{1/2}} \cdot \sum_{l=0}^{\infty} \sum_{m=-1}^{1} \frac{i^l \kappa^{l-2}}{(2l+1)!!} Q_{lm}(f,i) Y_{lm}(\theta, \phi) \tag{51}$$

Now substitute (51) into (46) to get, where (46) is,

$$T(L, \overline{e}_x) = \frac{4\pi}{\hbar} V \frac{m\kappa\hbar}{(2\pi\hbar)^3} \int |H_{fi}|^2 d\Omega$$, and after using the

orthogonality and the normalization of Y_{lm},

$$T(L, \bar{e}_x) = 128\pi \frac{me^2}{\hbar^3 a^3} \cdot \sum_{l=0}^{\infty} \sum_{m=-1}^{1} \frac{\kappa^{2l-3}}{[(2l+1)!!]^2} |Q_{lm}(f,i)|^2$$

(52)

Equation (52) represents the probability $T(L, \bar{e}_x)$ per unit time, associated with the internal conversion process. Due to selection rules, only a certain value of l actually occurs in the sum (52). In practice, only the lowest possible value of l gives an appreciable contribution. This lowest value for l in the electric radiation is to get a rough value of (I_{cc}). The case of parity-favored transition, in which the electric multiple radiation predominantly occurs, is taken from the rank $l = l_{min}$. It is known that

$$T(L, \gamma) = \frac{8\pi(L+1)}{L[(2L+1)!!]^2} \frac{\kappa^{2L+1}}{\hbar} |Q_{LM}|^2$$

(53)

and $\dfrac{T(L, \bar{e}_x)}{T(L, \gamma)} = \alpha$ (conversion coefficient), $\vec{l} \equiv \vec{L}$

Therefore, from (52) and (53) for $T(L, \bar{e}_x)$ and $T(L, \gamma)$, α_κ is given by

$$\alpha_\kappa \cong 16 \frac{L}{(L+1)} \frac{Z^3}{a_o^4} \cdot \frac{\kappa^{2L-3}}{\kappa^{2L+1}} \cong 16 \frac{L}{(L+1)} \cdot \frac{Z^3}{a_o^4} \cdot \frac{1}{\kappa^4}$$

$$\therefore \left[\alpha_\kappa \cong \frac{16L}{(L+1)\kappa^4} \cdot \frac{Z^3}{a_o^4} \right]$$

(54)

$\kappa = \hbar^{-1} P$, Z = atomic number, a_o is Bohr radius = $\dfrac{e\hbar^2}{mc^2}$.

From (54), the following can be concluded:

(1) α_κ is independent of nuclear model.

(2) α_κ is L-dependent, especially for small energy.

(3) $\alpha_\kappa \propto Z^3$

(4) α_κ is large for small energy.

Problem given: $J_{lm}(\vec{\kappa}) = \int e^{i\vec{\kappa}\cdot\vec{r}} e^{-r/a} Y_{lm}(\theta,\phi) Y^{-(l+1)} r^2 dr \sin\theta d\theta d\phi$

Compute it.

Hint: $\kappa = \dfrac{2\pi}{\lambda} \gg \dfrac{1}{a} \rightarrow e^{-r/a} \rightarrow 1$

$e^{i\vec{\kappa}\cdot\vec{r}} = e^{i\kappa r \cos\theta} = \sum_{l=0}^{\infty} i^l (2l+2) j_l(\kappa r) P_l(\cos\theta)$

Use recursion formula for the Bessel function $j_l(\kappa r)$.

V-3-3—Oriented Nuclei

Sometimes, the nuclei are subjected to experimental conditions for certain projective objectives. The orientation of nuclei is one of these conditions. There are two types of orientation:

(a) Polarization such as ↑↑↑↓↓ ↓ ↗, where the population $P(m_i) \neq P(-m_i)$, m_i represents the levels given by $(2l + 1)$, where l is the angular momentum.

(b) Alignment ↑↓ ↗↘ then $P(m_i) = P(-m_i)$, if $P(m_i)$ can be distinguished from different (m_i); this means the degeneracy of nuclear states is removed, such as

Degenarated

Level $B^{\rightarrow} = 0$ $B^{\rightarrow} \neq 0$ —nondegenerate level (split)

$$\Delta E = \frac{B\mu}{I}; \quad E(m) = E_\circ - \frac{B\mu}{I} \cdot m$$

; m is leading to the polarization.

For gradient, $\vec{E}$; $E(m) = E_\circ + \Delta E_Q(m^2) \rightarrow$ alignment.

Now since the degeneracy is removed, that indicates that the population is different for different (m), which invites the statistics to play an important role. So the population $P(m_i)$ is given statistically by

$$P(m_i) = ae^{-\Delta E(m_i)/KT}$$

(significantly in equal, i.e., $\Delta E/KT \approx 1$)

K is Boltzmann constant; T is temperature in K° (Kelvin) degree.

T = 273 at room temperature.

Let $B^{\rightarrow} = 10,000$ gauss, then $KT \cong 1/30$ eV; at room temperature and $\Delta E \approx 10^{-8}$ eV, then $\Delta E/KT \cong 3 \times 10^{-7}$, which is so tiny that it is too difficult to be observed.

Now let (1) large fields $B^{\rightarrow}$, $grad$ $E^{\rightarrow}$; too low temperature, such as $B = 10^6$ gauss, $T = 10^{-4}$ K°. When this can be done by adiabatic magnetization, things will be quite different. The trick here is to use internal fields such as paramagnetic ions.

These brief notes are given for thinking extension over possible treatments of nuclei under study. For example, let us study the emission of multipole radiation from oriented nuclei (γ-rays). This study will deal with direction of emission, i.e., studying the angular distribution of γ-rays emitted by such nuclei. Let the emission of γ-rays from state $(I_i m_i)$ to state $(I_f m_f)$ such as

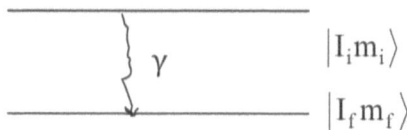

The radiation that is isotropic can be detected in all directions equally, where the orientation is random, which means the average of m_i is the same for any direction. Also, random orientation implies that $P(m_i)$ = constant and conversely. The angular distribution of this radiation can be determined by calculating the density angular distribution $\omega(\theta)$, which is given by

$$\omega(\theta) = \text{const.} \sum_{m_i m} P(m_i) \begin{pmatrix} I_f & L & I_i \\ -m_f & -M & m_i \end{pmatrix} \left| \vec{A}_{LM}(\theta, \phi) \right|^2 \tag{55}$$

To do that, consider the following schematic diagrams:

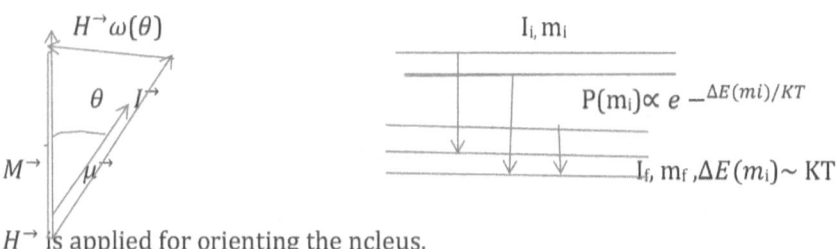

$\vec{H}$ is applied for orienting the ncleus.

The transition probability should be computed. Hence,

$$T(L, m_i \rightarrow m_f) = const. \begin{pmatrix} I_f & L & I_i \\ -m_f & -M & m_i \end{pmatrix}^2 \left| \vec{A}_{LM}(\theta, \phi) \right|^2 \tag{56}$$

Hence, $\omega(\theta)$ is given by (55),

where $P(m_i)$ is given by experiment, and "3j"-symbols are a weighing factor, and $| A_{L,M} |^2$ is angular distribution. The angle ϕ is used only if there is a polarization.

From previous sections, it is known that

$$A_{LM}(m) = h_L^{(1)}(\kappa r) \vec{Y}_{LL}^M(\theta, \phi) \qquad \text{and} \qquad \vec{Y}_{LL}^M \qquad \text{(gradient}$$

$$\text{formula)} \quad = \quad \frac{\vec{L} Y_{LM}(\theta, \phi)}{\sqrt{L(L+1)}} \;; \quad \text{thus,} \quad \text{it} \quad \text{follows} \quad \text{that}$$

$$\vec{A}_{L,M}(m) = \left[h_L^{(1)}(\kappa r) \frac{\vec{L}}{\sqrt{L(L+1)}} \right] \vec{Y}_{L,M}(\theta, \phi) = b_L \vec{Y}_{L,M}(\theta, \phi) \tag{57}$$

By rotation, as it is known, one has

$$Y_{L\mu}(\theta', \phi') = \sum_{\mu} D_{\mu\mu}^L(\theta, \phi, 0) Y_{L\mu}(\theta, \phi) = \sum_{\mu} D_{\mu}^L(\theta, \phi, 0) Y_{L\mu}(\theta, \phi)$$

Now operating by $\vec{L}$ leads to

$$\vec{L} Y_{L\mu}(\theta', \phi') = \sum_{\mu} D_{\mu\mu}^L(\theta, \phi, 0) \left(\vec{L} Y_{L\mu}(\theta, \phi) \right)_{\theta = \phi' = 0}$$

Now take the z-component that leads to

$$\left(L_z Y_{L\mu}(\theta, \phi)_{\theta = \phi' = 0} = \mu \cdot Y_{L\mu}(0,0) = \mu \sqrt{\frac{2L+1}{4\pi}} \delta_{\mu 0} = 0 \right)_{, \text{since}}$$

$\mu \neq 0$, then $\delta_{\mu 0} = 0$; hence, z-component is zero.

Take x and y components; here $L_\pm$ is to be applied, where $L_\pm = L_x \pm iL_y$, so

$$\left(\vec{L}_\pm Y_{L\mu}\right)_0 = \left(\sqrt{(L \mp \mu)(L \mp \mu \mp 1)}\, Y_{L,\mu\pm 1}\right)_0 = \sqrt{L(L+1)}\sqrt{\frac{2L+1}{4\pi}}\,\delta_{\mu,\mp 1}.$$

Now if the radiation is along z-axis such as

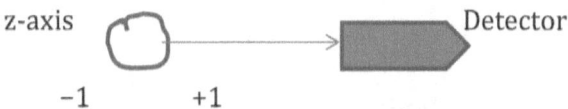

It is observed here as either positive circle polarization or negative circle polarization. Now $\left|A_{L,M}\right|^2$ can be written as

$$\left|A_{L,M}\right|^2 = \vec{A}_{LM}\vec{A}_{LM}^* = b_L^2\left(L_- Y_{LM}(\theta,\phi)\right)\left(L_+ Y_{LM}^+(\theta,\phi)\right) + \ldots\ldots$$

$$= \frac{1}{2}b_L^2\left\{\left(L_+ Y_{LM}\right)\left(L_+ Y_{LM}^+\right) + \left(L_- Y_{LM}\right)\left(L_- Y_{LM}\right)^* + 2\left(L_z Y_{LM}\right)\left(L_z Y_{LM}\right)^*\right\}$$

$$\Rightarrow \left|A_{L,M}\right|^2 = \frac{1}{2}\left\{\sum_{\mu=\pm 1} D_{\mu,M}^L(\theta,\phi,0)\cdot D_{-\mu,-M}^L(\theta,\phi,0)(-1)^{M-\mu}\right\}$$

By applying Clebsch-Gordan coefficients (CGC), one gets

$$\left|\vec{A}_{LM}\right|^2 = \frac{1}{4\pi\kappa^2 r^2}(2L+1)\sum_{\mu=\pm 1}\sum_{\kappa=0}^{\mu}(-1)^M(2\kappa+1)\begin{pmatrix} L & L & \kappa \\ \mu & -\mu & 0 \end{pmatrix}\begin{pmatrix} L & L & \kappa \\ M & -M & 0 \end{pmatrix}D_{00}^\kappa(0,0,0)$$

This is only for L, i.e., pure multipole radiations.

Now, $\displaystyle\sum_{\mu=\pm} = \begin{pmatrix} L & L & \kappa \\ L & -1 & 0 \end{pmatrix} + \begin{pmatrix} L & L & \kappa \\ -1 & 1 & 0 \end{pmatrix}$

But $\begin{pmatrix} L & L & \kappa \\ \mu & -\mu & 0 \end{pmatrix} = (-1)^{2L+\kappa}\begin{pmatrix} L & L & \kappa \\ \mu & -\mu & 0 \end{pmatrix}$

$$\Rightarrow \sum_{\mu} = 0$$

If κ is odd,

Therefore, $\sum_{\mu}$ survives only for even κ.

Also, it is known that

$$D_{00}^{\kappa}(0,0,0) = P_{\kappa}(\cos\theta)$$

From equation (55),

$$\left[\omega(\theta) = \text{const.} \sum_{m_i,M,\kappa} P(m_i) \begin{pmatrix} I_f & L & I_i \\ m_f & M & m_i \end{pmatrix}^2 \begin{pmatrix} L & L & \kappa \\ 1 & -1 & 0 \end{pmatrix} \begin{pmatrix} L & L & \kappa \\ M & -1 & 0 \end{pmatrix} (2\kappa+1)(-1)^M P_{\kappa} \right]$$

$$(58)$$

The following remarkable points should be observed:

(1) κ is even.

(2) $\kappa \le 2L$.

(3) If $P(m_i)$ is constant, then,

$$P(m_i) = \frac{1}{2I_i + 1}, \text{ which implies that } \omega(\theta) \text{ is constant.}$$

[As an exercise, prove (3).]

V-3-3-1—Angular Correlation of Multipole Radiations

As stated before, studying the angular correlations between emitting γ-radiations in cascade are very important for being acquainted with many kinds of nuclear structural properties such as the angular momentum, parity, and type of radiation emitted. Let us consider the following schematic diagram:

$(I_i m_i)$ is initial state,

$(I m)$ is intermediate state,

$(I_f m_{fi})$ is final state.

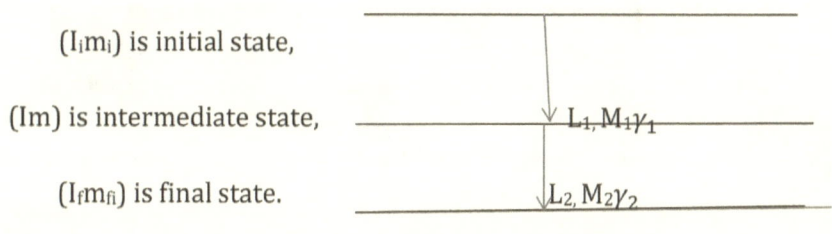

$L_1, M_1 \gamma_1$

$L_2, M_2 \gamma_2$

γ-Radiations in cascade

The experimental steps are as follows:

(1) Prepare P(m) ≡ f(m), so randum orientation is obtained, which

leads to
$$P(m_i) = \left(\frac{1}{2I_i + 1} \right),$$

where $(2I_i + 1)$ gives the possible values of (m_i).

(2) The orientation takes the z-direction, then the probability of forming (m) states is given as

$$P(m) = \text{const}(C). \sum_{m_i, \kappa} T(m_i \rightarrow m_f ; \vec{\kappa}_1) \left(\frac{1}{2I_i + 1} \right) \quad K_1^{\rightarrow} = Z^{\rightarrow} \gamma_1, \theta = 0$$

$$\left[= C. \sum_{\theta=0} \frac{1}{(2I_i + 1)} \begin{pmatrix} I_f & L & I_i \\ -m_f & M_1 & m_i \end{pmatrix} \begin{pmatrix} L & L & \kappa_1 \\ 1 & -1 & 0 \end{pmatrix} \begin{pmatrix} L & L & \kappa_1 \\ M_1 & -M_1 & 0 \end{pmatrix} (2\kappa + 1)(-1)^M \left(P_\kappa (\cos 0)_{\theta=0} \right) \right]$$

(3) ω (θ) is a direction correlation function. The experiment to measure this is designed as

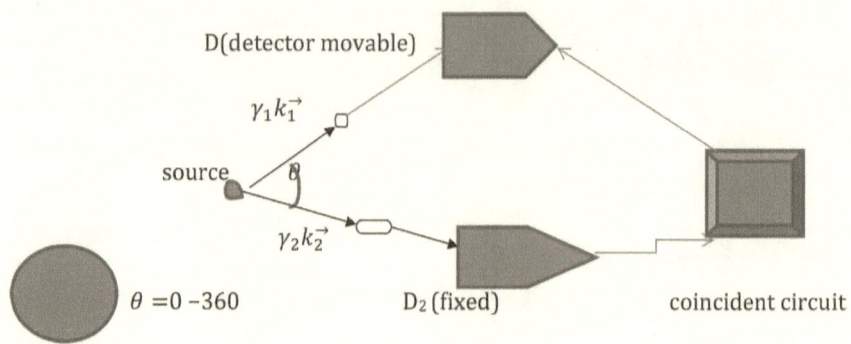

D(detector movable)

$\gamma_1 k_1^{\rightarrow}$

source

θ

$\gamma_2 k_2^{\rightarrow}$

$\theta = 0 - 360$ D_2 (fixed)   coincident circuit

Here, $\omega(\theta)$ is given by the following formula→:

$$\omega(\theta) = C.\sum_{\kappa_1} (2\kappa_1 + 1) \begin{pmatrix} I_f & L_1 & I_i \\ -m & -M_1 & m_i \end{pmatrix} \begin{pmatrix} L_1 & L_1 & \kappa_1 \\ M_1 & -M_1 & 0 \end{pmatrix} \begin{pmatrix} L_1 & L_1 & \kappa_1 \\ 1 & -1 & 0 \end{pmatrix} \times$$

$$(-1)^{M_1} \times (-1)^{M_2} (2\kappa_2 + 1) \begin{pmatrix} I_f & L_2 & I_i \\ -m_f & -M_2 & m_i \end{pmatrix}^2 \begin{pmatrix} L_2 & L_2 & \kappa_2 \\ M_2 & -M_2 & 0 \end{pmatrix} \begin{pmatrix} L_2 & L_2 & \kappa_1 \\ 1 & -1 & 0 \end{pmatrix}$$

$$\times P_{\kappa_2}(\cos\theta)$$

Remember that unless $\kappa_1 = \kappa_2 = \kappa$, $\omega(\theta) = 0 =$

$$= C.\sum_{\kappa} (2\kappa_1 + 1) \begin{pmatrix} L_1 & L_1 & \kappa \\ 1 & -1 & 0 \end{pmatrix} \begin{Bmatrix} I & I & \kappa \\ L_1 & L_1 & I_i \end{Bmatrix} \begin{pmatrix} L_2 & L_2 & \kappa \\ 1 & -1 & 0 \end{pmatrix} \begin{Bmatrix} I & I & \kappa \\ L_2 & L_2 & I_f \end{Bmatrix} P_{\kappa}$$

Therefore,

$$\omega(\theta) = \text{const.} \sum_{\kappa-\text{even}} (2\kappa_1 + 1) A_\kappa(1) A_\kappa(2) P_\kappa(\cos\theta)$$

where $A_\kappa(1) = \begin{pmatrix} L_1 & L_1 & \kappa \\ 1 & -1 & 0 \end{pmatrix} \begin{Bmatrix} I & I & \kappa \\ L_1 & L_1 & I_i \end{Bmatrix} \equiv "3j" \times "6j"$

and $A_\kappa(2) = \begin{pmatrix} L_2 & L_2 & \kappa \\ 1 & -1 & 0 \end{pmatrix} \begin{Bmatrix} I & I & \kappa \\ L_2 & L_2 & I_f \end{Bmatrix} \equiv "3j" \times "6j"$

Hence, for $\kappa > 2$ to κ_{max}, $\omega(\theta)$ is

$$\omega(\theta) = 1 + \sum_{\kappa \rangle 2}^{\kappa_{max}} A_\kappa(1) A_\kappa(2) P_\kappa(\cos\theta)$$

(59)

Note: $A_\kappa(1)$ is related to γ_1 (first radiation), and $A_\kappa(2)$ is related to γ_2 (second after γ_1, in cascade). Hence, γ_1 and γ_2 are emitted in cascade, as shown in the diagram above. As in the schematic experimental apparatus, $A_\kappa(1)$ contains only the γ_1 quantum number (or related to γ_1). And $A_\kappa(2)$ is of γ_2 in the same meaning. There is a well-defined function known as F-function; it is so useful here. It is given as

$$F = F_\kappa(L_1\bar{L}_1, \bar{\Pi}) = \Gamma(-1)^{\bar{I}+I-1}\left[(2\kappa+1)(2L_1+1)(2\bar{L}_1+1)(2\bar{I}+1)\right]^{1/2}\begin{pmatrix} I & I & \kappa \\ 1 & -1 & 0 \end{pmatrix}\begin{Bmatrix} I & I & \kappa \\ L_1 & \bar{L}_1 & \bar{I} \end{Bmatrix}$$

Then use it in $\omega(\theta)$ to abbreviate the expression before (59),

where I is referred to as the intermediate state;

$\bar{I}$ is referred to as either the initial state or the final state.

These F-functions are tabulated and can be used in the calculations according to the values of κ, L_1, $\bar{L}_1$, I, $\bar{I}$. Also, "3j" and "6j" are tabulated and can be calculated according to these quantum numbers.

Chapter Six

Nuclear Beta Decay

VI-1—Introduction

So far, we have discussed the electromagnetic radiation, sometimes called as -decay, which physically means a nucleus decay is the emission of γ-radiation to go back to its more stable state or to its ground state. As shown clearly in chapter 4 and 5, the other decay is the so-called α-decay, which is quite well treated in some undergraduate books in nuclear physics. Also, one has to be acquainted with the facts that β and α are particles with defined masses and sizes, while γ is an electromagnetic radiation, as very well given in the preceding chapter. The old physics books (up to 1960) referred to α, β, and γ as rays, because they were detected early (1897–98) as rays. But in fact, this is misleading now. So one must get familiar with the particle status of α and β and the electromagnetic radiation status of γ. Although the waviness character is associated with moving β and α according to de-Broglie principle $\lambda = \frac{h}{P}$, where λ is the wavelength of moving particle, p is the momentum of this particle, and h is Planck constant about 6.63 $\times$ 10^{-34} joule $\cdot$ sec. still, a differentiation between particles (with mass and size) and radiation of no mass and defined size, such as γ-rays or photon, is quite a necessity in physics.

Nuclear β-decay is classified among the weak nuclear interactions. If the force responsible for this kind of interaction is compared with the strong nuclear force, one finds that its strength is only about 10^{-13} of that of nuclear force and about 10^{-11} of that of the electromagnetic force. But actually, it is of strength about 10^{26} as strong as the strength of the gravitational force.

VI-2—β-decay Processes

The historical story of discovering -decay is well-known and demonstrated in details in many books of atomic and nuclear physics for undergraduate students as well as for graduates, where the proposed neutrino by Pauli was a necessity to work for conservation laws of energy and momentum, and the same is for the antineutrino. But here, we would like to clarify the theory of β-decay in more advanced details, including some elements of relativistic quantum mechanics, which is necessary for the treatment of β-decay theory.

In general, β-decay processes are given by

$$n \rightarrow p + \beta^- + \bar{\nu} \tag{1}$$

neutron decay by emitting β^-

$$p \rightarrow n + \beta^+ + \nu \tag{2}$$

proton decays by emitting β^+

$$p + \bar{e} \rightarrow n + \nu \tag{3}$$

electron capture by the proton

The decay (1) occurs inside the nucleus and also when the neutron is free, because it is an unstable nucleon with a half-life about 10^3 seconds. But the decays (2) and (3) occur only inside the nucleus, where an interaction takes place; the $\bar{e}$ is a stable lepton and till now considered an elementary particle of no structure, i.e., undividable. The proton is stable as free nucleon, or baryon, but it is not an elementary particle because it has a structure and it is made of quarks (down and up).

There are recent theoretical studies predicting a half-life of proton as in the range of 10^{31}–10^{32} years! The beta energy is about 1 MeV.

The mass of the neutrino and the antineutrino is about zero, i.e., $m_\nu = m_{\nu_-} \sim 0$. But there is experimental data claiming that these masses are much less than (200) electron volts, i.e., about 10^{-4} MeV = 2×10^{-4} $m_e \cong 2 \times 10^{-30}$gm. So the neutrino and the antineutrino, ν and $\overline{\nu}$, respectively, are relativistic particles. Therefore, a relativistic mechanics has to be used in tackling these particles, which are also leptons. As we all know, the β^+ is the antielectron, called a positron, which has the same mass (≈ 0.511 MeV) and spin (S = 1/2 $\hbar$, = 1/2, considering $\hbar$ = 1) and opposite charge, i.e., e$^+$.

VI-2-1—Simple Theory of β-decay

β-decay occurs due to a weak nuclear interaction, as stated before. For example, in γ-decay emission processes, the energy is E$\geq$1 MeV, and the average lifetime is about 10^{-13} sec. While in β-decay, the energy is E$\leq$1 MeV, and the average lifetime is about 10^3 sec (mentioned lifetime). So it is clear that when the force of interaction is strong, the energy and lifetime are different from the case of weak interaction. In strong interaction, the energy is equal, or more than one million electron volts (MeV), and the lifetime is quite short ($\approx 10^{-13}$ sec), while in the weak interaction, the maximum energy is about (1 MeV), and lifetime is large (10^3 sec). More differentiating concepts between γ-decay and β-decay are (1) in γ-emission, we know the Maxwell equation and the electromagnetic fields (chapter 4), (2) in Maxwell equations with quantum theory, one can calculate γ-decay rate. These may be represented schematically, as the photon is the messenger particle of interaction.

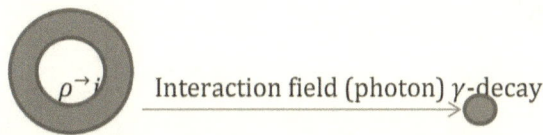

Interaction field (photon) γ-decay

Source of E and B fields, Maxwell equations

In β-decay, things are quite different, which might be represented as

Nucleus (n, p) interaction ⟶ ▌β-decay

Lepton field

In the beginning of the β-decay theory, the known data is that β-decay actually occurs, and p and n are the sources of β-emission from some elements. After that comes fermi, and a way is worked out to connect between β-decay and electromagnetism theory in introducing that $\left(\vec{\rho}, \vec{j}, e\right)$ are to be combined into four-vector potential field $S^{\to(4)} \equiv \left(\dfrac{1}{ic} j, \to \rho \to\right)$ (see chapter 4), where $\vec{S}^{(4)}$ satisfies the continuity equation $\sum_{\mu} \partial_{\mu} S_{\mu} = 0$, which implies that particles cannot be created or destroyed. The field consists of a vector potential field $\vec{A}$ and scalar φ. It is denoted as four vector $F^{(4)} \equiv f\left(\vec{A}, i\phi\right)$, where $\sum_{\mu} \partial_{\mu} F_{\mu} = 0$, μ = 1, 2, 3, 4.

The Hamiltonian of the interaction between the source and the field, is density is

$$H = \rho\phi - \dfrac{1}{c}\vec{j} \cdot \vec{A} = \vec{S}^{(4)} \vec{F}^{(4)}$$

(4)

$$\rho = e\rho_{\circ}........charge \quad density$$
$$\vec{j} = e \cdot \vec{V} = \quad current, \quad \vec{V} = the\ velocity.$$
$$\vec{S}^{(4)} = e\vec{S}^{(4)}_{\circ} \tag{5}$$

Hence, $H = e \cdot \vec{S}^{(4)} \vec{F}^{(4)}$ (6)

e—electron charge (coupling constant of E&M field)

$\vec{S}^{(4)}$ —the source

$F^{(4)}$ is four vector field, a function of $(\vec{A}, i\phi)$ as defined before.

Therefore, H= e $(\rho_0\phi - \frac{1}{C}\vec{V}.\vec{A})$ (from 4 and 5)   (6)

Now if $\vec{V}$ /C $\square$ 1, then 1/C $\vec{V} \cdot \vec{A}$ (relativistic effect term) is neglected.

VI-2-2—Fermi Theory of β-Decay

In β-decay, the source of the field is (n, p), lepton field. Now, it is necessary to construct ψ_n and ψ_p as wave functions for the neutron and the proton, respectively. As mentioned in the preceding chapters, n and p are two states for a nucleon, i.e.,

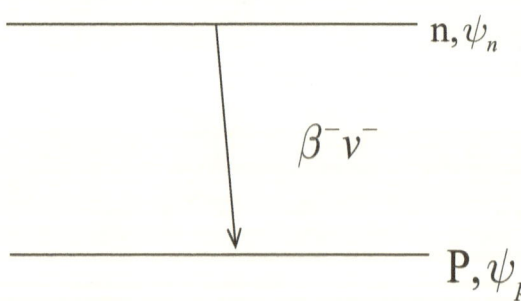

We construct four vector $\vec{S}^{(4)}$, $S^{(4)\rightarrow}$ and $F^{(4)\rightarrow}$ transform as four vector. Hence,

$$S_\mu^{(4)} = \vec{\Psi}(p)\gamma_\mu\vec{\Psi}(n),$$

where γ_μ is Dirac matrix.

Hamiltonian (H) is a scalar quantity.

The lepton field is given as

$$\vec{F}_\mu^{(4)} = \vec{\Psi}(\beta)\gamma_\mu\vec{\Psi}(v) \tag{7}$$

where $\vec{\Psi}(v)$ is the initial state of (v) and $\vec{\Psi}(\beta)$ is the final state of (β)

Hence,

$$H(\beta) = g_v\vec{S}_{(n,p)}^{(4)} \cdot \vec{F}_{(\beta,v)}^{(4)} \tag{8}$$

where g_v is a characteristic of the strength of the interaction. Therefore,

$$H(\beta) = g_v\sum_\mu(\vec{\Psi}(p)\gamma_\mu\vec{\Psi}(n))(\vec{\Psi}(\beta)\gamma_\mu\vec{\Psi}(v)) \tag{9}$$

The transition probability of β-decay can be calculated by using the fermi golden rule no. 2, i.e.,

$$N(E) = \frac{2\pi}{\hbar}\left|H_{fi}(\beta)\right|^2 \cdot dn/dE =$$ —number of particles emitted

within the energy range (E, E + dE)

or

$$N(E) = \frac{2\pi}{\hbar}\left|\langle f|H_v(\beta)|i\rangle\right|^2 dn/dE \tag{10}$$

Now

$$\langle f|H_v(\beta)|i\rangle = \int_{\substack{nucleus\\lepton}} H_v(\beta)\delta(\vec{r}_L - \vec{r}_n)dV_n dV_L$$

(11)

where n indicates the nucleus (n, p), and L indicates lepton (β, v). So equation (11) is reduced to

$$\langle f|H_v(\beta)|i\rangle = \int_{nucleus} H_v(\beta)dn$$

(12)

By substituting (9) for $H_v(\beta)$, one obtains

$$\langle f|H_v(\beta)|i\rangle = g_v \int_{nucleus} \sum_\mu (\vec{\Psi}(p)\gamma_\mu\vec{\Psi}(n))\cdot(\vec{\Psi}(\beta)\gamma_\mu\vec{\Psi}(v))dV_n$$

(13)

$\vec{\Psi}(v)$ is plane wave (in free space), but $\vec{\Psi}(\beta)$ is not well-known. Therefore, one has to solve Dirac equation (relativistic) to find

$$\vec{\Psi} = \begin{pmatrix}\Psi_1\\\Psi_2\\\Psi_3\\\Psi_4\end{pmatrix} = \begin{pmatrix}\Psi_1\\\Psi_2\\\Psi_3\\\Psi_4\end{pmatrix}$$ —the particle exists in four states (spinors),

where $\vec{\Psi}^* = (\Psi_1^*, \Psi_2^*, \Psi_3^*, \Psi_4^*)$ complex conjugate of $\vec{\Psi}$, but the present purpose may assume light nuclei, where Z is small. On the basis of this proposition, $\vec{\Psi}(\beta)$ may be considered as plane wave. Hence, $\vec{\Psi}(\beta)$ will be used as plane wave too. Now

$$\vec{\Psi}(p)\gamma_\mu\vec{\Psi}(n) = \vec{\Psi}^*(p)\gamma_4\gamma_\mu\vec{\Psi}(n),$$

where $\gamma_4 \gamma_\kappa = \beta \gamma_\kappa = -i\alpha_\kappa \,/\, if \dfrac{v}{c} \to 0 \Rightarrow \alpha_\kappa = 0$, also $\gamma_4 = \gamma_4^*$

—Hermitian; thus, $\gamma_4 \gamma_4 = 1$ —unitary. $\gamma_\kappa, \beta, \alpha$ are Dirac matrices in relativistic quantum mechanics, as we all know. It is also recognized that the speed of nucleon as inside the nucleus is about of one-tenth of light speed, i.e., v/c ≤ 0.1. Hence, all terms of 1/10 order are to be neglected, then β-decay is allowed.

Now under all these considerations and approximations, we try to solve the problem:

1) Neglecting the relativistic nuclear effect, i.e., $\alpha_k \to 0$, then we can write

$$\int_{nucleus} \left(\vec{\Psi}^*(p) \cdot \vec{\Psi}(n) \right)\left(\Psi^*(\beta)\Psi(v) \right) dV = \int \left(\vec{\Psi}^*(p) \cdot \vec{\Psi}(n) \right) \cdot D_\beta D_v e^{-i\vec{p}\cdot\vec{r}} e^{i\vec{q}\cdot\vec{r}} dV$$

where

$\vec{\Psi}^*(\beta) = D_\beta e^{-i\vec{p}\cdot\vec{r}}$ and $\vec{\Psi}(v) = D_v e^{i\vec{q}\vec{r}}$. D_β involves spin orientation;

$\vec{p}$ is the momentum of the β particle. D_v involves spin orientation, and $\vec{q}$ is the momentum of v.

We consider $\hbar$ = m = c = 1 in atomic units.

Therefore, we have

$\vec{\Psi}^*(\beta) = D_\beta e^{-i\vec{p}\cdot\vec{r}}$ as plane waves

$\vec{\Psi}(v) = D_v e^{i\vec{q}\vec{r}}$

The relativistic energy is given by $E^2 = \left(p^2 c^2\right) + m_o^2 c^4$ or $E^2 = p^2 c^2 + (m_0 c^2)^2$. Consider $(E/m_o c^2)$ = total energy/ ground state energy

Hence,

$$\omega^2 = \frac{E^2}{(m_o c^2)^2} = 1 + \frac{p^2 c^2}{(m_o c^2)^2} = 1 + P^2, \text{ where } P = p/m_0 c.$$

As it is known, the length unit is measured by $\hbar / m_0 c$. Also, we know that R (nucleus radius) $\cong 1.2 \times 10^{-13} A^{1/3}$; therefore, PR≤1/10, but here,

PR $\cong 2.8 \times 10^{-13} = \langle\langle\langle 1$. Therefore, $e^{-(\vec{p}+\vec{q})\vec{r}} \cong 1$.

Before going in details, let us consider allowing β-decay, assuming that $\hbar = m_o = c = 1$. Therefore, $N_\beta(E)$ is given by

$$N_\beta(\omega) = 2\pi \cdot \left|\langle f | H(\beta) | i \rangle\right|^2 dn/d\omega \tag{15}$$

where

$$\omega_e = \frac{E}{m_o c^2}; \quad E_t = (pc)^2 + \left(m_o c^2\right)^2 \text{ and that for electron, } \omega_e^2 = 1 + p^2$$

(as shown before).

For the neutrino $E_\kappa = E_{tot} = (pc)^2$, rest mass $m_0^v \approx 0$, which implies that $\omega_v^2 = q^2$, so, $\omega_v = q = (pc) = P$ on the basis that c = 1, by making more assumptions that $\frac{v}{c} \square 1$ so α_k =0 $\rightarrow \gamma_5$ =0 for the nucleon part. Hence, equation (14) is reduced to $\int \Psi^*(p)\Psi(n) dV_n = \int 1 = \langle 1 \rangle$ nuclear matrix element.

2) Now, take the lepton part, assuming $z \cong \ll 1$ or close to zero. So one has free particles to use a plane wave, as shown before, i.e.,

$$\Psi(\beta) = D_\beta e^{i\vec{p}\cdot\vec{r}} \text{ and } \Psi^*(\beta) = D_\beta^* e^{-i\vec{p}\cdot\vec{r}}, \ \Psi(v) = D_v e^{i\vec{q}\cdot\vec{r}}$$

Hence,

$$\left(\Psi^*(\beta), \Psi(v)\right)_R = D_\beta^* D_v e^{-i(\vec{p}+\vec{q})\cdot\vec{r}}$$

$$D_\beta^* D_v \approx \left(1 + \frac{r}{c}\cos\theta_{\beta v}\right); \quad \Lambda = +1$$

(16)

One can expand $e^{-i(\vec{p}+\vec{q})\cdot\vec{r}}$ at R (nucleus radius) such as

$$e^{-(\vec{p}+\vec{q})\cdot\vec{r}}\Big|_R = 1 - \frac{i(\vec{p}+\vec{q})\cdot\vec{r}}{1!} - \frac{(\vec{p}+\vec{q})^2 r^2}{2!}$$

(17)

This is used for approximation use (remember).

Now with the approximations mentioned before, then

$$\langle f|H(v)|i\rangle_{all} \cong g_v \langle I\rangle \text{ —equation (14)}$$

To use the fermi golden rule, the second, it is necessary to find the number of available states, i.e.,

$$dn = \frac{1}{(2\pi\hbar)^3}Pdp \cdot \frac{1}{(2\pi\hbar)^3}q^2 dq \text{, for state (1).}$$

Therefore,

$$\frac{1}{(2\pi\hbar)^3}q^2 dq \neq 0$$

only if $\omega_v = \omega_\circ - \omega$, which implies the use of Dirac delta function $\delta(\omega_v - (\omega_\circ - \omega)); \quad \omega_v \equiv q$.

Hence,

$$dn = \frac{1}{(2\pi)^6} \cdot p^2 dp \int q^2 dq \delta \left(\omega_v - (\omega_\circ - \omega) \right)$$

$$= \frac{1}{(2\pi)^6} p^2 dp (\omega - \omega_\circ)^2$$

(As an exercise, show this.)

Now,

$$\frac{dn}{d\omega} = \frac{1}{(2\pi)^6} p^2 dp \left((\omega - \omega_\circ)^2 \right) = \frac{1}{(2\pi)^6} \cdot p^2 dp (\omega - \omega_\circ)^2 \cdot (\omega / p dp)$$

From $\omega^2 = 1 + p^2$, it is obtained that

$$dp = \frac{\omega}{p} \cdot d\omega$$

, which was substituted for dp.

Therefore,

$$\frac{dn}{d\omega} = \frac{1}{(2\pi)^6} p \cdot \omega (\omega - \omega_\circ)^2$$

since

$$N_\beta(\omega) = 2\pi \cdot \left| \langle f | H(\beta) | \right|^2 \frac{dn}{d\omega} \tag{15}$$

Hence,

$$N_\beta(\omega) = (2\pi)^{-5} g_v^2 \cdot \langle I \rangle \cdot \left(\omega \cdot p / (\omega - \omega_\circ)^2 \right) \cdot F(z, \omega) \tag{18}$$

We can construct β-spectrum from equation (18), which represents an allowed β-decay formula.

Therefore, one can write

1) $\left(N_\beta(\omega)/p \cdot \omega F\right)^{1/2} = cons \tan t.\left((\omega_\circ - \omega)\right) = P$

If $\omega = 0 \Rightarrow P = cons\,\omega_\circ$, P = 0 if ω=ω$_\circ$; this can be shown graphically as below:

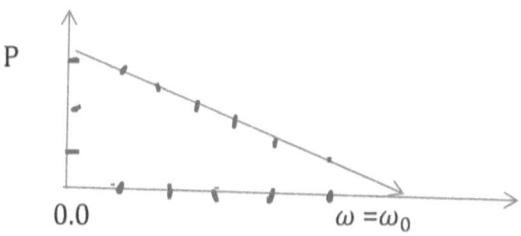

0.0 $\omega = \omega_0$

Fermi-Kurie plot

This is a Fermi-Kurie plot, which is in a perfect agreement with the experiment. The second plot is the N$_\beta$ distribution as a function of ω, as shown in equation (18).

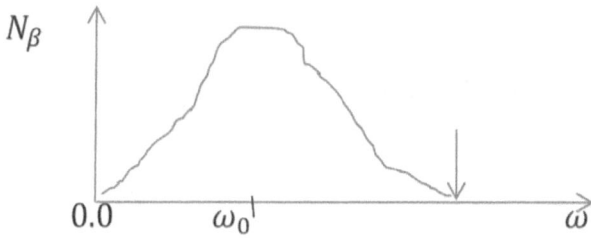

0.0 ω_0 $\tilde{\omega}$

N$_\beta$ versus ω

2) *Selection Rules*

In equation (18), < I > is scalar operator; ΔI = 0, Δπ=, no change. In this case, lepton field is zero, which indicates that oo transition, which is observed; such transition is called Fermi transition. But in ⁶⁰Co, as shown below, its decay scheme and the selection rules of transition are Gamow-Teller (G-T).

$$\Delta I = 1; (5-4) = 1, \Delta \pi = no., + \rightarrow + , \text{—this is (G-T) transition.}$$

As shown in the following decay scheme for $_{27}^{60}C$,

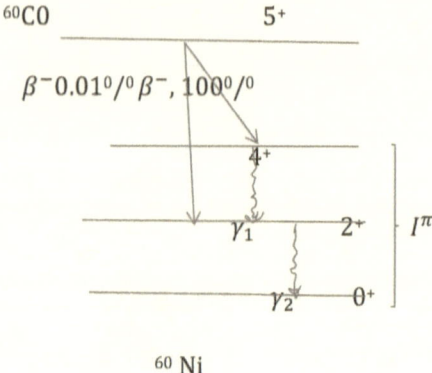

^{60}CO   5+

$\beta^- 0.01^o/^o \beta^-, 100^o/^o$

4+

γ_1 2+ I^π

γ_2 0+

60 Ni

Anyway,

$$H(x) = g_x \left(\vec{\Psi}(p) \cdot \Omega_x \vec{\Psi}(n) \right) \times \left(\vec{\Psi}(\beta) \Omega_x \vec{\Psi}(v) \right)$$

and $H(v) = g_v$ (source) (lepton field), where H(v) must be scalar. For H(x), therefore, there are five ways of the type H(x), which are relativistic invariant ways, as shown in table 1 below.

Table 1: Coupling Operator and Bilinear Covariant

Type	Ω_x	Bilinear Ccovariant	Number of Componet	ΔI	$\Delta \pi$	$\Lambda = D_\beta^+ \Omega D_v$
Scalar	1	$\widehat{\bar{\Psi}}\widehat{\Psi}$	1	0	No	−1
Vector	γ_μ	$\widehat{\bar{\Psi}}\gamma_\mu\widehat{\Psi}$	4	0	No	+1
Tensor	$\gamma_\mu\gamma_v$	$\widehat{\bar{\Psi}}\sigma_{\mu v}\widehat{\Psi}$	6	0, ±1 0→0	No	+1/3
Axial V.	$\gamma_\mu\gamma_v\gamma_\lambda$ $=\gamma_5\gamma_\mu$	$\widehat{\bar{\Psi}}\gamma_\mu\gamma_5\widehat{\Psi}$	4	0, ±1 0→0		−1/3
Pseudo-Scalar	$\gamma_\mu\gamma_v\gamma_\lambda\gamma_\kappa$ $=\gamma_5$	$\widehat{\bar{\Psi}}\gamma_5\widehat{\Psi}$	1	0	No	+1

Let us take the axial vector coupling,

$$H(A)=g_A\left(\widehat{\bar{\Psi}}(p)\gamma_4\gamma_\kappa\gamma_5\widehat{\Psi}(n)\right)\left(\widehat{\bar{\Psi}}(\beta)\gamma_4\widehat{\otimes}\gamma_5\widehat{\Psi}(v)\right)$$

Also, one must note that

$$\gamma_4\gamma_\kappa\gamma_5=\gamma_5\gamma_4\gamma_\kappa=\gamma_5\gamma_\beta\gamma_\kappa=-\gamma_5(i\alpha_\kappa)=i(-\gamma_5\alpha_\kappa)=i\sigma_\kappa^{(4)}$$

, for allowed β-decay v/c→0, then α→0 and γ_5→0; therefore, $H_{all}(A)$ is given by

$$H_{all}(A)=-g_A\left(\widehat{\bar{\Psi}}^*(p)\widehat{\sigma}\widehat{\Psi}(n)\right)\left(\widehat{\bar{\Psi}}^+(\beta)\widehat{\sigma}\widehat{\Psi}(v)\right)$$. Using equation

(14) form, one finds $N_\beta(\omega)=(2\pi)^{-5}g_A^2\langle\vec{\sigma}\rangle^2 p\cdot\omega(\omega_\circ-\omega)^2$.

Here, nuclear matrix elements $<\vec{\sigma}>$, vector, and AV by W-E theorem, one finds $\Delta I = 0, \pm 1$. For nonrelativistic limit, G-T transition $\vec{\sigma}_v \downarrow\downarrow \vec{\sigma}_\beta$, but for F-transition, $\vec{\sigma}_v \uparrow\downarrow \vec{\sigma}_\beta$.

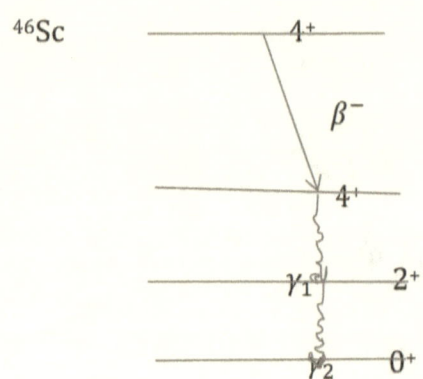

Decay scheme of ^{46}Sc

$$n \rightarrow p + \overline{\beta} + \overline{v}$$

Neutron→proton + electron + antineutrino: m_p=938 MeV, m_e=0.511 MeV, and $m_v \ll 200$ eV.

$$S_p = S_\beta = S_{\overline{v}} = \frac{1}{2}\hbar = S_n$$

$$e_n = 0, e_p = 1, e_{\overline{\beta}} = -1, e_v = 0, S_\beta \uparrow\downarrow S_{\overline{v}}$$

Here, the laws of consideration are valid.

As worked in equations (14) and (18), the number of electrons $N_\beta(\omega)$ is given by $N_\beta(\omega) = (2\pi)^{-5}(g_v^2\langle I\rangle + g_A^2\langle\vec{\sigma}\rangle^2)p\cdot\omega(\omega_o - \omega)^2$ (19)

where in (19), there is a mixture of vector and axial vector coupling in ^{46}Sc decay.

Let us take another example; take $^{14}_{6}O_8 \rightarrow {}^{14}_{7}N_7 + \beta^+ + \nu$, where

$p \rightarrow n + \beta^+ + \nu$, proton $\rightarrow$ neutron + positron + neutrino.

It is found that $<1> = 2 \rightarrow g_v = 3.0 \times 10^{-12}$ in units of $\hbar = c = 1$ for neutron

$<1>^2 = 1$, $\quad \downarrow \langle \sigma \rangle^2 = \sigma_x^2 + \sigma_y^2 + \sigma_z^2 = 3$. Hence,

$$\left| \frac{g_A}{g_v} \right| = 1.19 \pm 0.04$$

In the case of the electromagnetic coupling, $e^2 = 1/137$ (fine structure).

Let us go back to some important physical remarks and theoretical concepts. The beta decay was equationally demonstrated in equations (1), (2), and (3), which can be rewritten in the following forms:

(a) $(Z, N) \rightarrow (Z+1, N-1) + e^- + \overline{\nu}_e$ —electron emission

(b) $(Z, N) \rightarrow (Z-1, N+1) + e^+ + \nu_e$ —positron emission

(c) $(Z, N) + e^- \rightarrow (Z-1, N+1) + \nu_e$ —atomic electron captures

The notations ν_e and $\overline{\nu}_e$ are related to the emitted neutrino and the antineutrino accompanying the emission of the electron (negative and positive—that is, β^- and β^+). This is so because the neutrino and the antineutrino are also emitted by μ-decaying, i.e., there are ν_μ and $\overline{\nu}_\mu$, (a) and (b) are clearly explaining themselves well. But (c) needs a brief clarification. The electron-capture phenomenon is summarized as follows: an electron in atomic shell (say, K, L, M) might be captured by the nucleus of the atom itself. If the captured electron is from K-shell, it is called K-electron capture.

The β-decay processes are of a physical importance, which obeys the weak interaction (nuclear), as clarified previously. Therefore, scientists, such as Abdus Salam and Weinberg, have worked in the mid sixties on extensive theoretical tackling to unify this type of interaction (nuclear weak interaction) with the electromagnetic interaction. This theory of unifying (NWI) with (E&MI) predicts that the exchange particles or the messenger particles in nuclear weak interaction are heavy ones. Their masses are larger than the proton mass such as (80–90) m_p. These particles transmit this interaction, which is a vector field of spin (1) particles. In 1983, these particles were discovered by a research group at CERN laboratory in Switzerland; they were denoted as $Z°$ (neutral) and $ω^±$ (charged). The mass of $Z°$ was estimated by experimental measurement to be ninety times the mass of the proton, which is about 8.542×10^4 MeV, and the mass of $ω^±$ particles are eighty-five times the proton mass, which is about 7.504×10^4 MeV. These particles with spin (1) are particles of a vector field, also the photon of the electromagnetic field with spin (1). So these particles are comparable with the photon. Therefore, they are sometimes called heavy photons.

Hence, as the photon is considered as a carrier to the electromagnetic interaction, in analogy, $Z°$, $ω^±$ are considered as carriers to the nuclear weak interaction. Accordingly, the discovery of the photon-like particles ($Z°$, $ω^±$) has led to the unification of the nuclear weak interaction with the electromagnetic interactions. Therefore, after unification of the electric and magnetic forces (or fields), the unification of the electromagnetic fields with the nuclear weak interaction fields had come to existence. This new unified force of field is called "electro weak" interaction.

Some important implications based on β-decay can be summarized by the following

1. Energy release (E_o) in equation (a) is given as

$$E_o = \left[M(Z, N) - M(Z+1, N-1)\right]c^2 \quad (e^--\text{emission}) \qquad (20)$$

2. Energy release in equation (b), positron emission is given by

$$E_o = \left[M(Z, N) - M(Z-1, N+1) - 2m_e\right]c^2 \qquad (21)$$

3. The energy balance in atomic electron capture gives the energy release as

$$E_o = \left[M(Z, N) - M(Z-1, N+1)\right]c^2 \qquad (22)$$

where M = atomic mass, m = electron mass.

Comparing (21) with (22), one can notice there is a range of atomic mass differences for which atomic electron capture is possible, but positron emission is not possible. One can notice that the energy released in β-decay may be as low as 20 KeV as in the case of ^{3}H or as much as 20 MeV in the decay of ^{8}B.

Some important information to be mentioned are as follows:

a. The wavelength of the moving electron and neutrino for the electron λ is given by $\lambda = \dfrac{386}{\sqrt{\varepsilon^2 - 1}} fm$, where $\lambda = \dfrac{\lambda}{2\pi}$ and ε is (e^-) total energy including the rest mass.

b. An electron (e^-) with E_{kin} of $2m_o c^2$ (1.022 MeV) has a wavelength $2\pi\lambda$ of 137fm>>R (nuclear radius). Also, λ for neutrino is a good approximation since $\left(\Psi(e^-), \Psi(v)\right)$ is constant over nuclear volume.

Recoil energy (E_R) is a physical fact of any system that *emits* a particle; hence, there is a recoil energy (E_R) for the nucleus under decay processes. Therefore, the recoil energy of the residual nucleus is given by

$$E_R = \frac{E_\circ^2}{2MAc^2} = \left(\frac{e^2}{3672A}\right) \cdot m_\circ c^2$$

(for large electron energy as a limit)   (24)

[*As an exercise*, verify equation (24).]

It is also

$$E_R = \frac{1}{2MA}(P_e + P_v)^2$$

(25)

[Also verify equation (25).]

For a nucleus of mass number A = 100, where $2m_\circ c^2 = 1.022$ MeV, $E_R \cong 13 eV$.

Let us go back to equation (18) and write ω_{fi} as

$$\omega_{fi} = \frac{mc^2}{\hbar} \cdot \frac{\Gamma^2}{2m^2}|M_{fi}|^2 \, F(Z,R,\varepsilon_\circ) \text{ , where}$$

$$\Gamma = \frac{g}{mc^2}\left(\frac{mc}{\hbar}\right)^2$$

and

$$F_\circ(Z=0,R,\varepsilon_\circ) = (\varepsilon_\circ^2-1)^{1/2}\left[\left[\frac{\varepsilon_\circ^4}{30} - \frac{3\varepsilon_\circ^2}{20} - \frac{2}{15}\right] + \frac{\varepsilon_\circ}{4}\log\left(\varepsilon_\circ + \sqrt{\varepsilon_\circ^2-1}\right)\right]_{\varepsilon_\circ \gg 1} = \left[\frac{5}{30}\right]$$

(26)

Also,

$$|H_{fi}|^2 = g^2\rho(Z,R,\varepsilon_\circ)|M_{fi}|^2; \quad M_{fi} \neq 0$$

$$t_{\frac{1}{2}} = \frac{ln2}{\omega fi} = \text{the half-life, } \omega_{fi} \text{ is the transition rate.}$$

Let us define "ft."-value as

$$ft = \frac{\hbar}{mc^2} \cdot \frac{2\pi^3}{\Gamma^2} \cdot \frac{\ln 2}{|M_{fi}|^2} \qquad (27)$$

where F is a known function as in equation (26), where in general,

$$F(Z, R, \varepsilon_\circ) = \int_t^{\varepsilon_\circ} d\varepsilon \rho(Z, R, \varepsilon)(\varepsilon_\circ - \varepsilon)^2 \varepsilon(\varepsilon^2 - 1)^{1/2} \qquad (28)$$

and $t_{1/2}$ can be measured experimentally.

Hence, it is possible to determine the squared matrix elements $|M_{fi}|^2$ for each nuclear transition. Γ^2 value can be obtained through examining transitions in which $|M_{fi}|^2$ is considered known or sought to be known. $|M_{fi}|^2$, in fact, carries the nuclear information provided by β-decay through lifetime researches. Ft. values play a big role in determining the kind of transition (superallowed, allowed, degree of forbiddenness, etc.) so we know from undergraduate studies that ft. values are taken in terms of their logs for mathematical simplification. The cases of transition groups, which are classified according to the $(ft_{1/2})$ values, are shown in the graph below:

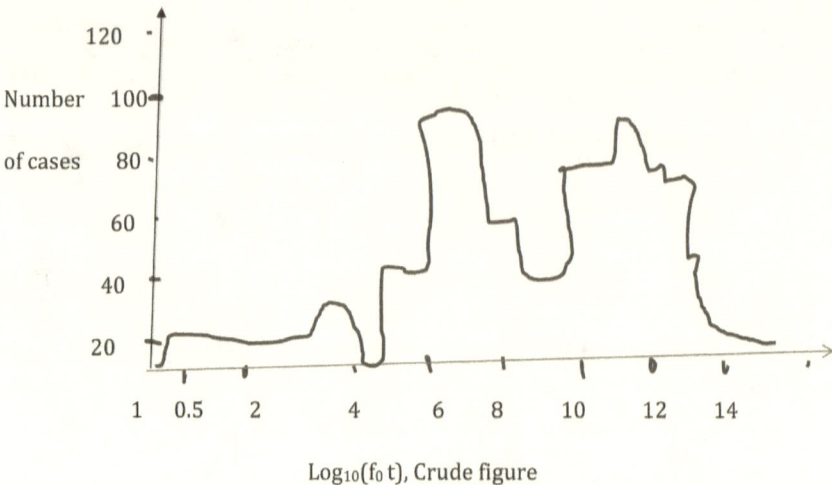

$\mathrm{Log}_{10}(f_0\,t)$, Crude figure

For small ft.-values, the smallest leads to allowed transitions. These allowed transitions are of two types (groups):

1) 1st group $\log_{10}\left(f_0 t_{1/2}\right) \cong 3.5$

2) 2nd group $\log_{10}\left(f_0 t_{1/2}\right) \cong 3.7 \pm 1.1$ $\Big\}$ superallowed transitions

The $\log f_0 t_{1/2} \geq 5.7 \pm 1.1$ are forbidden transitions of higher order. This leads to $\left|M_{fi}\right|^2 \to 0$. This implies one has to go to $\left|H_{fi}\right|^2$, considering $\Psi(\beta)$ and $\Psi(\nu)$ are no longer constant within the nucleus, but there are other terms to be considered in the calculation. Hence, one must have a complete description of (H_{fi}) before going further. To deal with this problem, the following has to be taken into consideration:

1. For allowed transitions, the character of $\left|M_{fi}\right|^2$ can be determined from general principles, such as assuming the operators involved in $\left|H_{fi}\right|$ are, to a first approximation, single-body operators. This assumption is reasonable because

a. free neutron decays ($t_{1/2} \approx 10^3$ sec);

b. this lifetime $t_{1/2} \approx 10^3$ sec is consistent with the decay in nuclei such as ^{3}H;

c. $\vec{K}_e$, $\vec{K}_v$, as well as the spin orientation of e^- and v, has been averaged over, and M_{fi} can only involve nuclear coordinates;

d. neutron at rest decays, and the operators in M_{fi} are nonzero, which implies these operators are not zero, even if the decaying nucleon is at rest;

e. the neutron is decaying and translational invariance, and the transition operators are independent of special coordinates;

f. there are only three operators: I, σ_{ij}, and τ_{ij}—unit operator, spin operator, and isospin operator, respectively;

g. finally, the operators must transform a neutron into a proton for β^--decay and a proton to a neutron for β^+-decay.

This transformation can be accomplished by the isospin operator (τ_{ij}) such as

$$\left.\begin{aligned}
\tau^{(-)}|p\rangle &= \frac{1}{2}\left(\tau_1 - i\tau_2\right)|p\rangle = |n\rangle \\
\tau^{(-)}|n\rangle &= 0 \\
\tau^{(+)}|p\rangle &= \frac{1}{2}\left(\tau_1 + i\tau_2\right)|p\rangle \\
\tau^{(+)}|n\rangle &= |p\rangle
\end{aligned}\right\} \tag{29}$$

M_{fi} must involve $\tau^{(-)}$ and $\tau^{(+)}$ through β^+ emission and β^- emission, respectively. The dependence on σ_{ij} is fixed by the requirement that $|M_{fi}|^2$ is rotationally invariant. Therefore, under these conditions, $|M_{fi}|^2$ can take the form

$$|M_{fi}|^2 = \frac{|C_F|^2}{2J_i+1}\sum_{f,i}\left|\langle f|\sum_\kappa \tau^{(-)}(\kappa)|i\rangle\right|^2 + \frac{|C_{GT}|^2}{2J_i+1}\sum_{f,i}\left|\langle f|\sum_\kappa \tau^{(-)}(\kappa)\sigma(\kappa)|i\rangle\right|^2$$

(30)

C_F is Fermi constant.

C_{GT} is Gamow-Teller constant.

J_i is the spin of initial state.

Also,

$$\left|\langle f|\sum_\kappa \tau^{(-)}\sigma(\kappa)|i\rangle\right|^2 = \sum_i\left|\langle f|\sum_\kappa \tau^{(-)}(\kappa)\sigma_j(\kappa)|i\rangle\right|^2$$

Here, j refers to the different components of the vector $\vec{\sigma}(\kappa)$. The sum over κ is a sum over different nucleons of the decaying nucleus. For simplification, equation (30) can be separated into Fermi matrix and Gamow-Teller matrix as follows:

$$|M_{fi}|^2 = |C_F|^2|M_F|^2 + |C_{GT}|^2|M_{GT}|^2$$

(31)

where

$$|M_F|^2 = \frac{1}{2J_i+1}\sum_{f,i}\left|\langle f|\sum_\kappa \tau^{(\pm)}(\kappa)|i\rangle\right|^2$$

(32)

and

$$|M_{GT}|^2 = \frac{1}{2J_i+1}\sum_{f,i}\left|\langle f|\sum_\kappa \tau^{(\pm)}(\kappa)\sigma(\kappa)|i\rangle\right|^2$$

(33)

For β^- emission, $\tau^{(+)}$ is used.

For β^+ emission, $\tau^{(-)}$ is used.

The (32) represents Fermi matrix elements.

The (33) represents Gamow-Teller matrix elements.

Selection Rules

1. Fermi, $\Delta J = 0$, no change in parity i.e. $\Delta \pi = 0$, no $\Delta T = 0$, $\Delta T_3 = 1$,
    e^- emission, =-1, e^+ emission   (34)

2. Gamow-Teller

$$\left. \begin{aligned} \Delta J &= \pm 1, 0, \Delta \pi = \text{no.0} \neq \longrightarrow 0 \\ \Delta T &= \pm 1, 0, \left(0 \neq \longrightarrow 0 \right) \quad \Delta T_3 = \pm 1 \end{aligned} \right\}$$ (35)

with the notice that

a. the Fermi selection rules follow because Fermi M_f is a special scalar and due to the fact that

$$\sum_\kappa \tau^{(\pm)} (\kappa) = T^{(\pm)}$$

$$\left[T^2, T^{(\pm)} \right] = 0 \ ;$$

b. the Gamow-Teller selection rules follow because $\sum \tau^{(+)}(k) \sigma^{\rightarrow}(k)$ transforms as a vector in isospin space.

These two types of matrix elements and selection rules are experimentally confirmed. Let us see this by some examples:

1. Decay of $^{34}_{17}Cl$ by β^+ emission,

$\Delta J = 0$, $\Delta \pi = $ no., This is a Fermi type.

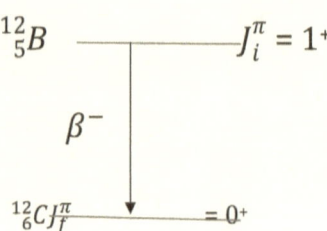

2. Decay of $^{12}_{5}B$ by e^- emission,

$\Delta J = 1$, $\Delta \pi = $ no.,

$$T_3(B) = 1, \; T_3(C) = 0$$

$$\therefore \Delta T_3 = 1$$

This is a G-T type.

In the decay of $^{1}_{0}n \rightarrow ^{1}_{1}p + e^{-1} + \bar{v}$, the contribution to M_{fi} comes from both F and G-T matrix elements.

The selection rules in (34) and (35) indicate that

a. in Fermi transitions of e^- and $\bar{v}$, it carries off a net zero angular momentum;

b. while in G-T transitions of e^- and $\bar{v}$, it carries off a unit angular momentum ($J = 1$).

VI-2-3—Superallowed Transitions (SAT)

When selection rules provide a superallowed transition, that indicates that M_f and $M_{G.T.}$ have their maximum values. For M_f this can occur for light elements; Ψ_i and Ψ_f (eigenstates) are members of an isospin multiplet. This is due to the fact that under these circumstances, a maximum overlap between the initial and the final wave functions occurs. So β⁻-decay (transition) is replacing a neutron in an orbital by a proton in the identical orbital. For example, in the transition between mirror nuclei, with T = 1/2, from $T_3 = -1/2$ to $T_{3=1/2}$, the wave function generated by the transition is shown in the schematic plot below. It is identical with $T_3 = 1/2$ state of the final nucleus (after β–decay).

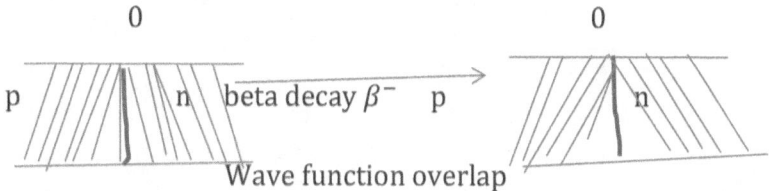

Now if $|i\rangle$ and $|f\rangle$ are members of an isospin multiplet, the Fermi matrix element can easily be obtained as follows:

$$\left.\begin{aligned}\left|M_f\right|^2 &= \left|\langle f|T^{(+)}|i\rangle\right|^2 = \left[T(T+1) - T_3(T_3+1)\right]...\beta^- - \text{decay} \\ &= \left|\langle f|T^{(-)}|i\rangle\right|^2 = \left[T(T+1) - T_3(T_3+1)\right]...\beta^+ - \text{decay}\end{aligned}\right\} \quad (36)$$

As it is known, $T_3 = (1/2)[Z - N]$ for the initial nucleus. So for the transition,

$$^{34}_{17}\text{Cl}_{17} \rightarrow\,^{34}_{16}\text{S}_{18} + e^+ + \nu, \quad p \rightarrow n + e^+ + \nu$$

T = 1, T_3 = 0; hence,

$$\left|M_f\right|^2 (\text{eq.36}) = 2....\beta^+ - \text{decay}.$$

For the decay $_0^1 n \to _1^1 p + \beta^- + \bar{v}$, where

$$T = \frac{1}{2}; \quad T_3 = \frac{-1}{2}$$

then,

$$|M_f|^2 = 1, \beta^- - \text{decay}.$$

For Gamow-Teller matrix element, it is not so easy to do, except for

the case of neutron decay ($n \to p + \beta^- + \bar{v}$), where

$$|M_{G.T}|^2 = \frac{3}{2} \sum_{m_i m_f} |\langle m_f | \sigma_\kappa | m_i \rangle|^2 = 3 \qquad \text{(as an exercise, verify this.)};$$

therefore,

$$|M_{fi}|^2 = |C_f|^2 + 3|C_{G.T}|^2 \quad \text{(n-decay)} \tag{37}$$

One might notice, as in the case of the electromagnetic transition, it is possible to obtain a "single-particle" model value for M_{fi}, which provides an order of magnitude estimate, where the jj shell model is assumed. It is assumed also that a single nucleon changes its isospin when β-decay occurs. The β-decay transition operators are independent of space coordinates. Hence, orbital angular momentum of the transformed particle stayed unchanged. Therefore, in the single-particle model ΔL = 0, the single-particle value assumes that it is possible to satisfy this requirement (ΔL = 0). But this estimate might be wrong. It might be a case that the appropriate orbital for the product particle might be occupied so that ΔL≠0, or the jj shell model cannot provide a good description to the nuclear states. For example, the core might excite many particles, so this excitation may make a significant contribution to the wave function. On this basis, the matrix elements are considerably reduced, compared with the single-particle estimate, as it well be shown next. In the single-particle, the isospin operators ($\tau(i)$) described in (33) might be replaced by unity. Now let us consider ($\vec{\sigma}$) and in terms of spherical tensors and write Gamow-Teller in the following form:

$$| M_{GT}|^2 = \frac{1}{2J+1} \sum_{k1, \, mf, \, mi} (-)^k (T_k^{(\sigma)})(T_k^{(\sigma)})^* \tag{38}$$

Where $T_k^{(\sigma)} = < \ ^- \ J_f m_f \, 1 \, T_k^{(\sigma)} I \frac{1}{2} l \ J_i \, m_i> \tag{39}$

In equation (39), (l) is the orbital angular momentum of the decaying nucleon. Since (σ) is independent of spacial coordinates (its space is spin space); hence, (1 - orbital) does not change during the transition. To find the most explicit form to equation (39), the Wigner-Eckart theorem might be used; hence, equation (39) takes the form

$$<\frac{1}{2} l \ J_f m_f 1 T_k^{(\sigma)} \ 1 \frac{1}{2} l \ J_i \, m_i> = (-1)^{Ji-Jf} \ 3j(J_f, l, J_i \, /-m_f, \, k, \, m_i) \times$$

$<1/2 \ l \ J_f\| \ T^{(\sigma)} \ l1/2 \ l \ J_i>.$ $< ll \ ll>$ are reduced matrix elements of $T_k^{(\sigma)}$ tensor operator. These reduced matrix elements can be written in terms of 6j-symbols as

$$< \ 1/2 l \ J_f\| \ T^{(\sigma)} \| \ 1/2 l J_i>= \frac{(-)^{\frac{1}{2}+l+J+I}}{\sqrt{(2J_f+1)(2J_i+1)}} \begin{Bmatrix} \frac{1}{2} & \frac{1}{2} & 1 \\ J_i & J_f & l \end{Bmatrix} < \frac{1}{2}\| T^{(\sigma)} \| \frac{1}{2}> \tag{40}$$

It is easy to show that $<j\| J\| j> = \sqrt{j(j+1)(2j+1)} \tag{41}$

So for j = 1/2, then, $<1/2 \| T^{(\sigma)} \| 1/2 > = \sqrt{\frac{3}{2}}$.

By the same way, it can be found that

$\left(\int T_{-k}^{(\sigma)}\right)^{*}$ = [equation (39)} [As an exercise, find it, then combine it with that of equation (40).] You find that

$$1M_{GT}1^{2} = 6(2J_{f} + 1) \begin{bmatrix} 1/2 & 1/2 & 1 \\ J_{I} & J_{f} & \ell \end{bmatrix} \sum \begin{bmatrix} J_{f} & 1 & J_{i} \\ -m_{f} & k & m_{i} \end{bmatrix} \begin{bmatrix} J_{i} & 1 & J_{f} \\ -m_{i} & -k & m_{f} \end{bmatrix}$$

$$\qquad\qquad\qquad\qquad\qquad 6j \qquad\qquad\qquad 3j \qquad\qquad\qquad 3j$$

Remember the orthogonal properties of the 3j-symbols, then,

$$1M_{GT}1^{2} = 6(2J_{f} + 1)\{6j\} \qquad\qquad\qquad (42)$$

The 6j-symbols are tabulated in certain tables according to J-values. Table 1 gives the values of 6j for SPA (superallowed transition) G-T ME.

Table 1

J_f	j_i	$1M_{GT}1^2$
$\ell + 1/2\ell + 1/2$ $(J_f + 1)/J_f$		
$\ell - 1/2$ $2J_f + 1/J_f$		
$\ell - 1/2\ell + 1/2$		$2J_f + 1/J_{f+1}$
$\ell - 1/2$	$J_f/J_f + 1$	

Substitute these values into equation (42), the values for $1M_{GT}1^{2}$, for the superallowed transition, then substitute that into equation (37) to get $1M_{fi}1^{2}$, which is reasonable for light nuclei, but it fails for heavy nuclei, where the single-particle model is not well satisfied, and $\Delta \ell = 0$ does not work. Table 2 gives values of superallowed G-T ME $|M_{GT}|^2$ for some light nuclei.

Table 2: $|M_{GT}|^2$ Values for SPAT

Decay	$j_i = j_f \ell_i$		$|M_{GT}|^2$	Configuration
$^{11}C \rightarrow ^{11}B$	$3/2$	$P_{3/2}$	$5/3$	$S_{1/2}\ P_{3/2}$
$^{15}O \rightarrow ^{15}N$	$1/2$	$P_{1/2}$	$1/3$	$S_{1/2}\ P_{3/2}\ P_{1/2}$
$^{17}F \rightarrow ^{17}O$	$5/2$	$d_{1/2}$	$7/2$	$S_{1/2}\ P3/2\ P_{1/2}\ D_{5/2}$
$^{31}S \rightarrow ^{31}P$	$1/2$	$s_{1/2}$	3	$S_{1/2}\ P_{3/2}\ P_{1/2}\ D_{5/2}\ 2\ S_{1/2}$
$^{33}Cl \rightarrow ^{33}S$	$3/2$	$d_{3/2}$	$3/5$	$S_{1/2}\ P_{3/2}\ P_{1/2}\ D_{5/2}\ 2S_{1/2}\ D_{3/2}$
$^{41}Sc \rightarrow ^{41}Ca$	$7/2$	$f_{7/2}$	$9/2$	$S_{1/2}\ P_{1/3}\ P_{1/2}\ D_{5/2}\ 2S_{1/2}\ D_{3/2}\ F_{7/2}$

VI-2-4—Antineutrino Absorption and Characters

The neutrino is an amazing particle that was proposed by Pauli to find a solution to the problem of nonconservation energy and angular momentum in the process of β-decay.

It is with a mass $\simeq 10^{-4}$ of the electron mass (0.511 MeV).

Its speed *is nearly equal to the speed of light*. Its interaction with matter is almost undetectable, has no charge, and is created during the decay of the nucleon-carrying energy and spin 1/2. It was proposed and became experimentally a fact. It is possible that it interacts with the proton and the neutron; this is called the inverse interaction, because it is itself the production of the decaying of these nucleons inside the nucleus, as seen early in the previous sections. This inverse interaction is represented by

$$p + \bar{v_e} \rightarrow n + e^+ \, (\beta^+) \tag{43}$$

$$n + v_e \rightarrow p + e^- \, (\beta^-) \tag{44}$$

Equation (43) is experimentally possible if the target is hydrogen ($^1H_1$) and the antineutrino (v^-) energy is quite high. Also, equation (44) is experimentally possible too, if the target is a nucleus. On this basis, equations (43) and (44) may take the forms

$$(Z, N) + \bar{v_e} \rightarrow (Z - 1, N + 1) + e^+ (\beta^+) \tag{45}$$

$$(Z, N) + v_e \rightarrow (Z + 1, N - 1) + e^- (\beta^-) \tag{46}$$

It is known today that $\bar{v_e}$ is produced in the nuclear reactors in huge amounts. The principal beta decays usually come from neutron rich elements, so that the relevant decay process is

$$(Z, N) \rightarrow (Z + 1, N - 1) + e^- + \bar{v_e} \tag{46}^*$$

There is another source for the antineutrino and the neutrino itself. These are the μ — *meson* and the τ — *meson decays.* μ — *decay is given by*

$$\mu^- \rightarrow e^- + \bar{v_e} + v_\mu \text{ or } \mu^+ \rightarrow e^+ + v_e + \bar{v_\mu} \tag{47}$$

The $\mu^\pm$-mesons have the same characteristics of the electron, but with heavier masses, about 270 $m_e \approx 138 MeV$. v_μ and $\bar{v_\mu}$, the neutrino and the antineutrino, respectively, associated with the μ — *meson* particle decay.

It is important to note that this particle is not a fundamental one; it has a structure and size.

The absorption of the antineutrino is observed experimentally. Cadmium chloride $(CdCl_2)$ was used as a target; the source of the antineutrino was Savannah River High Flux Reactor in the United States of America. The absorption cross section was measured in 1966; it was found to be $\sigma_{ext.} = (0.94\pm0.13) \times 10^{-43}$ cm^2 = 10^{-18} barns.

Problem: Consider beam of (v_e^-) with energy $E_{v_e^-}$ is absorbed in a certain target; show that the absorption cross section is given by

$$\sigma = (\frac{\hbar}{mc})^2 \frac{\Gamma function}{\pi}|Mfi|^2 \cdot E_{v_E^-}(E_{v_p^-}^2 -1)E_{v_e^-} -1]^{1/2},$$ where $E_{v_e^-}$ is in unit of mc^2.

In the beginning, it was thought that the neutrino and the antineutrino are identical particles. On this base, Majorana developed an experiment to study their behaviors, supposing if that is true, then the following inverse reactions should be induced by any of these particles:

^{37}Al + e^- = ^{37}Cl + v ,^{37}Cl + υ = ^{37}Al + e^-

This hypothesis was tested by subjecting ^{37}Cl to a reactor flux of antineutrino to find out whether it will absorb the antineutrino or not. The result of the test showed that the interaction cross section is about $(0.1\pm0.06) \times 10^{-45}$ barns. This is very much less than $(0.94\pm0.13) \times 10^{-43}$ barns. This is within the background errors coming from the neutrinos developed from the decay of the cosmic rays mesons ($\mu^\pm$). This result indicated that the neutrino and the antineutrino are not quite identical, but they differ. This conclusion was clearly confirmed by double beta decay. In addition to that, the difference in the helicity property, where it is (+1) for the antineutrino and (−1) for the neutrino, will be clarified later on.

VI-2-5—Violation of Parity Conservation (in Weak Interaction)

This physical phenomenon was so important in developing the theory of physics in general and in nuclear physics specially; it was thoroughly studied theoretically and experimentally in the fifties. In 1956, it was experimentally found that two pairs of particles are θ^+ and τ^+; their characteristics are summarized as follows:

Particle	Mass (m_e)	Lifetime (in Sec)
θ^+	966.7±2.0	$(1.21\pm0.2)\times10^{-8}$
τ^+	966.3±2.1	$(1.19\pm0.05)\times10^{-8}$

This result shows that these particles are quite identical, but why are they considered physically different? Because they were found to decay into different products, such as

$$\tau^+ \rightarrow \pi^+ + \pi^- + \pi^+ \tag{48}$$

$$\theta^+ \rightarrow \pi^+ + \pi^0 \tag{49}$$

As it is known, the pi-meson (π) is of odd parity, i.e., its parity differs from that of the nucleon. Take (48):

$$\tau^+\left(p=-1\right) \rightarrow w.i \rightarrow \pi^+ + \pi^- + \pi^+ \Rightarrow p|f\rangle = -|f\rangle,\ p = -1$$

$$\theta^+\ (p = +1) \rightarrow w.i. \rightarrow \pi^+ + \pi^0 \Rightarrow p\,|f\rangle = |f\rangle,\ p = +1$$

(+, −, 0) indicate the charge.

The parity, as it was briefly explained in the previous chapter, is an important physical dynamic operator. It describes the state of a physical system in spacial space and when it is reflected through the origin of its spacial coordinates. So the knowledge of whether it is conserved or not is quite necessary. In the electromagnetic interaction, it is confirmed experimentally as conserved, and in the nuclear reaction, it is proved to be conserved too. But in the nuclear weak interaction (W I), the conservation of parity was not experimentally confirmed; for this reason, the following was proposed before 1956:

1. The nuclear weak interaction (nuclei-decays) is not parity conserved.

2. An experimental test is necessary to check claim (1).

In the sixties, it was discovered that the particles τ^+ and θ^+ are the kaon (k^+), which has a mixed state of two states l 1+ > + l – 1>. The important step is to test that the parity is not really conserved. The well-known fact is that the nucleus constituents (nucleons) are bounded by a strong force, which is what remains of a very strong force that combines the quarks inside the nucleons. The process of beta decay is composed of a nucleon decay (inside the nucleus) and beta ($e^\pm$) and neutrino, following after the decay takes place. The nuclear force is stronger than the nuclear force by a factor of 10^{13}. For example, take that the state is mixed of even and odd parity, as shown in the diagram below:

$$----------\pi_i \rightarrow |\pi_i > +f| - \pi_i >. \quad f = 10^{-6}$$

$$--------- -\gamma - rays$$

$$---------- \pi_f \rightarrow \pi_f + f|-\pi_f >$$

So the mixed state can be written as

$1 \; \pi_{i>} + f1 - \pi_i > \rightarrow 1 \; \pi_{i, n.f} > + |\pi_{i}, wi> + 1 - \pi_i, wi.>$

Therfore, the Hamiltonian of the interaction consists of two parts:

$$- H_\beta = H + H^{'} \tag{50}$$

where H is even parity and $H^{'}$ is odd parity, i.e., = (+1) and (−1), respectively. Some books claim that (H) is conserved parity and ($H^{'}$) is not parity conserved. This claim is just nonsense, since = $(H_\beta)^2 = H^2 + H^{'2} + 2H \; H^{'}$ (interference term), which is clearly indicating that the only term that is non-conserved parity is the interference term. Hence, H is the sum of a vector part and an axial vector part, such as

$$H = g[C_v (\Psi_p^{'} \Upsilon_\mu \Psi_n^{'})^*(\Psi_{\beta}^{'} \cdot \gamma_\mu \Psi_v) + C_A (\Psi_p^{'} \gamma_\mu \gamma_5 \Psi_n^{'})^*(\Psi_{\beta}^{'} \gamma_\mu \gamma_{5\Psi_v}) \tag{51}$$

as it can be seen that Hamiltonian has nuclear parts, the nuclear part denoted by (*), which is a definite parities.

To elaborate more on this point, let $C_v^{'}$ and $C_\beta^{'}$ be considered of odd parity, so $H^{'}$ can be written as

$$H^{'} = g[C_v^{'}(\Psi_p^{'} \gamma_\mu \Psi_n^{'})(\Psi_{\beta}^{'} \gamma_\mu \gamma_5 \Psi_v)] + [C_A^{'}(\Psi_p^{'} \gamma_\mu \gamma_5 \Psi_n^{'}) \times$$

$$(\Psi_{\beta}^{'} \gamma_\mu \gamma_5 \Psi_v)] \tag{52}$$

Now let (Q) be defined as a parity effect parameter, such that

$$Q = [\frac{2CC^-}{C^2 + C^{-(2)}}] = \frac{1 + 2\frac{C^-}{C}}{1 + (\frac{C^-}{C})^2} \tag{53}$$

If Q is plotted versus (c'/c), the following diagram will be obtained

Q versus $\frac{C^-}{C}$

From the diagram, it is clear that if $Q = 1 \Rightarrow C'/C = 1$, $= 0$ if $Q = 0$ means no parity effect. This implies no odd-parity effect. Experimentally, it is found that

$Q_A = (1.00\pm.02)C_A$, $Q_A' = (1.002\pm0.2)C_A'$—this implies that $c_i' = c_i$; hence, H_β can be written as

$$H_\beta = g[C_v(\Psi_p'\gamma_\mu\Psi_n)(\Psi_\beta'\gamma_\mu(1 + \gamma_5)\Psi_v) + C_A(\Psi_p'\gamma_\mu\gamma_5\Psi_n) \times$$

$$(\Psi_\beta'\gamma_\mu\gamma_5(1 + \gamma_5)\Psi_v) \tag{54}$$

where $(1 + \gamma_5)$ is projection operator given by

$$1 + \gamma_5 = \begin{pmatrix} 1 & 0 & -1 & 0 \\ 0 & 1 & 0 & -1 \\ 0 & -1 & 0 & 1 \end{pmatrix}, \text{ therefore,}$$

$$1+\gamma_5\psi_b = \begin{pmatrix} 1 & 0 & -1 & 0 \\ 0 & 1 & 0 & -1 \\ 0 & -1 & 0 & 1 \end{pmatrix} \begin{pmatrix} \psi_1 & -\psi_1 & \psi_2 \\ \psi_2 & -\psi_2 & -\psi_4 \\ \psi_3 & -\psi_1 & +\psi_3 \\ \psi_4 & -\psi_2 & +\psi_4 \end{pmatrix}$$

(55)

VI-2-6—Two Components Theory of the Neutrino (c = c)

This theory proposes that the neutrino (v) has its momentum direction in parallel with its spin direction, i.e., $\vec{q} \uparrow\uparrow \vec{\sigma}$, where the vector (q) is the momentum and the vector (σ) *is the spin*, while for the antineutrino, the spin represented by Pauli matrix (σ) is antiparallel with the direction of its momentum (q), i.e., $\vec{q} \uparrow\downarrow \vec{\sigma}$. Let these two pictures be called A and B, respectively. Now how is it possible to distinguish between the two pictures practically? The answer is given by a suitable experiment with defining a physical concept related to the neutrino and the antineutrino plorization This concept is termed as helicity, denoted by the symbol $(\mathcal{H})$. It is a pseudoscalar type of screw sense. It is defined as

$$H = \frac{\vec{\sigma} \cdot \vec{P}}{|\vec{\sigma} \cdot \vec{P}|} = +1 \text{ or } -1, [\uparrow\uparrow \text{ or } \uparrow\downarrow \text{ respectively})$$

(56)

So each of both particles has a definite helicity. Experiment measurements showed that the helicity of the right-hand side is a mirror picture of the left-hand side and conversely, as shown in the schematic diagram below:

$$\text{-----------------------------R.H.S,} \mathcal{H} = +1$$

Mirror--

$$\text{--------------------L.H.S,} \mathcal{H} = -1$$

So by experiment, $\mathcal{H}_\nu = -1$

$$\mathcal{H}_{\bar\nu} = +1$$

Hence, the neutrino and the antineutrino have different helicity. This means that the two particles are distinguishable, which implies they are not identical particles, so they do not obey the Pauli exclusion principle.

Now let us go back to the coefficients C_ν', C_A', C_V, and C_A, where the experimental data showed that

$C_V' = C_V$, $C_A = C_A'$; also, $C_A = C_V$, $\lambda = -1.19 \pm 0.03$.

Introduce that g = C_V = 0.03; the Hamiltonian H_β is given by

$$H_\beta = \text{g} \sum \{\Psi_p' , \gamma_\mu (1 + \gamma_5) \, \Psi_n)\} + \{\Psi_\beta' \gamma_\mu (1 + \gamma_5) \, \Psi_\upsilon \}^* \qquad (56)$$

which is an axial-vector type of interaction (AV).

Hence, H_β = g . $B^{\to(4)}$. $L^{(4)}$ $\qquad\qquad\qquad\qquad\qquad$ (57)

where B = { }$^+$, baryon current $\qquad\qquad\qquad\qquad\qquad$ (58)

and L = { }$^{*\prime}$ lepton current $\qquad\qquad\qquad\qquad\qquad$ (59)

Therefore, $[H_\beta$ = g . B^4 . $L^4]$ –for WI, $\qquad\qquad\qquad\qquad$ (60)

compared with $[H_{E\&M}$ = e . S^4 . $A^4]$ $\qquad\qquad\qquad\qquad$ (61)

VI-2-7—Experimental Evidences for Parity Nonconservation in Weak Interaction

As it has been pointed out before, the particles θ^+ and τ^+ are look-alikes, except their decays product is different (equations 48, 49). Also, the product pi (π) is of odd parity, where that of the nucleon is even parity. Furthermore, the kaon particle was found to be of two components: short lifetime one, denoted by k_s, and long lifetime one, denoted as k_l. This kaon particle is neutral, and decays according to its components are as follows:

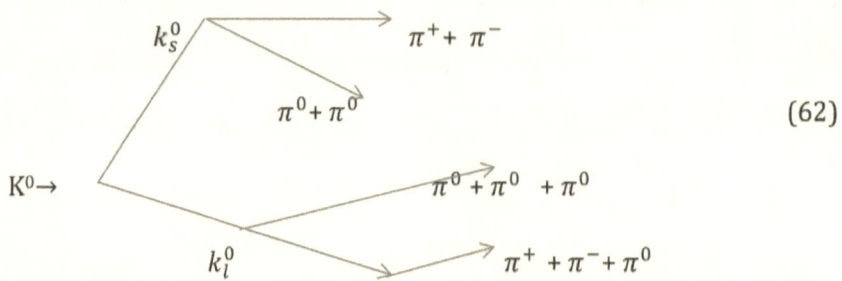

$$(62)$$

The two final states (k_s^0 , k_l^0) were found to be of different parities: $\pi^+ + \pi^-$ and $\pi^0 + \pi^0$ in k_s^0 state, with intrinsic negative parity. This suggested that conservation of parity needs to be experimentally checked—on the assumption that if the parity is not conserved in kaon decay, then it is not conserved in beta decay too. As it was pointed out in previous chapters, if it is operated on $\psi(r)$, which is eigenfunction or state of Hamiltonian (H) by inversion operator R(R^{-1}XR = −1), a new eigenstate Rψ is obtained, which is eigenstate to(H) too. This leads to the commutation between H and R so that [R, H] = 0, and R is a constant of the motion. In addition to that, the transition operator ($\Im$), with its matrix elements between <f l and l i> eigenstates representing the transition amplitude, also commute with R, i.e., [R, $\Im$] = 0 or R$^{-1}\Im R$ =

$\Im$. As it is known, $\Im$ is a function of $-\vec{p}$ and $\vec{\sigma}$, i.e., $\Im(\vec{p},\vec{\sigma})$; therefore,

$$R^{-1}\Im(\vec{p}\vec{\sigma})R = \Im(-\vec{p},\vec{\sigma}) = \Im(\vec{p},\vec{\sigma}) \tag{A}$$

Since $\vec{p}$ is (r) dependent, $R^{-1}\vec{p}R = -\vec{p}$, but $\vec{\sigma}$ is independent of $(\vec{r})$ but spin dependent, so $R^{-1}\vec{\sigma}R = \vec{\sigma}$.

Hence, equation (A) holds if and only if parity is conserved. An experiment done by Goldhaber et al. confirmed there is no conservation of parity, but how was it done? It is known that from quantum mechanics, describing a spin state emitting a neutrino is by

the expectation value of $\dfrac{\vec{\sigma} \cdot \vec{P}}{|\vec{\sigma} \cdot \vec{P}|}$, i.e., $<\dfrac{\vec{\sigma} \cdot \vec{P}}{|\vec{\sigma} \cdot \vec{P}|}> \dfrac{M}{g}$ (B)

$$<\psi_f |\psi i > = < I\psi i |\frac{M}{g}\Im | \psi i > =$$

$$<\psi_i | \Im^* \frac{M}{g}\psi_f \frac{M}{g}\Im|\psi_i> \text{ [remember equation (B) for } (\frac{M}{g})] \tag{C}$$

This holds for a given initial state. By averaging over the initial state, the result is

$$\frac{<I\psi i \dfrac{M}{g}\Im I\psi i >}{\Sigma <\psi i |I\psi i >} = T_r(\Im^*\left(\frac{M}{g}\right)\Im)/T_r(1) \tag{D}$$

This is the expected value for $<\dfrac{M}{g}>$, T_r for tracing.

Now if $\Im$ is replaced by $(R^{-1}\Im R)$ in equation (D), assuming the parity is conserved, the following result will be obtained:

$$-T_r\{\Im^*(\frac{M}{g})\Im\} \tag{E}$$

(E) is obtained using the mathematical facts that

$$T_r AB = T_r BA, \quad R^{-1} \vec{p}\, R = -\vec{p}, \quad R^{-1} \vec{\sigma}\, R = \vec{\sigma}$$

Therefore, the average of $\dfrac{M}{g}$ = 0, if the parity is conserved, i.e., any operator $0\,(\vec{p}, \vec{\sigma}) = 0\,(-\vec{p}, \vec{\sigma})$ means the parity is conserved.

Goldhaber et al. experimentally found that the $\dfrac{M}{g}$ value of the emitted neutrino is not zero but negative. This, as pointed before, indicates that parity is not conserved. On this basis, if the expectation value of a pseudoscalar quantity is not zero, it indicates that parity is not conserved. This usually can be demonstrated experimentally where the original picture is quite different from its corresponding picture in the mirror, as it is shown in the following illustrating schematic diagram:

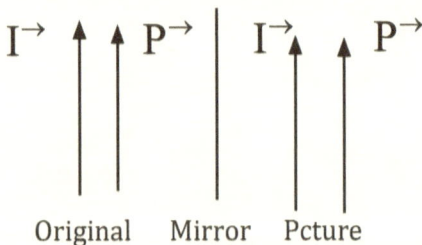

Original   Mirror   Pcture

This is pseudoscalar observable $\vec{p} \cdot \vec{I} = -\vec{p} \cdot \vec{I}$

Now $\vec{p} \cdot \vec{I} = pI \cos\theta = pI \cos(180 + \theta) = -pI \cos\theta$
$= -\vec{p} \cdot \vec{I}$, it is proved.

$$H = +1, -1$$

In another way, C. Wu et al. studied experimentally the theoretical concept of T. D. Lee in the mid fifties that the parity in beta decay is not conserved. It is based on the angular distribution function $\omega(\Theta)$ of beta decay with a momentum of $\vec{p}$. The simple form of this function (as shown in chapter 5) is given by $\omega(\theta) = 1 + A\cos\phi$

To study such case, a polarized atom is needed. That requires applying a strong magnetic field to the atom under low temperature such as

$B^{\rightarrow}$ ϕ $w(\phi)$

ATOM

This experiment was performed by C. Wu et al. They investigated the pseudoscalar quantity $(P^{\rightarrow} \cdot \mathfrak{I}^{\rightarrow})$, where P is the electron momentum emitted by the nucleus of the studied atom, and $\mathfrak{I}$ is the spin of the emitting nucleus subjected to polarization under the effect of the applied magnetic field. C^{60} was used in this experiment, which was done in 1957, decays by emitting β^{-}, as shown in the following schematic diagrams:

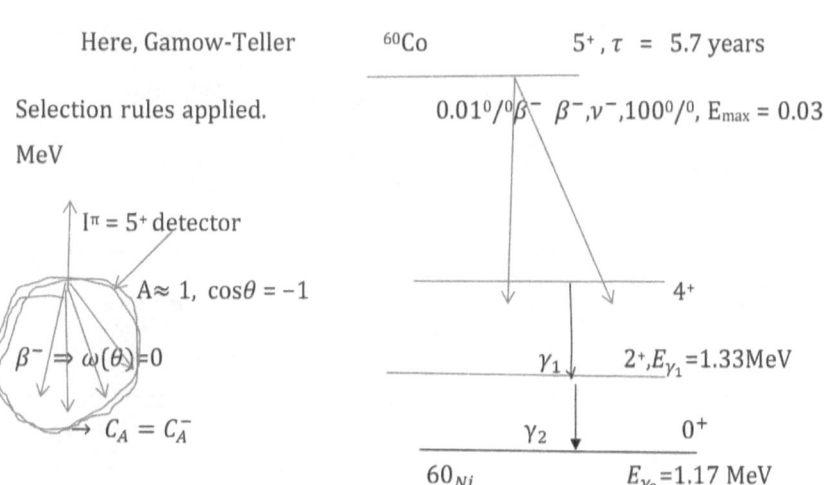

Here, Gamow-Teller

Selection rules applied.

MeV

$I^{\pi} = 5^{+}$ detector

$A \approx 1$, $\cos\theta = -1$

$\beta^{-} \Rightarrow \omega(\theta) = 0$

$\hookrightarrow C_A = C_A^{-}$

^{60}Co $5^{+}, \tau = 5.7$ years

$0.01^{0}/^{0}\beta^{-}$ $\beta^{-}, v^{-}, 100^{0}/^{0}$, $E_{max} = 0.03$

4^{+}

γ_{1} $2^{+}, E_{\gamma_{1}} = 1.33$MeV

γ_{2} 0^{+}

^{60}Ni $E_{\gamma_{2}} = 1.17$ MeV

Co^{60} is under strong magnetic field $B^{\rightarrow}$,   Decay scheme of ^{60}Co

So the nucleus of ^{60}Co will be polarized, i.e., its radiations will be aligned in a direction under the effect of the magnetic field, due to the crystal the atom situated in. This crystal is made of $[2C_e (NO_3)_3; M_g (NO_3)_2; 24H_2o]$. In that work, it was found that $\vec{P}.\vec{\mathfrak{I}}$ is negative, which implies that there is a preferential emission of the electrons (β^-) in the opposite direction of the spin direction of the nucleus, which confirmed the theory of Lee and Yang of the non-parity conservation in the beta decay processes, which take place under the nuclear weak interaction. This result can be understood in terms of the angular momentum balance, where the helicity ($\mathcal{H}$) of the antineutrino is positive, and it is negative for the electron. From the decay scheme of Co, as shown before, it is clear that the transition from $J^\pi = 5^+$ to $J^\pi = 4^+$ implies that v^- and β^- must have a unit of angular momentum in the direction of the ^{60}Co nucleus spin I. This is shown in the schematic diagram below:

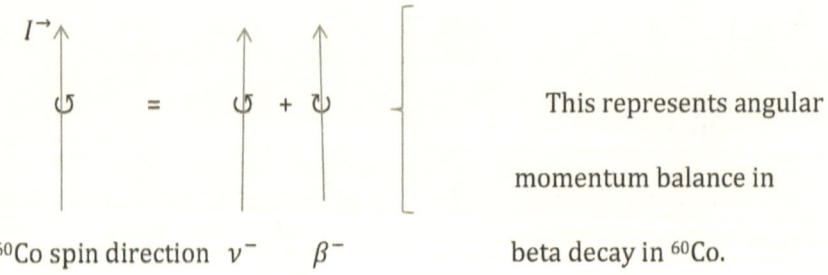

$I^\rightarrow$ ↑

$\mathfrak{I}$ = $\mathfrak{I}$ + $\mathfrak{v}$ [This represents angular

momentum balance in

^{60}Co spin direction v^- β^- beta decay in ^{60}Co.

In order to obtain a total unit of spin, the antineutrino must travel in a direction parallel to the spin direction of the nucleus, but the spin direction of the electron must be opposite to the direction of the nucleus spin. As mentioned before, the angular correlation function in its simplest form is $\omega(\theta) = 1 + A\cos\phi$, if $A = 1 \Rightarrow \cos\phi = -1 \Rightarrow \omega(\phi) = 0$.

For G-T , I= 5.

$$A = \pm < \frac{I_z}{I} > \cdot \frac{2C_A^- C_A}{[C_A^2 + C_A^{(-)2}]} \cdot \frac{P}{\omega}$$

$$= [< I_z > /I] \cdot Q \cdot (P/\omega), P/\omega = V/C \text{ for } \beta^{\pm}$$

where Q is the parity effect, as defined before. From that, it is easy to tabulate that

1) $\mathcal{H}(\beta^-) = -V/C$

2) $\mathcal{H}(\beta^+) = +V/C$ ^{60}Co ^{60}Ni β

$I^\pi = 5+ \longrightarrow I^\pi = 4+$

$\sigma^{\rightarrow}$ $\beta^- \leftarrow$ $p^{\rightarrow}$

$\beta^{+\rightarrow} \longrightarrow p^{\rightarrow}$

3) $\mathcal{H} = (\dfrac{p^{\rightarrow} \cdot \sigma^{\rightarrow}}{\downarrow P \square \sigma \downarrow})$

$C_A^- \approx C_A$ longitudinal polarization of $\beta^{\pm}$

Now, the polarization of the final state of ^{60}Co after β^- -decay and releasing ν^- might be represented a

$\nu^- \sigma^{\rightarrow}$ ^{60}Co $\sigma^{\rightarrow}$ $\beta^{\rightarrow}$ Z-axis, $\Delta m = -1$

The emitted gamma ray (γ) following beta decay is of circular polarization such as

γ $\beta^{\rightarrow}$

(C. P) $^{60}_{27}Co$

Hence, beta-gamma circular polarization correlation may be represented as

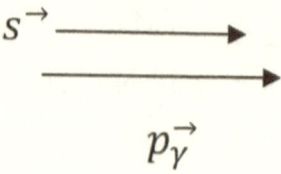

and the helicity, as defined before, is right-hand screw, i.e., $\mathcal{H}_y$ is (+1). Usually, the polarization for a given state of the electron is specified by the expectation value of the helicity of the state, as computed before. Now it is important to know the probability that the helicity is positive or negative. For that, let P_+ be considered as the probability that the helicity is positive and that P_- is the probability for negative helicity. If the helicity is negative, the expected value of $\vec{\sigma} \cdot \vec{P_e}$ is given by <

$$< \vec{\sigma} \cdot \vec{P_e} > = \frac{P_+ - P_-}{P_+ + P_-} \qquad (63)$$

P_+ is proportional to the transition probability of forming a state with helicity (+1), but this can be extracted from the probability of the transition with regard to the helicity ($\mathcal{H}$).

It can be performed by inserting or introducing a projection operator defined as $\hat{P}_+ = 1/2[1 + \vec{\sigma} \cdot \vec{P_e}]$, $\mathcal{H} = +1$ $\qquad$ (64)

$\hat{P}_- = 1/2[1 - \vec{\sigma} \cdot \vec{P_e}]$, for $\mathcal{H} = -1$ $\qquad$ (65)

where $[\hat{P}_\pm]^2 = [\hat{P}_\pm]$, so $\hat{P}_\pm = 1$ or 0, $\hat{P}_+ \hat{P}_- = 0$ $\qquad$ (66)

Under these physical hypotheses, it was found that polarization is given by $(-V_e/C)$ for unpolarized nucleus, for the case of β^- emission. And it is $(+V_e/C)$ for unpolarized nucleus, for the case of β^+ emission.

VI-2-8—Mu-Meson Decay, the Universal Weak Interaction

The weak interaction or the nuclear weak interaction, where the beta decay takes place under its action, is very much greater than the gravity force but too much smaller than the nuclear force that combines the nucleons inside the nucleus, and also, it is less than the electromagnetic force. This information is quite well explained in many textbooks in physics.

This type of interaction is found to have its effect in all nuclear particles as families, which decay in different processes, so beta decay is, in fact, a prototype to this universal nuclear weak interaction. Therefore, the μ-meson was found to decay under this interaction as follows:

$$\mu^{-} \longrightarrow e^{-} + V_{\mu} + V_{e}^{-} \ , \ \mu\text{-decay} \tag{67}$$

$$\mu^{-} + p \longrightarrow n + V_{\mu} \ , \ \mu\text{-capture} \tag{68}$$

From (67), one can conclude that μ^{-}-decay involves only leptons $\left(e^{-}, V_{\mu}, V_{e}^{-} \right)$, which is, for the time being, considered with the quarks (building block of the baryons' particles) as the fundamental particles of the matter. Also, the electromagnetic interaction, which is stronger compared with the nuclear weak interaction, can exist among these particles. In addition to that, the $\mu - capture$ (68) involves the strongly interacting baryons (n, p), a case similar to beta decay and the electro orbital capture, represented by

$$n \longrightarrow p + e^{-} + V_{e}^{-} \ , \qquad e^{-} + p \longrightarrow n + V_{e}$$

Also, it is important to note that there are two types of neutrinos: one associated with the electron emission (V_{e}) and one associated with μ-decay (V_μ). The μ^- itself is found to be a production of the pion (π)-decay, as shown:

$$\pi^- \text{ (pion)} \longrightarrow \mu^- + V_\mu^- \tag{69}$$

It was found that the helicity of V_μ is similar to that of V_e; i.e., $\mathcal{H} = -1$, which implies that the interaction is axial-vector type (A-V). This was experimentally proved by Dunby et al. In addition to this, it was observed that V_μ cannot replace V_e in the capture reaction. That means

$n + V_e \longrightarrow p + e^- \neq n + V_\mu \longrightarrow p + \mu^-$, which clearly shows the absorption of V_μ by the neutron leads to the emission of the muons. Hence, there are two leptons involved in the nuclear weak interactions:

1. The electron mode (e^-, V_e)

2. The muon mode (μ, V_μ)

It appears here that the description of electron capture is quite identical to the muon capture. The only difference is, the muon mass is larger than the mass of the electron. It is about 27o m_e (m_e =.511MeV). So , $m_\mu \cong$ 138 MeV. Equation (67) very clearly demonstrates the relation between the two leptons in (1, 2) above. Some results based on these facts were calculated. They were found to be

1) $C_{vg} = 1.4x10^{-49}$ erg.cm^3 for beta decay interaction strength;

2) The transition probability / decay time of the muon was the first approximation found to be

$$W = [\frac{G_\mu^2}{192\pi^3\hbar^7}]M_\mu^5 C^4 \tag{70}$$

G_u is analogous to C_{VG} of beta decay. Lifetime of the muon was experimentally measured to be $\tau_\mu = (2.2) \times 10^{-6}$ seconds. This leads to $G_u \cong (1.435) \times 10^{-49}$ erg.cm^3.

So $G_u / C_{Vg} \cong 2^0/^0$ larger than C_{Vg} of beta decay (reference: de-Shalit and Feshbach, *Theoretical Nuclear Physics*, vol. 1, 1977).

Chapter Seven

Nuclear Reactions

VII-1—Introduction

Nuclear reactions are types of scattering, elastic and nonelastic, where projectile of certain energy incidents on a nuclear target, or collision between nuclei, nucleons, or ions. This can happen between a fixed target at a definite state and moving projectile; the projectile particle is also called the bombarding particle. The different types of reaction depend on the energy carried by the bombarding particles. These types of reactions will be clarified when it is treated later. Also, the nuclear fission and the nuclear fusion are major sources of energy today for peaceful uses and harmful uses.

The scattering and the collision type of reactions occur in two different setup systems:

1) The laboratory system (lab S.) in whichis the target is fixed and the incident projectiles are moving with certain velocity, depending on the energy provided. This can be schematically demonstrated as follows (figure 1):

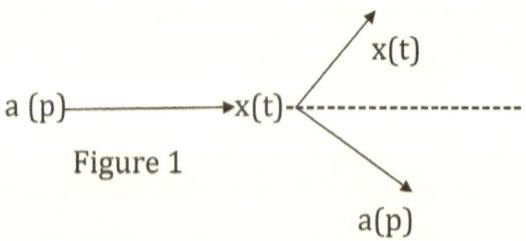

Figure 1

where a (p) is the projectile and x (t) is the target.

2) The center mass system (CMS) is where both the target and the bombarding particle are in motion with a certain energy toward each other (a kind of collision). The following diagram (figure 2) is a simple demonstration for it:

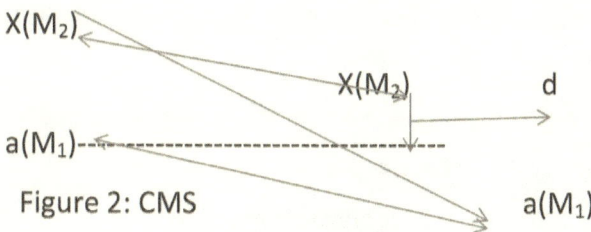

Figure 2: CMS

d is the impact parameter, a is the projectile, M_1 (its mass),X is the target, and M_2 is its mass. Usually, $M_2 \gg M_1$; therefore, the velocity of the bombarding particle $V_a \gg V_x$, V_x is the velocity of the target.

The following diagram is more explicit. Take a nucleus (X) as a target and (a) as a projectile; the (Y) is the product nucleus, and (b) is the ejected particle.

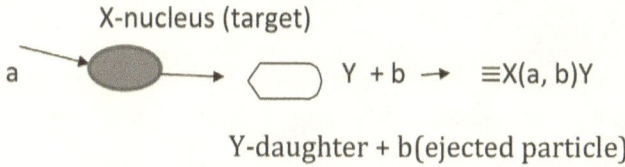

Y-daughter + b(ejected particle)

Or this can be written as

$$a + X \longrightarrow Y + b \equiv X(a, b)Y \qquad (1)$$

where X is the target nucleus, a is the bombarding particle (projectile), Y is the product nucleus, and b is the ejected particle due to the interaction. Sometimes a compound nucleus might be formed in excited state for a while, due to absorbing the incident particle and gain energy. Before the final productions take place, this will happen. The compound nucleus is, in fact, in unstable state, so it has to get rid of this excess energy by many ways to be in a stable state within a certain time. The ways of getting rid of this exciting energy might be (1) emitting it as radiations (2) or emitting the incident particle or particles, producing new nucleus or more. These possibilities can be shown as

$$a + X \rightarrow \begin{cases} Y + b \\ Y^- + b^- \\ Y^= + b^= \end{cases} \quad \begin{bmatrix} \text{These represent all possible} \\ \text{reactions allowed by energy} \\ \text{and selection rules.} \end{bmatrix}$$

Sometimes, the target nucleus captures the bombarding particle (a), where the interaction is weak, as in the following equation: $a + X \rightarrow W^* \rightarrow w + \hbar\omega$, which indicates that a compound nucleus is formed in excited state W^* then went back to its stable state (w) given by the exciting energy as radiation of energy ($\hbar\omega$). But if the interaction is strong, it will take the form

$$a + X \rightarrow \begin{cases} a + X & \text{elastic scattering} \\ a + X^* & \text{inelastic scattering}^*\text{—excited state} \\ b + Y \text{ or } c + Z, & \text{etc., ejected particles with new nuclei} \end{cases}$$

(products)

The energies of the colliding particles or the projectiles and the targets (CMS) play a big role in recovering the mystery of nature building and its behavior. The range of the energy used these days (2014) in the order of tera (up to 14 tera-megawatt), in Cern giant machine, where scientists are looking after the theory of big bang to understand well the real origin of matter and the source of the mass and other nature mysterious this huge energy is to be used . As it is well-known to nuclear particles physicists, the energy levels play an important role in revealing such particles, their properties, and interaction behaviors. On this basis, they are divided as families with common physical properties. This will be treated well in chapter 8. In addition to that, this chapter is dealing with nuclear reactions only up to 150 MeV, which is equivalent to the threshold energy for the production of the mesons. To study nuclear physics, the acquaintance with the basic general theories of the nucleus started in the forties and the fifties is quite important. These theories are based on two well-known concepts:

1. Liquid drop model

2. Shell model

As an example, the liquid drop model is the basic theory to understand the features of forming the compound nucleus. When the projectile incidents on the target nucleus, it interacts strongly with all the nucleons of the nucleus, quickly sharing with them its energy. But the compound nucleus formed decays in a manner that is so independent of the mode of its formation it does not remember its way of forming. In shell theory, the explanation of the compound nucleus takes the following steps. The incident particle is supposed to interact with the target nucleus through the potential created by the central force of the shell model structure (remember $F = -\Delta V$); therefore, the probability of it being absorbed by the compound nucleus is so small. Weisskopf unified these two concepts, proposing that any nuclear reaction takes a series of stages, as shown schematically in the following figure, when the incident particle reaches the edge of the potential. The first interaction will be a type of partial reflection of the wave function, which is called shape elastic scattering. Figure 3 shows that.

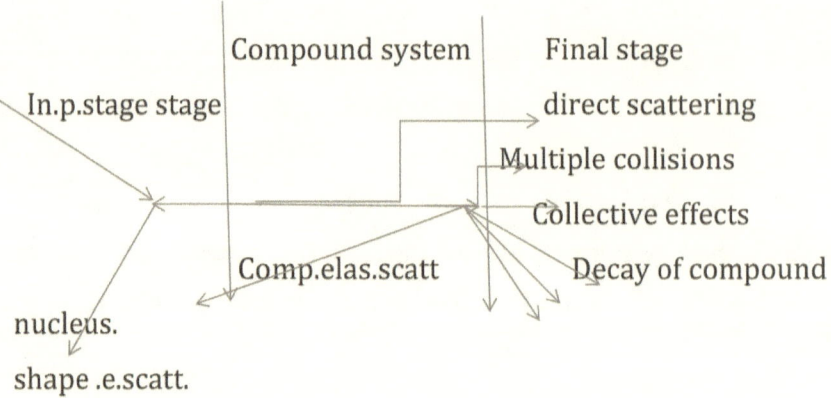

Figure 3: Weisskopf shape elastic scattering

Therefore, any potential discontinuity has a finite reflection coefficient for any incident wave. It is well-known from quantum mechanics that the wave part that enters the nucleus will be absorbed. In the first step of absorption, it is a kind of two-body collision, which can be stated as follows: If the projectile particle is nucleon, then it interacts with one of the target nucleus nucleons, which get excited to unfilled level. If the energy of the incident particle is quite enough, the nucleus nucleon will be ejected, leaving the nucleus. This reaction is called the direct reaction. In case the projectile energy is not enough to knock out a nucleon from the target nucleus, the interaction somehow will be complicated, like forming a compound nucleus. If the compound nucleus emits the incident particle or another one similar to it with the same energy of the incident particle, the scattering is called the compound elastic scattering. But if the energy of the leaving particle is less than the energy of the incident particle, the scattering is called compound inelastic scattering, which means some of the incident particle energy is taken away by certain reaction. Also, it is known that the compound nucleus can be formed by different ways. For example, the compound nucleus ^{14}N can be formed by the following interactions:

$$
\left.\begin{array}{l}
^{10}B + \alpha \rightarrow \\
^{12}C + d \rightarrow \\
^{13}C \rightarrow + p \dashrightarrow
\end{array}\right] \quad , \quad \left.^{14}N + \right\} \quad
\begin{array}{l}
\alpha\text{-particle (He-nucleus)} \\
d \text{ is the deuteron } (^{2}D_1). \\
p \text{ is the proton.}
\end{array}
$$

Usually, the lifetime of the compound nucleus is about $10^{-16\pm3}$ seconds, while the time for the proton to cross the nucleus (r = 10^{-13}cm.) is about 10^{-22} sec, which is too short, compared with 10^{-16}. The compound nucleus ^{14}N might decay through many ways such as n + ^{13}N particles leaving with α + ^{10}B, the energy of incident particle:

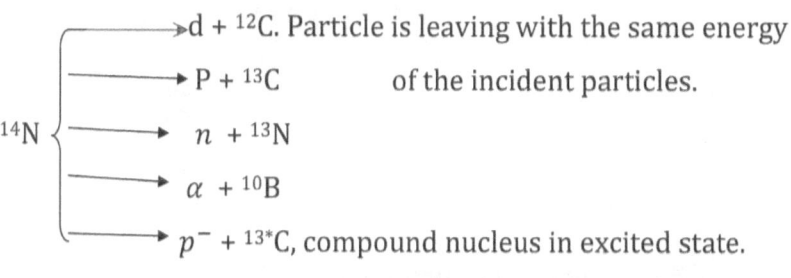

$$
^{14}N \left\{
\begin{array}{l}
\longrightarrow d + {}^{12}C. \text{ Particle is leaving with the same energy} \\
\longrightarrow P + {}^{13}C \qquad\qquad \text{of the incident particles.} \\
\longrightarrow n + {}^{13}N \\
\longrightarrow \alpha + {}^{10}B \\
\longrightarrow p^- + {}^{13*}C, \text{ compound nucleus in excited state.}
\end{array}
\right.
$$

The particle is leaving with less energy.

Sometimes, when the energy of the incident projectile E_i might be exactly corresponding to a virtual level of nuclear potential, the probability of finding the incident particle within the target nucleus is quite high, which leads to a nuclear interaction with a single particle or potential resonance (single-level resonance). This phenomenon is shown in figure 5.

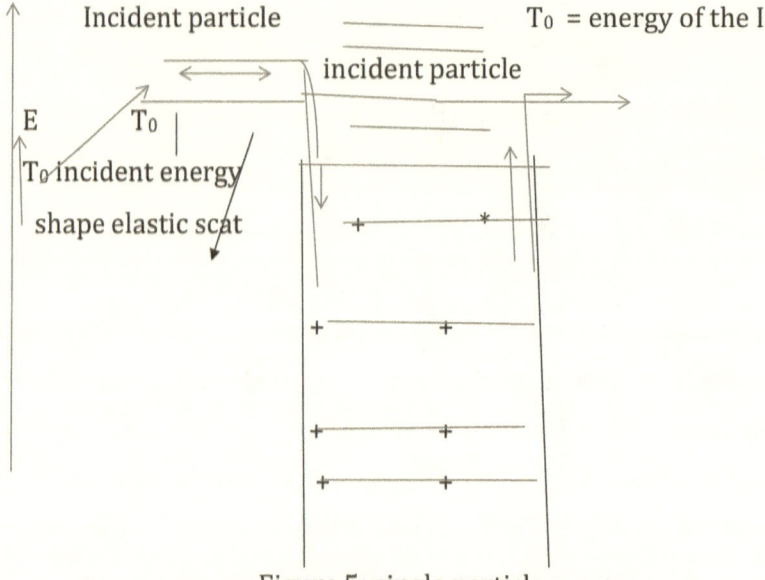

Figure 5: single particle

The conservation laws in nuclear reaction are well-known. They are summarized in the following:

1. The mass number A (Z + N) conservation

2. The angular momentum conservation

3. The linear momentum conservation

4. The charge conservation

5. The statistic conservation

6. The parity conservation

7. The isobaric spin conservation

8. The energy-mass conservation

The nuclear energy of disintegration or the so-called Q–value of the reaction is defined as the difference between the kinetic energies of the in and the out particles, which is given as

$Q = T_0(outs)-T_0(ins)$. This reaction can be represented schematically as follows (figure 6):

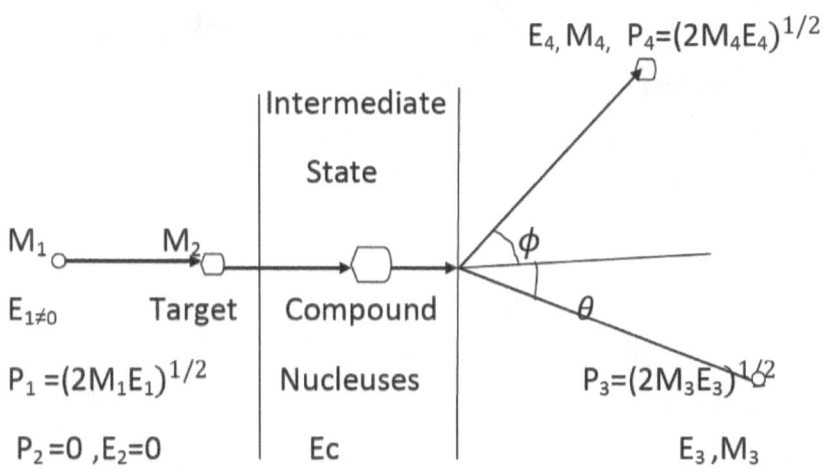

Figure 6

Therefore, $E_1 + Q = E_3 + E_4$ (mass-energy conservation)

$$[2M_3E_3]^{1/2} \sin\theta + [2M_4E_4]^{1/2} \sin\phi = 0$$

$$0 = [2M_3 E_3]^{1/2} \cos\theta + [2 M_4 E_4]^{1/2} \cos\phi$$

Solving for Q, it is given by

$$Q = E_3[1 + M_3/M_4] - E_1[1-M_1/M_4] =$$

$$Q = \frac{\sqrt{(2E_1 E_2 E_3 E_4)}}{M_4} \cos\theta \qquad (3)$$

Equation (3) is well-known as Q-value equation. Details of the endoergic and the exoergic reactions are very well tackled in some undergraduate standard textbooks (see Meyerhof).

VII-2—Wigner Channels of Interaction

Let α^-, β^-, and γ^- be the entering nuclear states, and α^*, β^*, and γ^* are the emerging states. Therefore, in general, the following demonstrates that

$$a_{\alpha^-} + X_{\alpha^-} \longrightarrow \begin{cases} X_{\alpha^-} + a_{\alpha^=} \\ X_{\beta^-} + \alpha_{\beta^=} & ------- (4) \\ Y_{\gamma^-} + b_{\gamma^=} \end{cases}$$

Here, the conservation laws restrict the channels β^- and $\beta^=$, γ^- and $\gamma^=$ reaction channels. So in the center of mass system, the total energy is given as

$E_{total} = \varepsilon_\alpha + E_{\alpha^-} + E_{\alpha^*}$, where ε_α is the kinetic energy of relative motion (channel energy), E_{α^-} is the internal energy of the target (X) (moving), E_{α^*} is the internal energy of the incident particle (a). The kinetic energy of the relative motion in the laboratory system can be found in terms of that of the center mass system to be

$$(\varepsilon_\alpha)_{lab} = (\frac{M_x + M_\alpha}{M_X}) \, \varepsilon_\alpha$$

(5)

Now let the initial and the final particles be different as

$a_{\alpha^*} + X_{\alpha^-} \rightarrow Y_{\beta^-} + b_{\beta^*}$, then the energy is

E = $\varepsilon_\beta + E_{\beta^-} + E_{\beta^*}$, where, $\varepsilon_\beta = \varepsilon_\alpha + Q_{\alpha\beta}$

By conservation of energy, it can be found that

$$Q_{\alpha\beta} = E_{\alpha^-} + E_{\alpha^*} - (E_{\beta^-} + E_{\beta^*})$$

(6)

($Q_{\alpha\beta}$) is the Q-value of the reaction channel ($\alpha\beta$).

It is important to notice that if it barely stated Q-value without specifying it to a certain state, it means it is in ground state. This implies that

$$Q_{\alpha\beta} = -Q_{\beta\alpha} = \Rightarrow Q_{\beta\alpha} = E_{\beta^-} + E_{\beta^=} - (E_{\alpha^-} + E_{\alpha^=})$$

which takes the reaction X (p, n) Y. This is represented as

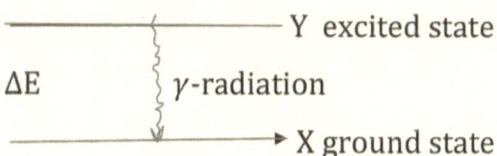

ΔE Y excited state

γ-radiation

X ground state

VII-2-1—Types of Nuclear Reactions

Nuclear reactions, as they have been seen briefly in the previous section, are of different physical processes. They are very important reactions helping in understanding the nature structures, because it is well-known that the first nuclear reaction experiment, which was a type of scattering between alpha particle and a thin foil of gold in 1911, discovered that the nucleus of the atom has positive charge and contains the major mass of the atom (99%) and that the radius of its size is about 1.4×10^{-13} cm. Following that is the study of the fine details of the nucleus structures and properties regarding the constituents of the nucleus (the protons and the neutrons), the force combining them together, the nuclear force. Studying the potential under which these nucleons (p, n) reside showed a difference between the potential of the protons and that of the neutrons. The types of the nuclear models and the characters of each one,also lead to understanding the nuclear levels and their quantum parameters for each states, the ground state and the excited states due to a type of these nuclear reactions.

Of course, for each type of reaction, a well-designed experiment is performed, taking into consideration all the required scientific conditions. In general, the reaction is a physical processes occurring between an incident particle (projectile) with a certain velocity (it depends on the energy) on a certain target (chosen according to the goal of the experiment); the particle energy is preferred to be homogenous. The projectile is usually a homogenous beam of particles impinging on a target to be studied. Due to the impact between the incident particle and the target, the reaction will take place, and new products might emerge. These are quantitatively studied by measuring the directions of motion and the speed of those new product particles, using a program that takes all the physical laws governing the physical phenomenon under investigation. These kinds of characters are called the dynamic of nuclear reactions, which, in fact, classify the nuclear reaction type occurring according to certain conditions. In the following, a very brief discussion will be given to these different types of nuclear reactions (some cited references are quite useful to refer to for a detailed discussion). These types of reactions are

A. Rutherford Scattering and Coulomb Excitation

This experiment, as mentioned before, was the first done by Rutherford to explore the secret structure of the atom in 1911, following the discovery of the division of the atom by J. J. Thomson, who discovered the electron as a part of the atom in 1897. Rutherford has had sent an alpha particle as a projectile with a certain energy toward a target made of gold foil. It is scattering in lab system. Alpha particle, as it is well-known, is a helium nucleus with positive charge; it is surrounded by electromagnetic field. So when it came close to the target, it deviated from its straight moving. The distance between the point of the deviation and the nucleus of the gold atom is called the impact distance; it is the distance at which the interaction between the alpha particle and the target has taken place. The analysis of the scattering experiment, as it is well-known to physicists, had led to the discovery of the nucleus, its mass, its approximate size, and its charge. As consequences, the nucleons were discovered. The interaction between the alpha particle and the nucleus is a type of scattering (repulsion) between two similar-charge-sign particles; the force acting here is a coulomb electric force, as it is well understood from the electric field theory. This was proof that the nucleus has a positive charge. In this type of scattering, which is, in fact, an elastic scattering (no loose of energy), the interaction is an electric field interaction under the effect of the coulomb force. Sometimes, the electric field associated with the charged projectile interact with the electric moments such as dipole, quadruple of the target nucleus, which excites the nucleus to higher energy states. At these states, the nucleus will be unstable. To go back to stable or ground state, it has to emit the exciting energy as gamma radiation. This process is known as coulomb excitation. It is quite useful to study the low-lying excited states of heavy nuclei.

B. Nuclear Potential Scattering

Any nucleus has its own potential due to the nuclear force, as mentioned in chapter 1, since any potential related to a force and vice versa, where $F = - \Delta V(r)$.

The nucleus potential is, in fact, the average of that of nucleons, where the neutrons' potential is deeper than the potential of the protons, which is due to the coulomb force with respect to the protons with positive charge. So when the incident particles encounter the target nucleus closely, they will be scattered due to the nuclear force that is related to the potential physically and mathematically, as shown above, but this happens in a relatively coherent way. It is a kind of elastic scattering in which neither the incident particle nor the target nucleus is in an excited state. In the beginning, the neutrons (no charge) were used as bombarding particle.

After that, the protons (which have charge) and other charged particles and ions are used.

C. Nuclear Surface Scattering

This type of scattering occurs when the incident bombarding particle is strongly interacting with surface nucleons of the target nucleus, giving part of its energy to these nucleons leaving the nucleus in excited state. It is inelastic scattering. It is somehow similar to coulomb excitation as an inelastic scattering. The difference is the type of force, and here, it is the nuclear force, which is one hundred times stronger than the electric coulomb force.

D. Surface Transmutation Interaction

Sometimes, the incident particle ejects one or more particles from the target nucleus by direct interaction between the bombarding particle and these ejected nucleons or particles. The incident particle is either scattered or captured by the target nucleus.

E. Compound Nucleus Formation

As it was mentioned in the introduction, most of the nuclear interactions or reactions take place into two steps, which are quite well defined for the kinetic energy of incident particle within 30 MeV. The first step is the interaction between the incident particle and the target nucleus, forming a compound nucleus. The mass of the compound nucleus is the sum of the masses of the incident particle and that of the target nucleus. Its charge is the sum of their charge. The incident particle is captured by the target nucleus; it shares its energy with the nucleons of the target nucleus. They mix together, forming the resulting compound nucleus. Therefore, at the same time, they are losing their individual identities. The second step is, after a certain time (lifetime), the formed compound nucleus, which is unstable, decays giving off particles and nuclei as products; therefore, its decay is independent of the mode of its forming. But it depends on its excitation. For more details about the physical features of the compound nucleus, one might refer to some books cited in the list of references at the end of the book. For historical reason, it is plausible to say the concept of the compound nucleus is related to Niels Bohr (1936).

F. Striping Reaction

It is clear from the title of the reaction that it is a reaction between two nuclei; one is a projectile passing by a target nucleus. When the projectile nucleus passes close to the target nucleus, some of its nucleons get striped by the target. So it is not forming a compound nucleus, but it is a kind of direct reaction similar to the surface reaction mentioned before. The mechanism of this reaction can be imagined qualitatively as such. At given energy and orbital angular momentum, there is a certain probability that one or more of the nucleons will be moving within the incident nucleus in such a way that as this nucleus is moving by close to the target nucleus, these nucleons are also moving in just the right way to be capturable in a certain orbit about the target nucleus. The striped nucleus recoils from the encounter with momentum change. It is approximately equal and opposite to the momentum carried by the missing nucleons. Due to the direct nature of the striping process, the center of mass angular distribution of the outgoing particles is not isotropic; it shows characteristic diffractive maxima and minima, predominantly in the forward direction.

G. Pickup Reaction

This reaction is quite inverse in its action to the striping reaction. The bombarding projectile, nucleus or nucleon, when passing by close to the target nucleus, might pick up one or more nucleons from the target nucleus. In this reaction, the center mass angular distribution also shows diffractive pattern. In fact, essentially, the same theory might be applied to the inverse cases: (d, p) and (p, d) or (d, n) and (n, d) reactions such as

$D + X \rightarrow Y + p$

$p + X \rightarrow Y + d$

$d + X \rightarrow Y + n$

$n + X \rightarrow Y + d$

H. Spallation Reactions

If the energy of the bombarding particle is in the range of no more than a few times of the average binding energy of the nucleon, which is about 8 MeV, it is in the low range of the required energy for nuclear reactions. At this low energy, the state of the reaction is quite different from the state where the energy of the reaction is within or more than 50 MeV. The mechanism of the reaction processes is totally indifferent from that of the low energy reaction. The basic features of the reactions of 50 MeV and higher are the primary interaction occurring more and more between individual nucleons of the bombarding particles and the nucleons of the target nucleus, not between the nuclei, the incident, and the target, as a whole, as might be expected, on the basis that both decreased the de- Broglie wavelength of the incident nucleons and the higher energy available to the system.

While in the lower energy range, there is seldom enough energy available to eject more than one nucleon from the target nucleus. Even this is on the expense of capturing the incident particle (nucleon). But in the higher range of energy, there is quite enough energy for the ejection of more than one nucleon, either single or as compound structure. These product particles may be of different sizes and masses. Hence, this type of reaction is called spallation reaction; the product is called spallation product.

I. Higher Energy Reactions

When the available energy for the nuclear reaction is in the range of 150 MeV and beyond, those new nuclear particles, in addition to the proton and the neutron, are produced, such as the pi-mesons, which are the first to be discovered in the cosmic rays. Theses π-mesons (π^+, π^-, π^0) play a main role in the nuclear force, which combines the nucleons inside the nucleus domain. During the development of the nuclear physics theory theoretically and experimentally, a new domain of high energy uses is created. Therefore, new hundreds of nuclear particles were discovered. So beyond 500 MeV, whole families of particles, the so-called strange particles, are discovered. For the energy 6000 MeV and beyond that, the antinucleons (p^-, n^-) are produced. The details of the nuclear particles physics will be treated in chapter 9.

VII-3—Cross Section, Definition, and Formalism

VII-3-1—Introduction

The cross section is a very important parameter in studying nuclear reactions. So it is a physical quantity too important to be measured. It is defined as the probability that the reaction occurs under given experimental conditions. Sometimes, this probability might exhibit very sharp beak at a certain energy, where the reaction is termed as resonance reaction. Also, this probability might be a rather smooth function of the energy, and usually, it is expressed in terms of an effective area (A), which is represented by the nucleus subjected to the bombarding beam of particles. Therefore, the number of the incident particles impinging the target within this area can be calculated easily using simple geometrical principles. These particles actually lead to the required reaction; hence, this, effective, is the cross section of the reaction or the interaction. The cross section, in general, depends on many geometrical factors and on the designed shape of incident beam. Therefore, it might be defined by several ways as follows:

1. The probability that an event will take place if a single target is subjected to beam of bombarding particles of total flux equals to one particle per unit area.

2. The probability that an event will occur when a single particle is incident perpendicularly at target is made of one nucleus per unit area.

3. If the extent of the bombarding beam is larger than the extent of the target, the cross section is the ratio of the number of the events observed to occur to the product of the integrated incident beam

flux in particles per unit area and the total number of eligible nuclei presented in the incident beam.

4. In a case inverse to (3), the target extent is larger than the extent of the bombarding beam; the cross section is the total number of events observed to occur to the product of the total integrated number of incident particles, plus the number of eligible nuclei per unit area of the target normal to the beam.

These descriptions to the possible definition of the cross section indicates the importance of being careful when you plan to study and calculate the cross section using a different experimental setup. Definitions (1) and (2) are quite easily visualized, but (3) and (4) are the most expected cases, where good attention and careful treatment are required. Usually, the total cross section is the sum of the scattering and the reaction cross sections; hence, if σ_t is the total cross section, σ_{sc} is the scattering cross section and σ_r is the reaction cross section, then,

$$\sigma_t = \sigma_{sc} + \sigma_r \qquad\qquad (5)$$

Where $\sigma_{sc} = \sigma_{el} + \sigma_{inel}$ $\qquad\qquad$ (6)

σ_{el} and σ_{inel} (elastic, inelastic) are termed as partial cross sections.

In elastic scattering, the processes of the scattering are not independent of each other, so two or more processes may exist, which may lead to a possibility of interference between them; therefore, it is not possible to specify or to write separately a partial cross section for such events. In order to pass this physical fact, the spin of the nucleus must be reoriented; otherwise, the elastic scattering by the coulomb field and by the nuclear forces will interfere. Such interference might be thought of as analogous to the interference effects in the optical diffractive scattering by obstacles, which are separated by distance comparable to the wavelength of the incident light. Therefore, all inelastic scattering processes are incoherent, so their cross sections are additive, i.e.,

$$\sigma_{ine} = \sigma_{i1} + \sigma_{i2} + \sigma_{i3} + \cdots + \sigma_{in} \tag{7}$$

Since various possible nuclear reactions may take place, in a similar way, as in the inelastic scattering cross section, the partial cross sections for nuclear interaction can be divided into component parts associated with these various possible nuclear reactions, so σ_{nr} is

$$\sigma_{nr} = \sigma_{n1} + \sigma_{n2} + \sigma_{3n} + \cdots + \sigma_{nn} \tag{9}$$

Here, it is important to notice that each partial cross section is a function of the energy of the incident particle.

Now if the cross section is a sharply peaked function of the energy, it is exhibiting a resonance. Knowing the energy dependence of each partial cross section is of great importance. One of the main objectives of the performed experiment is to determine this energy dependence of the partial cross section for particular nuclear reactions.

VII-3-2—Differential Cross Section

As a physical fact, the traveling of the particles emitted from the nuclear interactions depends on the angle between their direction and the direction of the incident particles. Therefore, studying the angular distribution is very important. It can be measured experimentally. The angular correlation function $\left[\omega(\theta)\right]$ is theoretically calculated to be

$\omega(\theta) = 1 + \sum A_i P_i \cos\theta)$, this function can be found experimentally.

It is the full information that helps in defining the characters of the nuclear forces and aids in identifying the quantum numbers designating the excited nuclear levels or states, such as the total spin I, the parity π, the lifetime τ, etc. Hence, the cross section on this basic fact can be described as function of the angles while it defines the number of particles emerging from the reaction per unit solid angle ($d\Omega$) at the coordinates (Θ, φ). This cross section is termed as the differential cross section and written as $\dfrac{d\sigma}{d\Omega}$, so the partial cross section for a given process is $\sigma = \int \dfrac{d\sigma}{d\Omega} d\Omega$ (9)

VII-3-3—Formation of Cross Section (σ)

From the previous definitions of the cross section, the general definition can be summarized as

$$\sigma == \frac{\textit{Number of events of given particles type per unit time per unit nucleucieus}}{\textit{Number of incident particles per unit area per unit time}}$$

Therefore, the total cross section for α-channel is given by $\sigma_T(\alpha) =$

$$\sigma_{sc}(\alpha) + \sigma_r(\alpha) \tag{10}$$

as in equation (5).

For simplification, let the spin of the nucleus be zero. In other words, the reaction is between even-even nucleus and aloha particle as a bombarding particle. This reaction is written as

$\alpha\ (_2^4\text{H}_2) + _Z^A\text{X}_N$ (even Z-even N).

Now the channel of reaction can be broken into sub-channels such as

$$\sigma_{sc}(\alpha) = \sum_{\ell=0}^{\infty} \sigma_{sc}(\alpha,\ell) = \sum_{\ell=0}^{\infty} \sigma_{sc,\ell} \tag{11}$$

This is for the scattering cross section.

For the reaction cross section, it is written as

$$\sigma_r(\alpha) = \sum_{\ell=0}^{\infty} \sigma_{r,\ell}(\alpha) \tag{12}$$

Equations (11) and (12) represent the partial cross sections for the given orbital angular momentum (ℓ).

This can be shown schematically as follows:

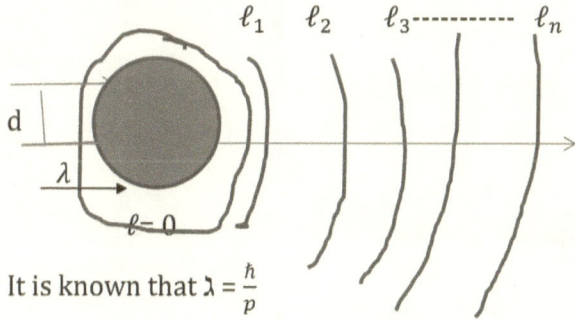

It is known that $\lambda = \dfrac{\hbar}{p}$

where λ is the wavelength λ divided by 2π and $\hbar = \dfrac{h}{2\pi}$, the Planck constant *divided by* $2\pi \cong 1.05x\ 10^{-35}$ J .s, p is the linear momentum of the projectile particle = mv, where m is the particle mass, and v is its velocity, d is the classical impact parameter.

From the rough diagram above, it is possible to conclude

1. $d < \lambda$ is the zero zone;

2. $\lambda < d < 2\lambda$ is the first zone;

3. $2\lambda < d < 3\lambda$ is the second zone;

4. $\ell\lambda\ < d < (l+1)\,\lambda$ is the ℓ th zone.

The general angular momentum (L) of the reacting particle is given as

$$mv\ \ell < L < mv(\ell+1).$$

Now the area of the cross section of the ℓ th -zone is given by

$$\pi(\ell+1)^2\lambda^2 -\quad \pi\ell^2\lambda^2 = \pi(2\ \ell + 1)\,\lambda^2.$$

Hence, the upper limit to the reaction cross section is given by

$$[\sigma_{r\ell} \leq \pi\,(2\,\ell + 1)\,\lambda^2]\tag{13}$$

For the reaction cross section, for the scattering cross section, consider the following diagram:

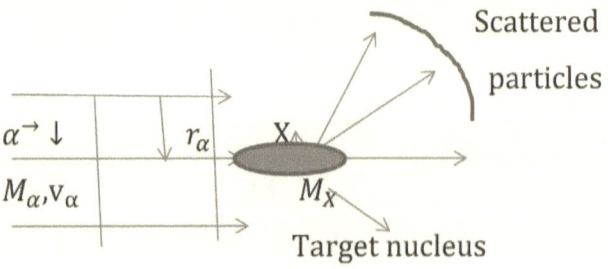

Scattered particles

Target nucleus

Consider the incoming alpha particle wave as plane wave, then $\psi \equiv e^{ik_\alpha.r_\alpha} = e^{ik_\alpha z}$ along z-direction

$$k_\alpha = \frac{m_\alpha v_\alpha}{\hbar} \; , \; P_\alpha = \hbar k_\alpha$$

$$m_\alpha = \frac{M_\alpha M_x}{M_\alpha + M_x} \; , \; \text{the reduced mass of the system}$$

$$\vec{V_\alpha} = \vec{r_\alpha^{(\cdot)}} = (\frac{dr}{dt}) \text{, the wave number } k_\alpha = \frac{1}{\lambda}$$

therefore, $k_\alpha = \dfrac{m_\alpha v_\alpha}{\hbar}$

So the problem here is a plane wave $e^{ik_\alpha z}$ incident on the target X. What is the scattering cross section? The answer is using the partial wave analysis method in quantum mechanics; the cross section will easily be found. Also, it is well-known that the plane wave can be expanded in terms of the physical harmonic function such as

$$e^{ik_\alpha z} = \sum_{\ell=0}^{\infty} A_\ell(r) \, Y_{\ell,0}(\Theta), \qquad \{z = r \cos \Theta\} \text{ and}$$

$$A_\ell(r) = \int Y_{\ell,0}^*(\Theta) e^{ik_\alpha r\cos\Theta} d\Omega = i^\ell [4\pi(2\ell+1)]^{1/2} j_\ell(kr)$$

Let $k_\alpha \equiv k$, where $j_\ell(kr)$ is the well-known spherical Bessel functions, a mathematical function well studied in quantum mechanics (see I. Schiff). Bessel functions have special properties of much interest in physics. One of these properties is $j_\ell(kr) = [\frac{\pi}{2kr}]^{1/2} j_{\ell+\frac{1}{2}}(kr)$. (Here,

$j_{\ell+\frac{1}{2}}$ is the ordinary Bessel function.) The asymptotic case where kr>>

is given by $j_\ell(\mathrm{kr}) = \dfrac{\sin(-\dfrac{\ell\pi}{2})}{kr}$

From complex variables, it is known that $\sin X = \dfrac{e^{ikX} - e^{-ikX}}{2i}$; therefore,

e^{ikz} can be written as

$$e^{ikz} = \frac{\sqrt{\pi}}{kr}\sum_{\ell=0}^{\infty}(2\ell+1)^{1/2}i^{\ell+1}\left[e^{-i(kr-\frac{\ell\pi}{2})} - e^{i(kr-\frac{\ell}{2}\pi)}\right]\, xY_{\ell,0}(\Theta)$$

$$(14)$$

$$\psi_{in} \qquad \psi_{out}$$

After the scattering takes place, the incident wave is not affected, but the out wave scatter, so $\psi_{sc} = \psi_{out}$. Hence, ψ_{out} is affected by the processes of the scattering by the potential of the target. Let η be the effect parameter of the scattering. So

$\psi(r) = e^{ikz} = e^{ikr\cos\Theta}$ can be written as

$$\psi(\mathrm{r}) = \frac{\sqrt{\pi}}{kr}\sum_{\ell=0}^{\infty}(2\ell+1)^{1/2}j^{\ell}+1_e - i(kr-\frac{\pi\ell}{2})-$$

$$\eta e^{i(kr-\frac{\ell\pi}{2})}y_{\ell,0}(\Theta)$$

Then, $\psi_{sc(r)} \cong \psi(r) \sim e^{ikz}$; hence,

$$\psi(r) = \frac{\sqrt{\pi}}{kr}\sum_{\ell=0}^{\infty}(2\ell+1)^{1/2}i^{\ell+1}\times(1-\eta_\ell)e^{i(kr-\frac{\ell\pi}{2})}y_{\ell,0}(\Theta)$$

for the case $kr \gg \ell$.

$$\sigma_{sc} = \frac{N_{SC}}{N_I} = \frac{\textit{number of scattered particle per unit area}}{\textit{number of incident particles per unit area}}$$

where $N_{sc} = \dfrac{\hbar}{2mi} \int [\dfrac{d\psi_{sc}}{dr}\psi_{sc}^* - \dfrac{d\psi_{sc}^*}{dr}\psi_{sc}]\, d\Omega$ (15)

By the same way, N_I can be found.

(As an exercise, show that $\sigma_{sc\ell} = \pi \lambdabar^2 (2\ell+1)\, |1-\eta_{\ell}|^2$.)

So $\sigma_{sc\ell} = \pi \lambdabar^2 (2\ell+1)|1-\eta_{\ell}|^2$ (16)

The reaction cross section $\sigma_{r,\ell}$ can be calculated if the number of incident particles N_I and the number of absorbed particles N_a are found. Then $\sigma_{r,\ell} = \dfrac{N_a}{N_I}$,

where $N_a = -\dfrac{\hbar}{2mi} \int [\nabla \psi\psi^* - \nabla \psi^*\psi]\, d\Omega$, over surface.

N_I can be easily calculated using the incident wave:

$$\sigma_{r,\ell} = \pi\lambdabar^2(2\ell+1)[1-|\eta_{\ell}|^2]$$ (17)

[As an exercise, verify equation (17).]

VII-3-3-1—Properties of (η)

This parameter is of important properties in the scattering theory; they are summarized briefly in the following:

1) $\eta \le 1$

2) $0 \le \eta_{\ell} \le 1$

Hence, $\sigma_{r,\ell} \leq \pi \lambda^2 (2\ell+1) \Rightarrow \sigma_{r,0} \leq \pi\lambda^2$

3) $\eta_\ell < 0 \Rightarrow \sigma_{sc,\ell} \leq 4\pi \lambda^2 (2\ell+1)$ —phase problem

The important conclusion here to be noticed is, the scattering can occur without interaction, but no interaction without scattering. For example, if $\eta_\ell = -1$, then, $1-|\eta_\ell|^2 = 0$, so $\sigma_{r,\ell} = 0$—no interaction, but there is scattering, where $\sigma_{sc} \leq 4\pi\lambda^2 (2\ell+1)$, $|1-\eta_\ell|^2 = 4$, so for $\eta_\ell = -1$, there is scattering but no interaction.

Now consider the so-called black nucleus. The radius of such nucleus $R \gg \lambda$, can crudely be represented by the following diagram: Everything has impact.

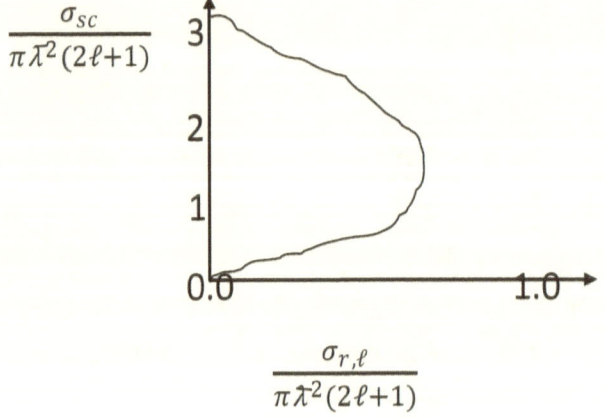

$$\frac{\sigma_{sc}}{\pi \lambda^2 (2\ell+1)}$$

$$\frac{\sigma_{r,\ell}}{\pi \lambda^2 (2\ell+1)}$$

To $\ell < R/\lambda$, where the nucleus is hit by the bombarding particles.

Now, if $\eta_\ell = 0$ for $\ell\lambda < R$, and equals 1 for $\ell\lambda > R$, this indicates that the incident particles did not impinge the nucleus,

where $\eta_\ell = 0$, then $\sigma_{sc} = \sigma_r = \sum_{\ell=0}^{\frac{R}{\lambda}} \pi \lambda^2 (2\ell + 1) \cong \pi R^2$.

Hence, $\sigma_T = \sigma_{SC} + \sigma_r = 2\pi R^2$, for the black nucleus. $\rightarrow$

$$[\sigma_{TBN} = 2\pi R^2] \tag{18}$$

In fact, this result is not surprising. It can be explained by the so-called shadow scattering, which is shown crudely in the schematic diagram below:

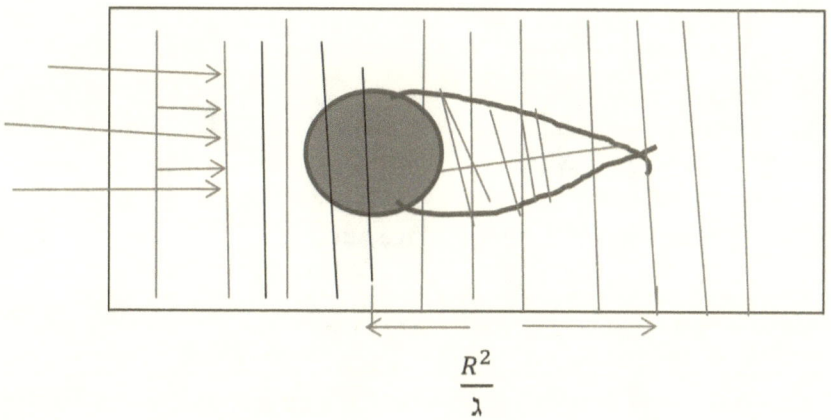

$$\frac{R^2}{\lambda}$$

Shadow scattering

$N_{SC} = N \pi R^2$ scattered particles; N is incident particles.

Hence, $\sigma_{sc} = \dfrac{N_{sc}}{N} = \pi R^2$, an area.

The basic thing in nuclear reactions is to relate η to the internal structure of the nucleus, so the main question is how to find η. To do that, consider the target X with radius r_x and the radius of the incident particle; say, α is r_α. Let the nucleus and the projectile just touch each other, as shown below:

The channel radius $R = r_x + r_\alpha = r$

If $r > R$, no interaction outside

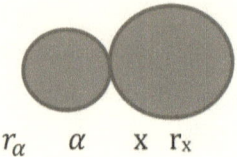

r_α α x r_x

If $r < R$, inside the nucleus, interaction will occur.

Potential $(V = 0)$

Therefore, the Schrodinger equation describing such system is $\nabla^2 \psi + k^2 \psi = 0$ for $r > R$, $V(r) = 0$

where k is the entrance channel wave number

$$\psi(\vec{r}) = \psi(r, \Theta) = \sum_{\ell=0}^{\infty} \frac{u_\ell}{r} Y_{\ell,0}(\Theta)$$

Consider $\ell = 0$, very slow, then, $\psi(\vec{r}) = \dfrac{u_0}{r}$, $Y_{00} = 1$

So the Schrodinger equation for this case takes the form

$$\frac{d^2 u_0}{dr^2} + k^2 u_0 = 0, \ r > R$$

From quantum mechanics, the solution is

1) $u_0(r) = a \ e^{-ikr} + b e^{ikr}$ $, r > R$

From equation (14) for $\ell = 0$, the asymptotic form of the solution is

$$\psi(r) = \frac{\sqrt{\pi}}{kr}[e^{-ikr} - \eta_0 e^{ikr}].$$

2) Let $a = \dfrac{i\sqrt{\pi}}{k}$, $b = \dfrac{-i\sqrt{\pi}}{k}\eta_0 = -a\eta_0$.

Using the log derivative, denoting by f_0, such that

3) $f_0 = [(\dfrac{du_0}{dr}) \cdot \dfrac{1}{u_0}]_{r=0} = [(\dfrac{du_0}{dr}) \cdot \dfrac{1}{u_0}]_{r=R}$

Suppose f_0 is known, how can η be obtained?. Now,

For $\ell = 0$, $f \rightarrow \eta \rightarrow \sigma$

Substitute 2 into 3; η_0 can be found as

$$\eta_0 = [\frac{f_0 + ikr}{f_0 - ikr}]e^{-2ikr} \qquad (19)$$

(Verify it as exercise.)

f_0 could be real or complex. If it is assumed as real quantity, then the absolute value of l η_0 l² = 1 [equation (19)]

This means the following:

If , $\ell = 0$, $\sigma_{r,0} = 0$, no interaction, as shown before.

Only scattering occurs, since l η_0 l² = 1 $\rightarrow \eta_0 = \pm 1$. As found before, l1 – η_0 l² is in the formula of $\sigma_{sc\ell}$, which gives 0 or 1 to its value. It indicates again that there is no interaction without scattering, but there can be scattering with no reaction, as stated previously. But if f_0 is complex quantity, then things are somehow quite complicated. Hence, f_0 can

take the form $\eta_0 = [\dfrac{Re\, f_0 + i\, Im\, f_0 + ikr}{Re\, f_0 + i\, Im\, f_0 - ikr}]$

Therefore, $= 1 - |\eta_0|^2 = |a|^2 + |b|^2 / |a|^2 - |b|^2 =$

$-4kR\,Imf_0 / |Ref_0|^2 + |Imf_0 - kr|^2 = \mathcal{M} ----(20)$

$-b = -i\,Imf_0 - ikr = -i\,(Imf_0 + ikR)$

So the reaction cross section is given as

$\sigma_{r,0} = \pi\lambda^2\,\mathcal{M}$ -------------------------------21

The scattering cross section $= \pi\lambda^2 |1 - \eta_0|^2 \Rightarrow$

$$\sigma_{sc,0} = \pi\lambda^2 \left| 1 - \frac{f_0 + ikR}{f_0 - ikR} e^{-2ikR} \right|^2 = \pi\lambda^2 \left| [(\frac{e^{-2ikR}}{f_0 - ikR}) + (e^{2ikR} - 1)] \right|^2$$

$$\Rightarrow \sigma_{sc,0} = \pi\lambda^2 \left| A_{isa} + A_{hsd} \right|^2 \qquad\qquad (21)$$

Here, A_{isa} is the internal scattering amplitude.

And A_{hsa} is the hard sphere amplitude.

Now, $u_0 \Big|_{r=R} = 0 \rightarrow f_0 \rightarrow \infty \Rightarrow A_{isa} \rightarrow 0$

This is called a hard sphere scattering case.

Hence, $\sigma_{sc,0} = \left| A_{hsa} \right|^2 = \pi\lambda^2 \left| e^{2ikr} - 1 \right|^2 \qquad\qquad (22)$

VII-3-4—Nuclear Fission

Introduction

From the energy-mass equivalence theory of Einstein, formulated as $E = mc^2$, E is the energy, m is the mass, and c is the velocity of light $\cong 3x10^8$ m/sec. Therefore, any change in the mass Δm is accompanied with energy Δm c^2. This principle is found of great use in the development of nuclear physics theory and in the particles physics since 1905. In addition to that, it played a great role in discovering the fission and the fusion theories as types of nuclear reactions. In 1939, the fission of some heavy nuclei was discovered, like uranium isotopes 238 and 235, if bombarded by thermal or fast neutrons. For example, if the uranium 235 is bombarded by thermal neutrons, it fissions to about two equal mass numbers or too close with the emission of gamma rays and some neutrons 2–3. This reaction, in addition to the release of these products, with an energy of about 180 MeV, is released too from the nucleus of the atom of this uranium isotope. It is quite a huge amount of energy if it is remembered that the molar gram contains about 10^{23} atom (Avogadro no.). The source of this fission energy is the binding energy mentioned in chapter 1. Figure 1 illustrates this idea quite well. It shows the relation between the binding energy per nucleon B/A and the mass number (A).

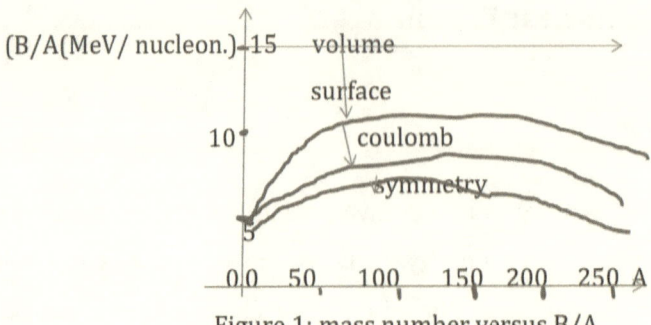

Figure 1: mass number versus B/A

Thus, when a heavy nucleus is fissioned into two medium mass nuclei, energy is released, as stated earlier; it is due to the energy that was binding the nucleons together inside the parent nucleus. Remember that for each nucleon, there is about 8 MeV to be combined to the parent nucleus. This will be released when the fission occurs. The total released energy for the fission is the difference between the binding energy of the parent nucleus and that of the daughter's nuclei. There are two types of fission: spontaneous and induced, as it will be shown next.

VII-3-4-1—Spontaneous and Induced Fission

As it is clear now, the fission of a heavy nucleus occurs if it is bombarded by projectile particles like the neutrons. The products of the fission are radiation, particles, and two daughter nuclei. The product nuclei are of a medium mass; let one of the mass be $M(A_1, Z_1)$ and the second mass be $M(A_2, Z_2)$, where A is the mass number and Z is the atomic number (protons). Hence, the required conditions for the fission to take place and the fission processes that would proceed can be quantified by examining the following processes:

$$M(A, Z) \implies M(A_1, Z_1) + M(A_2, Z_2) \tag{1}$$

Parent   Daughter 1   Daughter 2

The energy released after the fission is given by

$$Q = B\,(A_1, Z_1) + B(A_2, Z_2) - B(A, Z) \tag{2}$$

The binding energy of any nucleus can be estimated using the liquid drop model, assuming the terms of pairing and symmetry in the semiempirical mass formula are negligible, since $A = A_1 + A_2$; therefore, the difference in binding energy will involve only the surface and coulomb terms, as shown in figure 1. To simplify the calculation, let the nucleons (p, n), when the nucleus is split, be divided in the same ratio, i.e., <BR> is considered: $A_1/A = Z_1/Z = \eta_1$, $A_2/A = Z_2/Z = \eta_2$

$$\text{where } \eta_1 + \eta_2 = 1,\ \eta_2 = 1 - \eta_1 \tag{3}$$

Now, since the pairing and the symmetry terms are considered negligible, the released energy of the fission is only given by the surface (E_s) and the coulomb (E_c) terms of the original nucleus (parent) denoted by $M(A, Z)$.

Hence,

$$Q = E_s[1 - (\eta_1)^{2/3} - (\eta_2)^{2/3}] + E\,[1 - (\eta_1)^{5/3} - (\eta_2)^{5/3}] \tag{4}$$

To find Q_{max} set $dQ/d\eta_1 = 0$, where $\dfrac{d\eta_2}{d\eta_1} = -1$

This leads to Q_{max} occurring at $\eta_1 = \eta_2 = 1/2$; thus,

$$Q = 0.37\,E_c - 0.26E_s \tag{5}$$

Therefore, if the nucleus fissions into equal nuclei (so-called symmetric fission), the largest energy output (Q-value) is produced.

This process is exothermic ($Q > 0$) if $E_c/E_s > 0.7$. Often, this limit is specified in terms of the fission parameter ($\varkappa$), which is defined as,

$$\varkappa = E_c / 2 E_s = (\frac{Z^2}{A})/(2\, a_s/a_c) \sim (Z^2 / A)/50 > 0.35 \tag{6}$$

Remember, a_s and a_c are the coefficients that appear in the liquid drop model expression for the binding energy, or in the surface and coulomb terms, respectively, in the semiempirical mass formula. This indicates that all nuclei with (Z^2/A) > 18 (i.e., heavier than ^{90}Zr) release energy in undergoing symmetric fission. Nuclei that is well beyond this limit, which do not fall apart, however, might go for very long lifetimes against spontaneous fission. If the half-lives are plotted versus (Z^2/A), figure 2 will be obtained:

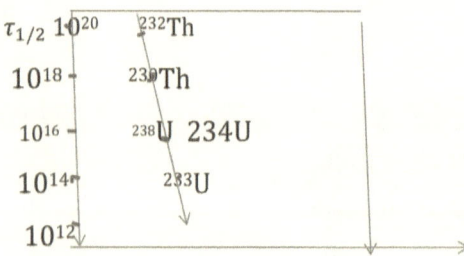

Figure 2: $\tau_{1/2}$ versus Z^2/A

The half-life of spontaneous fission against (Z^2/A) for heavier nuclei is shown in figure 3 below:

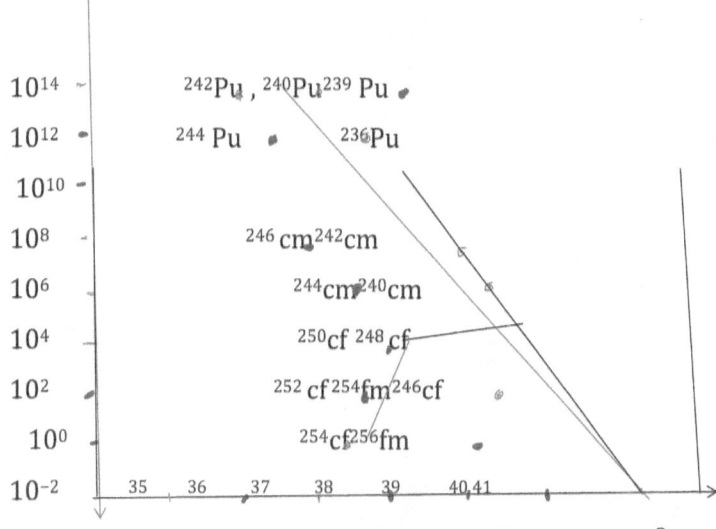

Figure 3: spontaneous fission half-life versus Z^2/A

Physically, it is well-known that the presence of the potential barrier in the nucleus has the effect of suppressing the fission processes. Figure 4 shows schematically the shape of the barrier as function of the deformation or fragment separation for different values of the atomic mass number.

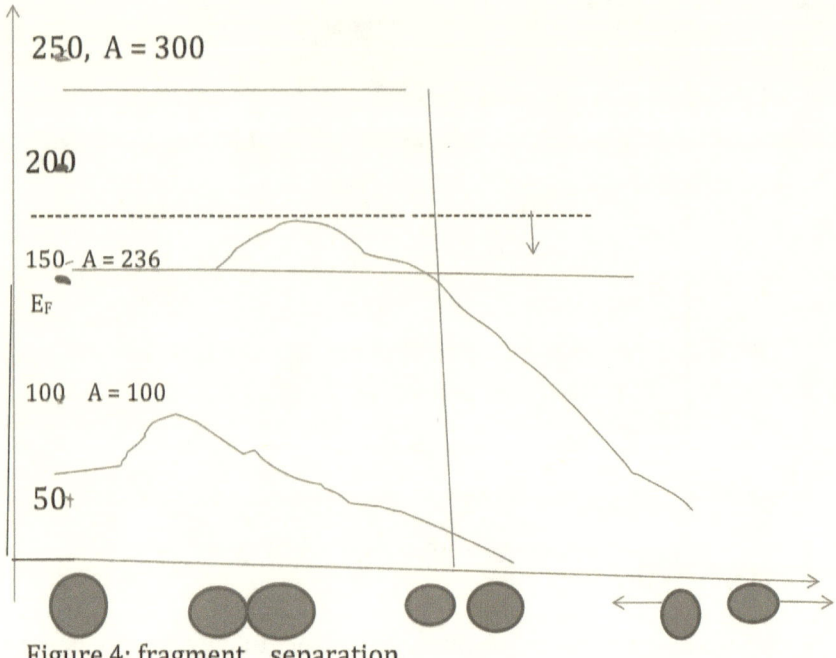

Figure 4: fragment   separation.

Figure 4 shows that at large separations, the barrier shape is determined by the coulomb potential between charges of the two nuclei. Competition between coulomb repulsion and surface tension at small distortion occurs where coulomb potential is separating the protons apart (same charge), while the surface tension is pulling the nucleus back to more spherical shape. The shape of the potential barrier can be determined with the aid of the liquid drop model. An example is the quadruple distortion, which is a measure of the deviation of the nucleus from the spherical shape. Figure 5 below (to be found and drawn) shows the energy of the deformation in unit of surface energy E_s for a spherical nucleus plotted versus the size of the quadruple distortion, for different values of the fission parameter $\varkappa$. When $\varkappa > 1$, the nuclei is completely unstable with respect to this distortion. For small values of $\varkappa$, fission can occur by barrier penetration. It is known from alpha-decay theory that the transmission factor is given by

$$T = \text{Exp.} (- \gamma); \text{ where}, \gamma = 2(\hbar)^{-1}(2m)^{1/2} \sqrt{\left[V(r)-E\right]} dr$$

It is very clear here that T is strongly depending on the mass. For alpha particle, as an example, the mass is considered too large compared with the mass of the electron. When the mass is very large, it means the transmission factor is very small, which indicates that the lifetimes are so long (lifetime $\tau = 1/T$.)

VII-3-4-2—Energy Released of the Fission

As stated before, the products of the fission are intermediate nuclei mass, particles, gamma radiations, and energy released. Also, the energy can be calculated, as pointed out before, from the liquid drop model. For example, this energy has been calculated for the fission of uranium 235. It is about 200 MeV, mostly from coulomb energy. The released energy is divided among the fission products as follows:

The Product	Energy MeV
Fragment kinetic energy	165
Prompt neutrons	5
Prompt gamma rays	7
Radioactive decay fragments	25

Because the fragment nuclei produced have excess neutrons, radioactive decays occur. Also, it is known that the heavy nuclei have more neutron per proton than the light nuclei. Therefore, when a heavy nucleus under fission splits into two nuclei, this product nuclei will either have to shed neutrons or transform them into protons in order to achieve a distribution of nucleons to come to the state of stability. Of course, there are different types of distribution that might lead to fission. The using of the liquid drop model helps in determining the actual type causing the fission, because during the processes of the fission operation, there are certain critical shapes at which a narrow neck between proto-fragments exists. This is known as the *scission point*.

In the fission processes, which are of spontaneous or low-energy excitation, the symmetric fission, or the equal masses fragments predicted by the liquid drop model, is not obeyed. But in the high-energy excitations, it is the dominant mode.

VII-3-4-3—Chain Reaction and Fission Reactor

The chain reaction is a process of reactions taking place consequently. It is used for different purposes. The source of this kind of reactions are, for example, uranium 235, uranium 238, and plutonium-239 as targets of nuclear foils and sources of neutrons, which are either thermal with energy of 0.025 ev or fast with energy of 1 MeV.

That depends on the type of the target foil and the purpose of the required reaction and the type of the used reactor. As mentioned previously, the neutrons are one of the fission products, about two to three neutrons, in each fissioned nucleus. When a neutron is absorbed by a nucleus, a compound nucleus is formed; it will be in an excited state. To go back to its ground state, the excitation energy will be released as products of fission such as radiations, fragments, and neutrons. This is illustrated in the following reaction, where the target is uranium 235 and the projectile is the thermal neutron

$$n + {}^{235}U \rightarrow {}^{236}U^* \rightarrow {}^{139}Xe + {}^{95}Sr + 2n \tag{7}$$

The ${}^{236}U^*$ is the compound nucleus; it fissions to the xenon nucleus, strontium nucleus, and two neutrons. Here, it is clear that the nucleons' number is conserved, 236 = 139 + 95 + 2. This equation can be schematically demonstrated as follows:

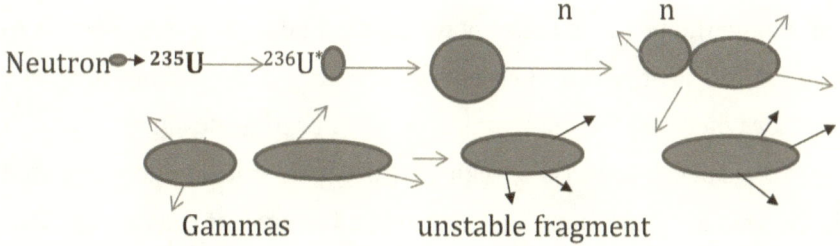

Unstable fragments nuclei usually emit beta, gamma, and delayed neutrons. The emission of neutrons (two to three) in the fission processes [equation (7)] leads to the successive fissions of nuclei in the target foil. This forms self-sustaining or chain reactions. The following is illustrating this fission process in ${}^{235}U$.

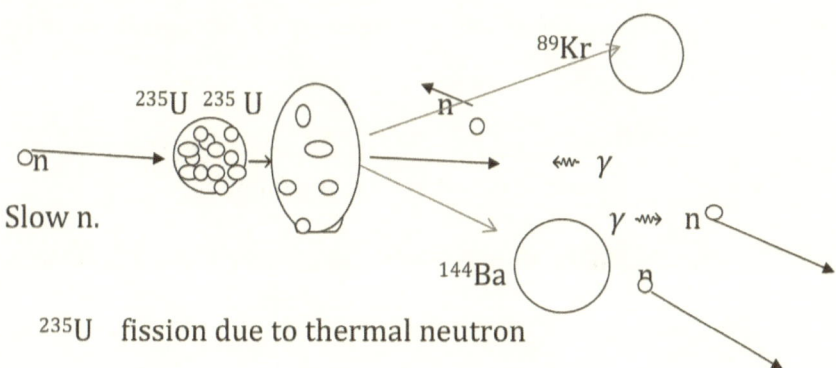

${}^{235}U$ fission due to thermal neutron

This type of reaction actually produces a huge amount of energy. Of course, a huge amount of heat will follow, which must be put under control to avoid huge explosion. The instrument that does that control is the reactor. There are different designs for this reactor according to its purpose. Just to imagine the hugeness of the produced energy after the fission takes place, take one gram of uranium 235; the produced energy from one process is 19.2 × 10^{12} joules, which is equivalent to the energy of burning 4.4 tons of coal or burning twenty barrels of oil. These values can easily be calculated. It is known today that the fuel of an ordinary reactor like a research type one is some few tons, then how huge the produced energy will be and how the necessary it is to control the heat out of such energy are very much required. So the reactor where the fission reaction occurs has to be designed and built up from special materials, taking into consideration such expected huge heat and very high temperature in the core of the reactor. This huge heat is used according to the purpose of the designed reactor. As it is known, there are different isotopes for uranium element in nature: uranium 238 represents 99.3%; and uranium 235 represents only 0.7%. The usual fuel is made of uranium 235 with enrichment not less than 4%. For this isotope fuel, thermal neutron is used as a bombarding particle. Its interaction cross section with uranium 235 is quite high. For uranium 238, the bombarding neutron is the fast neutron with energy not less than 1 MeV, while the energy of the thermal neutron is 0.025 eV.

Before going to some details into reactors physics, let some of the reactions terms be pointed out; these are the ollowing

1. $^{235}_{90}U$ fissions (n, f) at all energies of absorbed neutron, so it is fissile material.

2. The cross section for (n, f) is quite high compared with (n, γ) and (n, n) reactions. At thermal energy <0.1 eV, it is about ten times of (n, γ) and one hundred times of (n, n).

3. The general form of the nuclear reaction is X(a, b)Y $\Rightarrow$ a + X $\rightarrow$ Y + b, in which (a) is the projectile particle, (X) is the target, (Y) is the product nucleus, and (b) is the emitted particle, such as n, f, and γ.

4. $^{238}_{90}U$ —its threshold energy for fission (n, f) is 1 MeV. So with thermal neutron, no interaction will take place.

5. The difference between these isotopes of uranium is explained by the pairing term in the semiempirical mass formula, where only uranium 238 has this term.

6. For the energy range (10–100) eV, a strong resonant capture of neutrons (n, γ) clearly appears, especially for uranium 238, where the cross section reaches very high value >1000 barn and (100–1000) barns for uranium 235.This phenomenon can be illustrated in the following diagrams*:

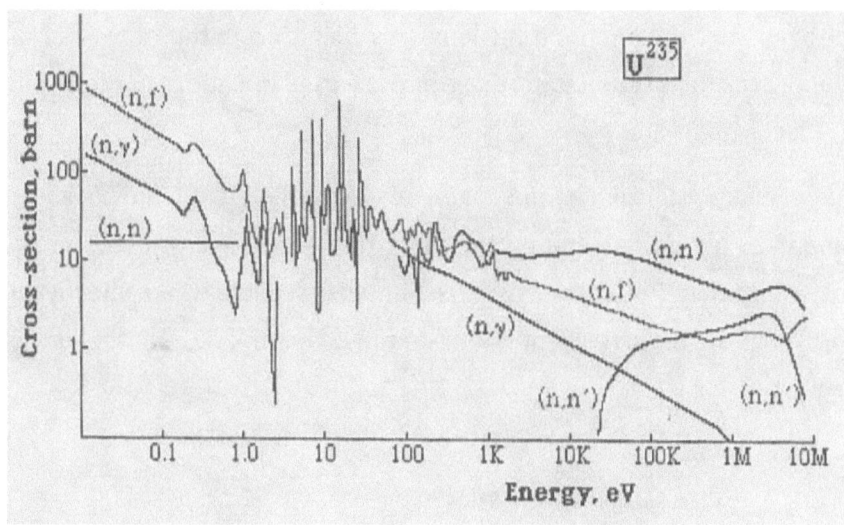

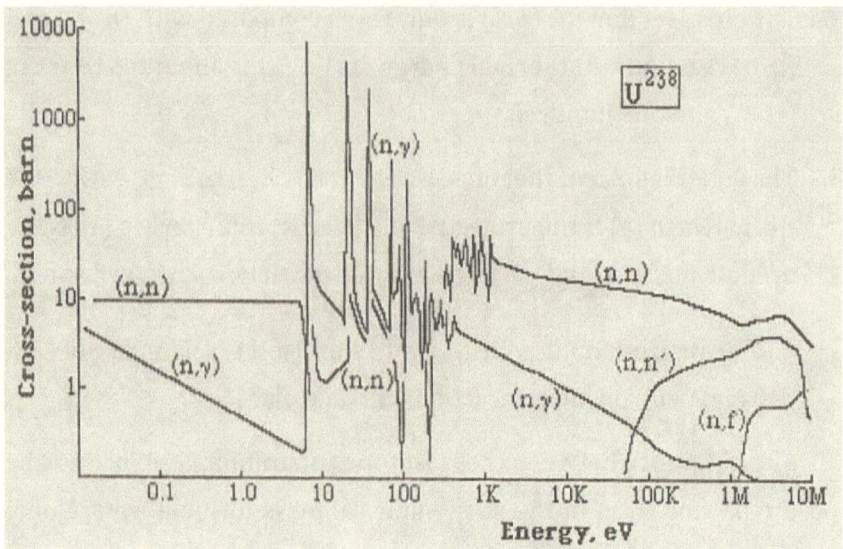

*Cross section versus energy of bombarding neutrons

From these graphs, it can be noticed that the fission of $^{235}_{90}U$ occurs in the thermal energy range (0.01–1) eV, and the largest cross section ($\cong$ 1000 barns) is at the energy value (0.025) eV. At this thermal energy, the neutrons are called thermal. For $^{238}_{90}U$, the fission starts at the energy (1–10) MeV. Other reactions, (n, n), (n, γ), and (n, n⁻), indicate a formation of compound nucleus, in which during the processes of going back to its stable state, these particles and radiations are emitted; it means no fission.

Usually, the energy spectrum of the fission neutron peaks at about 1 MeV. At this energy, the inelastic scattering cross section (n, n⁻) in uranium 238 exceeds the fission cross section, which effectively prevents the occurrence of the fission in uranium 238. See the figure below

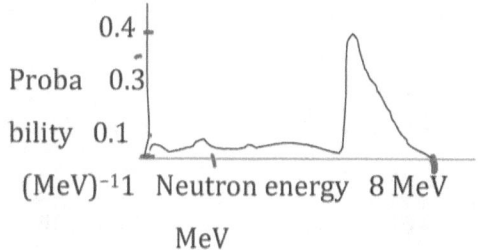

It is important to note the fact that the fission cross section is small at the energy of the fission neutrons, so in general, these neutrons are absorbed in other processes such as (n, n), (n, γ), and (n, n⁻). This implies that a lump of natural uranium cannot sustain a fission chain reaction. This might be more quantitatively clarified if it is examined. What might happen to the fission neutrons in the natural uranium lump? Start with one hundred fission neutrons; let ninety-eight be captured in uranium 238. Only eight of them results fission; the remaining two neutrons produce fission in uranium 235 (remember it is only 0.7% of the natural uranium). As shown before, each fission is associated with two or three neutrons; therefore, there will be only twenty-five neutrons in the second generation of the fission, which is clearly insufficient to sustain a chain reaction. As previously mentioned, a huge energy will be produced as a result of chain fission reaction, which is converted to a huge amount of heat of very high temperature in the reactor core. This heat is used to generate electricity to distilled salt water or for both purposes. The use of fission energy for generating power includes basically two ways surrounding this limitation:

a. The enrichment of uranium 235 is a very important step in order for a sufficient fission of this isotope to be obtained. If half of each isotope of uranium is mixed then used as fuel to the reactor, a chain reaction will sustain. Most of the fission events are from uranium 235, at range of energy (0.3–2) KeV. This is usually used in a specially designed reactor called fast reactor (FR).

b. When the designed reactor and the fuel contain certain materials that is moderate to the speed of the bombarding neutrons to the energy within the range of thermal neutrons, the cross section of the uranium 235 fission will be high enough, (10^3) barns, to sustain the fission that is mostly induced by those neutrons with thermal energy (0.025 ev). The reactor of such process of fission is called thermal reactor (TR).

VII-3-4-4—Nuclear Reactors

The nuclear reactor was first built to do scientific researches about the nuclear fission, which was discovered in 1939 in Germany. Its simplest structure is core made of special materials that resist very high temperatures, as mentioned before. Control rods, also from specific material resisting high temperatures, play the role of heat moderation within the bearing capability of the core to the expected high temperature, due to the huge amount of energy released from the fission processes, which is virtually converted to this heat. The fuel that is made as a rod too from some alloys made of certain elements has the highest resistance to the corrosion and the highest expected temperature containing the fuel material (enriched uranium 235 or 238 or plutonium-239). All this structure is contained in a container of high pressure resister, made of so solid materials. In addition to all that, the core is under cooling operation using cooling materials such as water, gas, and other possible cooling materials, according to the reactor design and use purpose. Since the fuel is the main material to the reactor, its description is important. In general, the fuel is packaged in the form of rods arranged in a regular lattice inside the reactor core moderator. The rods are typically about 2–3 cm in diameter, spaced by about 25 cm. The fuel is contained within a metal sheath or cladding, which is mostly a common stainless steel or alloys of zirconium. Such cladding is used to support the fuel mechanically and to prohibit releasing radioactive fission product into the stream of the coolant material and also to introduce extended surface contact with the coolant, to promote effective heat transfer. Periodically, within the lattice, there are control rods containing cadmium or boron elements (Cd, B). These elements are very good absorbers to the neutrons. These rods are designed to be movable up and down inside the core of the reactor. Their motion is controlled by remote operation so that the neutrons flux is controllable. The reactor core actually is surrounded by a reflector material, which is, in fact, a type of moderator without fuel elements; its effect is to keep the neutrons flux high during the active part of the core, thereby ensuring efficient fuel consumption. The core, as stated before, must be highly protected. Therefore, it is contained within a pressure vessel of quite well-welded steel. This vessel typically withstands a pressure of about 1.55×10^7pa or 153 bars. In addition to that, the whole vessel is surrounded from outside by a solid concrete biological shield. As examples, the following are typical types of nuclear reactors:

1. Pressurized water reactor (PWR), figure 1

2. Boiling water reactor (BWR), figure 2

3. Fast reactor (FR), figure 3

These reactors are common types that make use of the excellent properties of the water as coolant material and moderator. A light disadvantage of using ordinary water is absorbing the neutrons, and then the deuteron is formed as follows: $n + {}_1^1H \rightarrow {}_1^2D$. In this case, when the moderator is ordinary water, the fuel used has to be enriched with fissile materials (such as ^{235}UO, ^{239}Pu). The use of water as heat conveyor is too limited due to the fact that the critical temperature of water to stay in liquid state is 374 C^0. This fact has a direct effect on the efficiency of converting the produced energy into heat. From thermodynamic laws, it is known that the efficiency of the Carnot engine is given by $\eta_{th} = 1 - T_2 / T_1$. In the ideal case, the heat is received isothermally by the working fluid at temperature T_1 and is rejected isothermally at T_2; all processes are reversible. Also, to be quite clear, the Carnot engine is not an ideal case. Therefore, no power plant operates as such. It only illustrates that the higher T_1 is the higher efficiency (η), and T_2 cannot be lower than the earth atmosphere temperature or ocean.

For illustration in the PWR, the water is allowed to boil in the reactor core. The produced steam is then used to drive the turbines to generate electrical energy. The condensed water under the cooling of the remaining steam can be used as drinking water or for irrigation purposes. This is usually called nuclear plant of multiple purposes, which is more economically used. The boiling water reactor, as in the PWR fuel, must be enriched, but the operating pressure is lower; it is about 70 bars. In the case of the fast reactor (FR), in figure 3, no moderator is used; hence, its core is too small compared with PWR and BWR. The very high-powered density means that the liquid metals have to be used as coolants, such as the liquid sodium, which is most commonly used. The main disadvantage of its use is, it becomes radioactive due to the interaction ^{23}Na(n, γ)^{24}Na.

As well as the generating power, the fast reactors are used for breeding fissile materials through the reaction

$$^{238}\text{U (n, } \gamma \text{) } ^{239}_{90}U \rightarrow (\beta^-) \rightarrow ^{239}\text{Np} \rightarrow (\beta^-) \rightarrow ^{239}\text{Pu.}$$

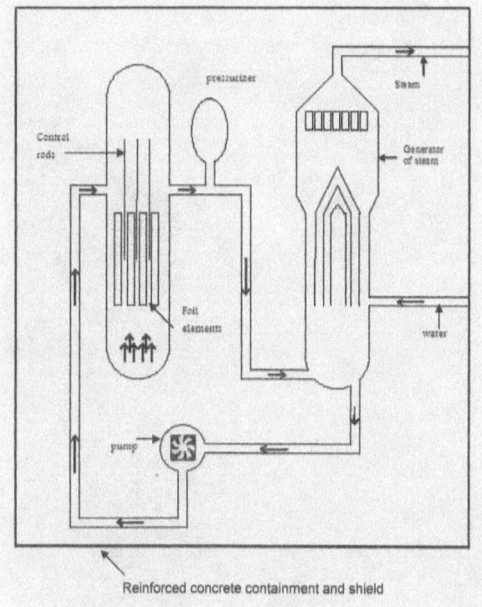

Figure 1: PWR

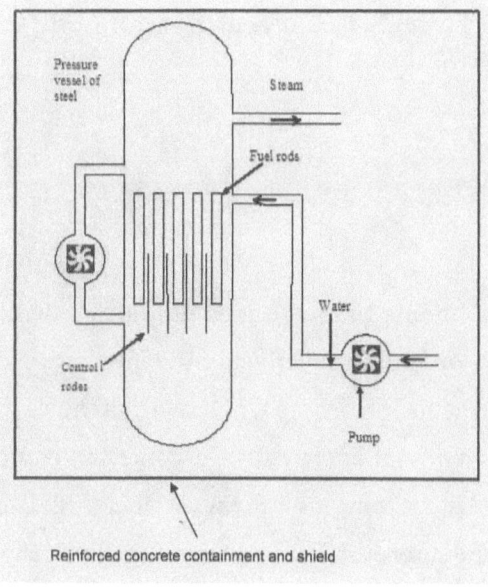

Figure 2: BWR

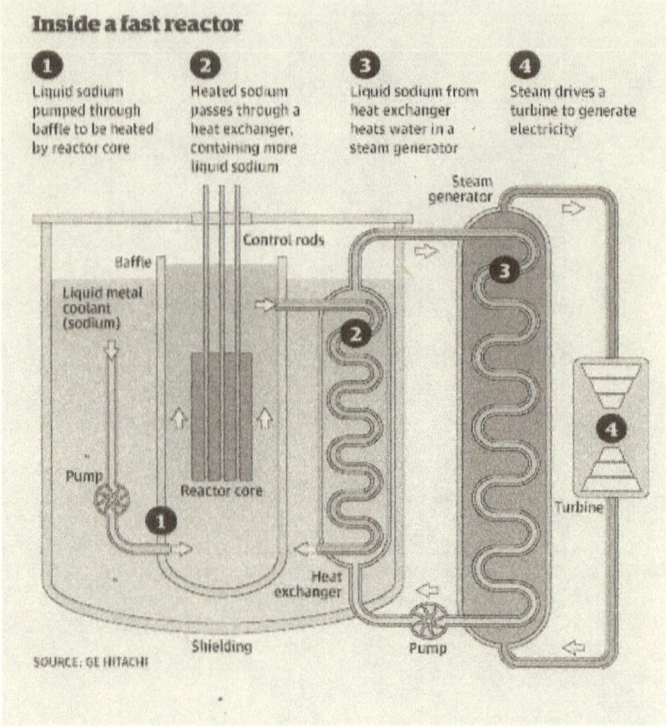

Inside a fast reactor

1 Liquid sodium pumped through baffle to be heated by reactor core

2 Heated sodium passes through a heat exchanger, containing more liquid sodium

3 Liquid sodium from heat exchanger heats water in a steam generator

4 Steam drives a turbine to generate electricity

SOURCE: GE HITACHI

Figure 3: Fast breeder reactor [FBR]

VII-3-5—Fusion

The fusion physically is the reverse process to the fission; it is a fusing of light elements to create medium nuclei, while the fission is heavy nuclei fission under the effect of the interaction of bombarding neutrons, while the fusion occurs when two light nuclei pushed toward each other with enough energy to overcome the coulomb force and to get them fused, forming new heavier nuclei like, as an example, tritium (^3T) and deuteron (^2D). The fusion is physically illustrated by the following configuration of the process compared to the fission process:

Fusion

Fission n

The fusion reaction is not easy to occur on the earth; it needs special reactors and high energy as kinetic energy to the light nuclei to be fused after overcoming the coulomb barrier. Naturally, it occurs in the sun and other stars in the universe, so the sun is a natural fusion reactor, where H and He are fused under very high temperature, high pressure, and high ions density. This fusion is a natural fusion.

VII-3-5-1—Fusion Process

The fusion, as stated before, naturally occurs in the stars, especially between hydrogen and hydrogen, or hydrogen and helium. For the hydrogen-hydrogen fusion, it means p.p fusion, but it is known that there is no two-proton-bound state, so this fusion must be associated with positron decay, where it converts one of the protons to neutron such as

$$^{1}H + {}^{1}H \rightarrow {}_{1}^{2}H + e^{+} (\beta^{+}) + n + Q \ (0.42 \ \text{MeV}) \tag{8}$$

This fusion interaction represents the first step of hydrogen burning in the sun. This process means an important physical concept related to the life of the sun and the evolution of life on the earth. Since this is positron decay, it indicates that it is a rare process, which, in turn, shows and determines the long life of the sun. Because according to this process, the consumption of the sun hydrogen is too slow. Therefore, the weakness of the weak interaction is important to human beings.

The other possible fusion reactions are those between hydrogen isotopes like the deuteron.

$$^2_1H + {}^2_1H \rightarrow {}^4_2He + \gamma + Q\ (\approx 23.8\ \text{MeV}) \tag{9}$$

It is clear that the energy released in this fusion reaction is more than the energy required to remove either proton or neutron from the helium (^4He). Other possible fusion processes are

$$^2H + {}^2H \rightarrow {}^3He + n + Q(3.3\ \text{MeV})$$

$$^2H + {}^2H \rightarrow {}^3H + p + Q\ (4.0\ \text{MeV}) \tag{10}$$

where ^3He is isotope of ^4He, and ^3H is isotope of hydrogen. One of the possible fusion reactions may occur between ^3H and ^2H such as

$$^2H + {}^3H \rightarrow {}^4He + n + Q\ (= 17.6\ \text{MeV}) \tag{11}$$

Equation (11) is quite important, where this fusion reaction is of great interest to the international quest for the future energy demands. Therefore, there are considerable experimental studies to develop a fusion reactor to control the energy released to be a great source of the future energy. This type of fusion reactors in nature, like our sun and other stars, even much bigger than the sun, in this huge universe, are responsible for the nucleosynthesis of elements up to iron or so. These scientific facts play main roles in the life of this expanding universe.

VII-3-5-2—Energy Production in the Stars

It is quite well-known that the stars are actually formed from clouds of gas, which, in fact, is collapsed under the mutual gravitational attraction effect, where the gravitational potential energy is converted into kinetic energy for the gas molecules. As a consequence, the gas will be heated up, and its density increases. As it is also known, the nuclei in the universe are 99.9% hydrogen or helium, but the dominant one is the hydrogen, about 92.5%, so the helium is 7.4%. The helium is known to be four times more massive than hydrogen; hence, helium accounts about 24% of the universe mass. Therefore, the stars' formation depends first on the fusion processes, which involve protons. Fundamentally, four protons are converted into alpha particle as the first step in the fusion processes. By simple calculations, it can be seen that 4H goes into ^{4}He, and the released energy is about 26.7 MeV.

As mentioned earlier, the coulomb potential barrier is working against the fusion process, so it is quite useful to understand the nature of this inhibiting effect using Maxwell-Boltzmann distribution. The spectrum of the effective kinetic energy of nuclei in the gas is given by the high-energy part of this distribution law where

$$N(E) \sim \exp(-E/KT) \tag{12}$$

This can be represented schematically as

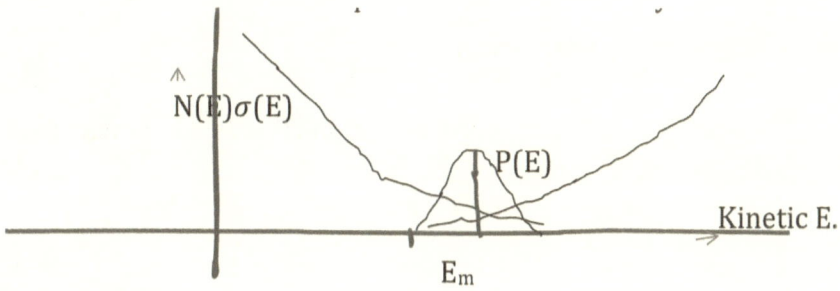

It is known from the tunneling effect (alpha decay) that the cross section is related to the barrier penetration probability such as

$$\sigma(E) = \exp.(-\gamma) \tag{13}$$

where $\gamma = 1/\hbar\,(2m/E)(\pi\,z_1\,z_2\,e^2)/4\pi E_0 \propto E^{-1/2}$ (14)

Hence, the probability of fusion event at kinetic energy (E) is proportional to the product of these two functions (12, 13), so it can be written as

$$P\,(E) \sim \exp.\,[-(\alpha\ E^{-1/2} + E/KT)] \tag{15}$$

Here, K is Boltzmann constant, and T is the temperature in kelvin unit. Equation (15) has the maximum at $E_m = [\alpha\dfrac{KT}{2}]^{2/3}$

$$P\,(E) \sim \exp.[-1.89\ \alpha^{2/3}(kT)^{-1/3}] \tag{16}$$

As an example, let the two-protons-fusion case be considered. Here, (α) is about $(21\ \text{kelvin})^{1/2}$, then P(E) becomes $P(E) \sim \exp.[-14(kT)^{-1/3}$, KT in electron volt, ev.

For T $\sim 10^6$ kelvin degree, kT $\cong 0.1$ keV, so for T $\sim 10^8$ K⁰, KT ~ 10 KeV. If this data used in (16) changing T by a factor 100, the fission rate changes by 10^{10}.

Fusion Reactions in the Stars

There are two different sequences of nuclear fusion reactions in the stars; they are as follows:

(a) The proton-proton chain reactions

1) $^1H + {}^1H \rightarrow {}^2H + e^+ + \nu$ (Q = 0.42 MeV)

2) $^2H + {}^1H \rightarrow {}^3He + \gamma$ (Q = 5.49 MeV)

3) $^3He + {}^3He \rightarrow {}^4He + 2\ (^1H)$ (Q = 12.86 MeV)

(b) The carbon-nitrogen cycle

1) $^{12}C + {}^1H \rightarrow {}^{13}N^* + \gamma$ (Q = 1.95MeV)

2) $^{13}N^* \rightarrow {}^{13}C + e^+ + \gamma$ (Q = 1.20 MeV)

3) $^{13}C + {}^1H \rightarrow {}^{14}N + \gamma$ (Q = 7.55 MeV)

4) $^{14}N + {}^1H \rightarrow {}^{15}O^* + \gamma$ (Q = 7.34 MeV)

5) $^{15}O^* \rightarrow {}^{15}N + e^+ + \nu$ (Q = 1.68 MeV)

6)$^{15}N + {}^1H \rightarrow {}^{12}C + {}^4He$ (Q = 4.96 MeV)

It is clear from the carbon-nitrogen cycle that four hydrogen nuclei are converted into helium nucleus, producing 24.68 MeV instead of 26.7 MeV, as expected from the six processes pointed clearly above. So the difference is 2.02 MeV. This is because the two positrons in the processes of the reactions have been annihilated with the electrons, so this energy of annihilation is not included, which is about (0.511 + 0.511) × 2 = 2.02 MeV.

It is quite important to note that the elements C, N, and O are, in reality, only a reaction catalyst, because the whole processes are a conversion of 4^1H into ^4He. From the big bang theory, it is known that these elements did not form after this event, within a reasonable time, but the protons were there after the quarks were being created as plasma soup, and very strong forces were combined to build up baryons and mesons, under very high temperature and pressure. Therefore, the first generation of stars is expected to build their statuses via the only route through burning the hydrogen by the proton-proton chain fusion reactions. Then other elements such as C, N, and O were formed, so the next stars' generations might be formed through both processes (a) and (b). The importance of each process in forming the star is relative, depending on the temperature, as demonstrated in the figure below:

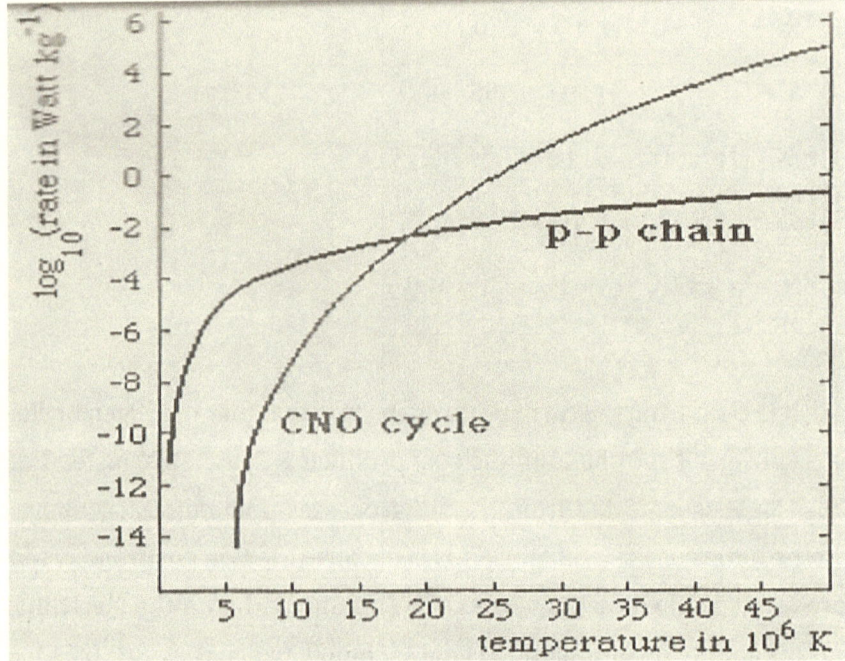

This figure shows the rate of energy production as a function of the temperature (T). The actual curves are for the densities in the sun: $10^5/m^3$ for protons and $10^3 kg/m^3$ for ^{12}C. The heat, generated by these fusion reactions, raises the temperature of the star core, and the pressure of the blackbody radiation is sufficient to counteract the gravitational collapse, but when the hydrogen in the central region of the star is exhausted, the gravitational collapse will continue, so the temperature rises due to the gravitational potential energy conversion into kinetic energy. When the temperature reaches $10^8 k^0$, the burning of helium starts up. This burning is as follows:

$$^4He + {}^4He \leftrightarrow {}^8Be, Q = -19.9 \text{ KeV} \tag{17}$$

$$^4He + {}^8Be \rightarrow {}^{12}C^* \rightarrow {}^{12}C+, Q = 190 \text{ KeV}$$

Equation (16) shows that both reactions are reversible. For example, in the first one, ^8Be exists as a resonant state; it decays within 10^{-16} sec. This resonant state requires a kinetic energy of 91.9 KeV, in the initial state. At a temperature of $10^8 k^0$, the fraction of helium nuclei, with at least this amount of energy (91.9 KeV), is given by Boltzmann factor exp $(-91.9/kT) = \exp(-91.9/8.6) \sim 2.2 \times 10^{-5}$, where 8.6 KeV is the mean thermal energy kT.

The second reaction in the equation (16), the branching ratio for the decay $^{12}C^* \rightarrow {}^{12}C + \gamma$ *is* about 4×10^{-4}, relative to the much more probable return to the initial state producing stable ^{12}C. Then absorbing the alpha particle leads to producing the elements ^{16}O, ^{20}Ne, and ^{24}Mg, due to the following reactions:

$$^{12}C + {}^4He \rightarrow {}^{16}O + \gamma , Q = 7.16 \text{ MeV} \tag{18 a}$$

$$^{16}O + {}^4He \rightarrow {}^{20}Ne + \gamma , Q = 4.73 \text{ MeV} \tag{18b}$$

$$^{20}Ne + {}^4He \rightarrow {}^{24}Mg + \gamma , Q = 9.3 \text{ MeV} \tag{18c}$$

Again, most of the helium in the core of the stars gets exhausted, which leads to more further collapsing, so also, the gravitational potential energy is converting into kinetic energy, which is raising the temperature to about 5×10^8 to 10^9 k^0, where at this level of temperature, C and O can be burned as follows:

$$^{12}C + {}^{12}C \rightarrow {}^{24}Mg + \gamma \,,\, Q = 13.9 \text{ MeV}$$

$$^{16}O + {}^{16}O \rightarrow {}^{32}S + \gamma \,,\, Q = 16.5 \text{ MeV}$$

$$(19)$$

Finally, elements up to iron are produced at temperature about $2 \times 10^5 \text{ k}^0$ in the so-called silicon-burning phase.

This is really an equilibrium process through which successive absorptions of alpha particles are balanced against the photodisintegration by thermal radiation.

At these temperatures, the thermal photons have an average energy of 170 MeV; their absorption can easily lead to breaking up the nuclei such as ^{28}Si into ^{4}He, so ^{4}He then is absorbed by other ^{28}Si nuclei, producing ^{56}Ni, etc. Beyond the iron, cobalt, and nickel region, there are no further exothermic fusion reactions, so there are no further elements building by this process.

VII-3-5-3—The Fusion Reactor

The problems of using the fusion reaction as sources of peaceful uses of energy are technological, like getting the energy output to be economically accepted and the reactor controlling the very hot plasma in certain core to get use of the generated huge heat. From equations (9) and (11), it is clear that the deuteron-deuteron interaction represents a continuous source of enegy; because the deuteron is found in the seas and occasions water, it is a promising source of energy. But as stated above, the problems are in controlling the plasma, supplying high heat about 10^6 k^0, so the fusion reaction occurs, and how can the effect of such temperatures be avoided? By also providing high density concentration. Both the fusion reactions shown in equations 9 and 11 are now of world concerns. The cross section of such reaction is shown in the figure below.

These reactions (D-D, D-T) are supported at the rate $R = n_1 n_2 < \sigma v >$ by heated and confined plasma, which actually contains these nuclei. n_1, n_2 are the ions densities (nuclei, usually because of the high temperature, are ionized); $< \sigma v >$ is the average value of the reaction cross section times the velocity of the particle, averaged over all particles, representing the transition probability. Now let Q be the output of the reaction energy, then the output per unit volume for the

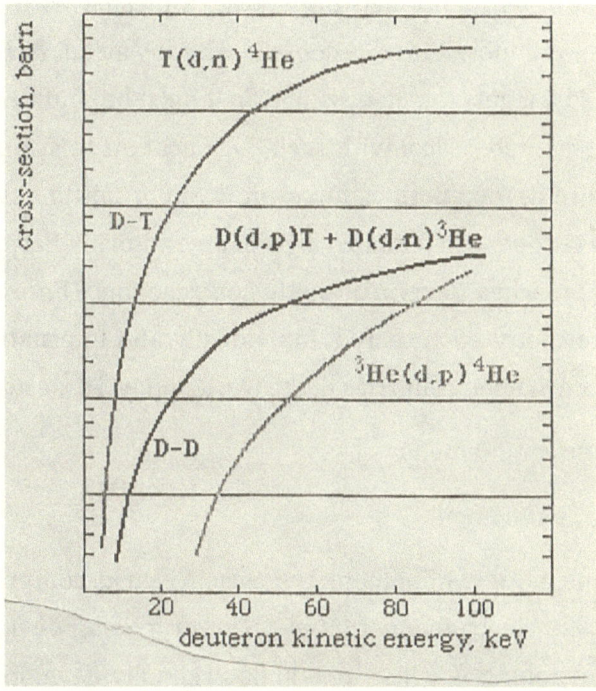

plasma contained for τ seconds is given by

$$w = n_1 n_2 < \sigma v > Q \tau \qquad (20)$$

From equation (11), Q = 17.6 MeV; therefore, if $n_1 \approx n_2 \sim 10^{21} /m^3$, then the output will be $w = 3 \times 10^8$ watt/m^3 (per/second) confinement. Note that the energy required to heat the plasma is $[n_1 + n_2]3kT/2 = (3/2)$ NKT, from kinetic theory of gases. So for net energy output, this has to be less than the calculated w above. Therefore,

$$n^2 < \sigma \ v > Q\tau > 3/2 \ n \ kT, \text{ or } n\tau > 3kT/2 \ / < \sigma v > Q \tag{21}$$

Now for an operation of reaction with temperature of $1.2 \times 10^8 \ k^0$, the KT = 10 keV and the D-T reaction with $<\sigma v > \sim 10^{-22} \ m^3/\text{sec}$. Hence, $n\tau$ $> 10^{20} \ s \ m^{-3}$ to generate power. This $n\tau$ is the product of the ions density with the plasma confinement time τ named as *Lawson criterion*.

This criterion is setting the target for successful design for the fusion reactor. As mentioned previously, a plasma with such high temperature is impossible to be contained by any metal. The metal will be vaporized if it gets touched with it. So it must be confined by other means that can safely deal with such huge heat. As it is so important for the future of the energy sources, it is important to develop new technology to cope with it financially and operationally. Basically, there are two approaches to control the fusion reaction. These are under detailed concentrated research, individually and internationally, by the greatly developed countries in the world today. These methods are

a. magnetic confinement;

b. inertial confinement.

In the magnetic confinement, the plasma is held, compressed, and heated by the electromagnetic field. The promising device of such type of functioning is called tokamak, originally developed by the previous Soviet Union. The figure shown below is the simple form of the tokamak:

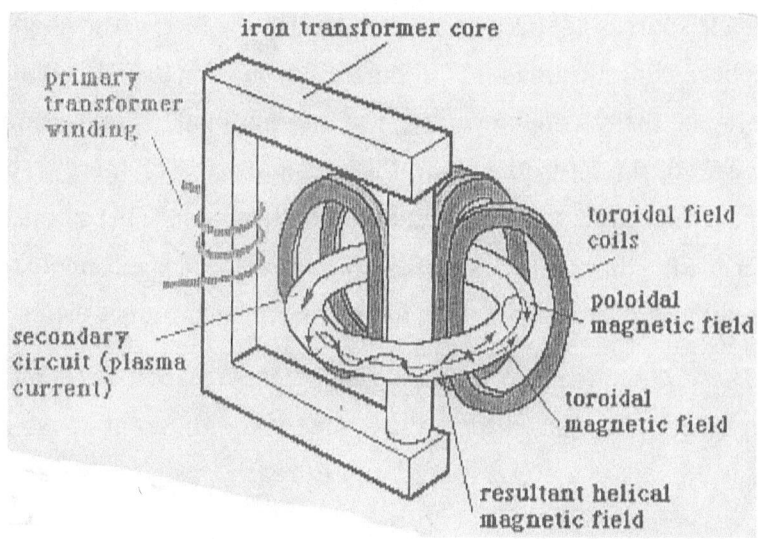

As shown in the figure, it combines toroidal field provided by coils wrapped around a ring-shaped tank and poloidal field, due to a large ring current induced in the plasma. This kind of fusion reactor is developing internationally, by a group of developed countries such as the European countries, United States, Russia, and Japan, under the support of the United Nations. This project is called Joint European Torus (JET), but it is still far away from the economical uses.

The technique of the inertial confinement involves the process of heating and compressing of a small pellet containing deuteron and tritium, using for that intense laser pulses. The temperature required to get KT ~ 10 KeV indicates that the pellet will fly apart in about a nanosecond. This, according to Lawson criterion, implies that the particles' density of 10^{20} m^{-3} must be reached. This density is still not reached yet. For example, in liquid hydrogen, it was found to be much less than this required density. The heating pellet within such a short time indicates that the laser must deliver an instantaneous power of 10^{14} watt or so. This shows the great difficulty opposing the development of fusion reactor to be economically acceptable compared to other available sources of energy. As stated earlier, the international efforts have been concentrated on building up a promising fusion reactor by developing the tokamak prototype to international fusion reactor to take care of the future needs to the expanding use of energy. These international concerns are to build the International Thermonuclear Experimental Reactor (ITER) in Cadarache, France. It is to demonstrate the feasibility of using sustained fusion reaction to produce electricity.

These attempts are still under development to reach the economical compatibility compared to the use now of the different sources of energy. However, there is a big hope to reach this goal in the near future.

Chapter Eight

High-Energy Physics: Nuclear Particles

VIII-1—Introduction

Physicists still concerned, theoretically and experimentally, with what is perhaps the most fundamental fact of the physical world, which might be stated briefly that most, if not all, of what is going on in the universe is ultimately resting upon interactions between the varieties of particles, of which all matter and radiation are composed. These particles, as it will be shown later, from different families or groups according to their properties, are governed by certain quantum numbers and a type of interaction, which creates them and decays them. The physicists started with the photon, electron, proton, and neutron, and numeral extended to tens of them during the interval (1936–1964), where the positron, neutrino, antiproton, antineutron, antineutrino, π-mesons, μ-mesons, anti-π-mesons, anti-μ-mesons, κ-mesons, anti-κ-mesons, $\overset{\circ}{\Lambda}$ -particle, sigma $\Sigma^{\pm^0}$ particles, Xi particles $\Xi^{\pm 0}$, omega-minus Ω^-, and resonance particles are discovered, and their physical properties and decays are studied. Also, they are categorized according to their type of interactions (strong, electromagnetic, or weak) and classified into families such as photon, lepton, meson, baryon, or hadron and no hadron, i.e., obey the strong interaction (hadron), and obey the weak interaction (no hadron). The particles κ-mesons and those following them such as Xi, sigma, lambda, and omega are called strange particles due to the fact that they are created by a strong interaction, but they decay under the weak interaction. As it will be clarified in the next sections, most of these particles are unstable.

Some of them decay under the effect of a strong interaction and electromagnetic interaction, and others decay by the weak interaction. Therefore, none of these strange particles are immutable, so all can go changing either spontaneously or under suitable external stimulus into other particles, which depends on certain conservation laws corresponding to the type of the particle and its processes of decay. Clearly, the understanding of the properties of these particles and their relative relations in such a way that they ought to be within the compass of the theory would help the physicists to understand the physical world as completel as possible. Therefore, the field of high-energy physics that concerns with the creation and decaying of these particles was the most active field in physics since the fifties of the twentieth century, theoretically and experimentally. The research was of two sides: experimental and theoretical. The experimental part was still concerned with the discovery of new particles, measuring their intrinsic properties and their interactions and decays. The theoretical part was still concentrating on a workable theoretical framework or theoretical model that can embody each property of these particles.

Studying these so-called nuclear particles or fundamentals and non-fundamentals is of great importance to understand the behavior of nature, its entity, its constituents, and its ultimate building blocks. This kind of research is thought to be started by Greek philosophers and scientists (500 BC); they proposed that matter elements were considered as components of all other materials. Later on, these thoughts were developed more scientifically to find out that the atom, the molecule, the element, and the compound then the matter. The atom was discovered to be dividable when the electron was discovered by J. J. Thomson (1897), followed by the discovery of the nucleus of the atom and that of the proton and the neutron, which gave the planet structure of the atom. Later on (1935–1964), a stream of particles and antiparticles was discovered in the cosmic rays and in the laboratory.

On the basis of such development of the nature and the properties of these particles, one can no longer be convinced that such particles can be elementary or fundamental particles. But rather, they are composites of a small number of basic particles called quarks, as it will clearly be shown in this chapter later on. This conclusion is based on the very well-known characters of these particles such as their decaying to other particles and their defined dimensions, which show that they are of structural entity and of dimensions, where the elementary particle is defined physically and mathematically as structureless and as a dimensionless point. On this basis, the only considered elementary particles yet are $\left(e^{-}, e^{+}, \overline{V}_{e}, V_{\mu}, \overline{V}_{\mu}\right)$ the so-called leptons and the proposed stable quarks particles (till the year 2013).

The development of certain accelerators around the world for a very high energy (200–2000 GeV) and for colliding different particles such as p-p, p-n, n-n, e-p, etc., might lead to new important information about the concept of the elementary entity, whether there is such a particle or not. Again, this conclusion is constrained by the ability of developing higher energy accelerator and the very advanced mathematical frame of work toward the theory of the unified fields. All such ideas about these nuclear and subnuclear particles will be described in some details in the following sections of this chapter. However, there is, today (2013), a great scientific effort to explore the deepest fine structure of the matter and its origin. For this aim, a huge collider is established at the common boundary between France and Switzerland. It is called Large Hadron Collider (LHC). It is designed for very high energy (14 tera eV). It used collisions between hadrons and between heavy ions to study the universe stage at 10^{-6} seconds after the big bang event, where the universe is a plasma of quarks and gluons, and also to investigate some proposed theory like Higgs boson particle and field, string theory, multiple dimensions up to eleven dimensions, supersymmetry particles. This collider started the research work in 2008. The first discovery, as announced in 2012, is the Higgs boson particle. This discovery was confirmed in 2013.

VIII-2—Force and Conservation Laws

Creation and Decay of Particles

VIII-2-1—Classification of Interactions

The studies of the elementary-particles physics, in fact, are partly studying the four fundamental forces acting among them. These fundamental forces are classified as

1. The strong nuclear force. This is responsible for the strong interaction between particles. Yukawa forces are such strong forces that are associated with π-meson exchange field charge denoted by "g," which has the characteristic coupling constant $g^2/\hbar c$, where its numerical value far exceeds that of other interactions, such as electromagnetic, gravitational, and weak interactions. In addition to the internuclear binding force, which is accounted for by the exchange of π-mesons (pions), the pion-nucleon scattering is also a type of strong interaction. As it will be shown, this strong interaction is responsible for kaon production such as

$$\bar{\pi} + p \,(\bar{\pi}\text{-meson bombarding a proton}) \rightarrow \bar{\Sigma}(sigma) + K^{+}(kaon)$$

This was discovered in the bubble chamber, as shown in the following schematic figure 1:

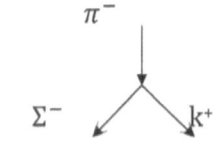

Figure 1: π^{-} decay

2. The electromagnetic interactions. This type of interaction is, in fact, much weaker than the nuclear interaction, due to the fact that this inverse square law coulomb forces may be recognized as due to the photon exchange between the fields of the interacting charged particles. The dimensionless coupling constant in terms of the electron charge is given as $\alpha = e^{2}/\hbar c = 1/137$, which represents the fine structure coefficient in atomic spectroscopy, because it has, in fact, a significance role in this spectroscopy. Therefore, in most instances, processes that proceed by the electromagnetic interaction involve one or more photons, where these photons are actually messengers between interacting electromagnetic fields. Hence, the photon exchange between these fields is responsible for the electromagnetic binding forces, so the photon capture can affect the production of mesons or hyperons by an electromagnetic interaction such as

$$\gamma(photon) + p(proton) \nearrow \pi^{\circ} + p \quad \text{(photo-pion production)}$$
$$\searrow \Lambda + \kappa^{+} \quad \text{(Associated two strange particles)}$$

These kinds of decay process appear as paradoxical at first sight, because uncharged particles such as (π^0, Λ^0) can also appear as subjected to electromagnetic interactions. So π^0 decaying into two photons such as $\pi^0 \rightarrow \gamma + \gamma$ occurs through electromagnetic interactions, with mean life $\tau \cong 5 \times 10^{-17}$ sec.

Now, how can such paradoxical phenomenon be resolved scientifically and acceptable? This can be resolved by introducing, as an intermediate step in the overall reaction, the virtual production of a nucleon-antinucleon pair (by strong interaction), using the uncertainty principle $\Delta E \, \Delta t \approx \hbar$ to borrow the energy within a very short time. This is not making a violation of conservation law or symmetry that can be sensed. Thus, $N - \bar{N}$ pair then decays into $\gamma + \gamma$ by an electromagnetic process. This overall decay can be written as

$$\pi^0 \xrightarrow[strong]{} \left(N + \bar{N} \right) \xrightarrow[E.M.]{} \gamma + \gamma \rightarrow \gamma + e^+ + e^-$$

$$\underset{virtual}{}$$

(one chance in eighty chances)

or $\gamma + \gamma \rightarrow e^+ + e^- + e^+ + e^-$ (one chance in 29,000 chances)

As we know, the inverse inhalation process can give further one or two electron pairs, which is, in fact, an electromagnetic interaction where no photon explicitly enters. As another example of such decay scheme, one might take the fast electromagnetic (strong) decay of the eta $\left(\eta^0 \right)$ meson, where

$$\eta^0 \rightarrow \pi^+ + \pi^- + \pi^0$$

with mean lifetime $\tau < 10^{-16}$ sec.

Another decay of such character is the neutral sigma $\left(\Sigma^{\circ}\right)$ decaying

into neutral lambda $\begin{pmatrix} \overset{\circ}{\Lambda} \end{pmatrix}$ and photon (γ), i.e., $\Sigma^{\circ} \rightarrow \overset{\circ}{\Lambda} + \gamma$, which again represent a neutral charge particle decaying under the effect of electromagnetic mode of interactions, which also can be explained on

the basis of intermediate formation of a virtual $N - \bar{N}$ pair such as

$$\Sigma^{\circ} \xrightarrow[strong]{} \overset{\circ}{\Lambda} + \underbrace{\left(N + \bar{N} \right)}_{virtual} \xrightarrow[E.M.]{} \overset{\circ}{\Lambda} + \gamma$$

As it will be shown later, Σ -particles are classified within the hyperon family, so this process of decay represents the only known instance of hyperon decay not occurring under the weak interaction effects.

3. The weak interaction force. In general, many of the interactions in particles physics found it is very much weaker than the force mentioned above. The most familiar weak interaction is the one under where beta decay occurs; this subject was fairly treated in chapter 6. Using Fermi theory, the numerical constant characterizing the identity of the weak interactions is obtained to be $g_f \sim 1.41 \times 10^{-49}$ er.cm^3. It is possible to construct a dimensionless constant from this constant. It is reasonable, in analogy with the expression for other coupling constants, to set g_f^2 in the nominator with balancing it by a denominator with dimension that accordingly must be$[\left(\hbar c \right)^2 \times (\text{length})^4]$.

The suitable quantity taken for the length found is the rationalized Compton wavelength for pi-meson—namely, $\dfrac{\lambda c}{(2\pi)_{pion}} \equiv \dfrac{\hbar}{cm_\pi} = 1.43$ fm. It is the same order of magnitude as the rationalized Compton wavelength of other nuclear particles, which is, in fact, the range of the nuclear force. By inserting this value, a dimensionless coupling constant for the weak interaction will be $\dfrac{g_f^2}{c^2\hbar^2}[\dfrac{M_\pi c}{\hbar}]^4 \cong 5\times10^{-14}$. Hence, the weak interaction is responsible for the radioactive beta decay for unstable nuclides, also for the beta decays of the muon, neutron, K-meson, and the moun decay to pi-meson—that is to say, to all those decays in which neutrinos are parts of the decay.

4. The weak interaction with about the same strength of that which is responsible for beta decay leads to the decay of strange particles. Statements 3 and 4, in fact, represent the universal weak interactions, as was pointed out in chapter 7. This universal weak interaction had been unified in the sixties by Abdus Salam and Weinberg with the electromagnetic interaction theoretically and had been proved experimentally in 1983. The unified force agent particles were discovered at CERN L arge Hadrons Collider. These particles are $W^\pm$ and Z^0; their masses are eighty-five times the proton mass and ninety times the proton mass, respectively. So these are heavy particles compared with nucleons, which indicates that the range of this weak interaction is quite short. This interaction does not appear to have the capacity for binding yet is responsible for the number of capture reactions such as μ^- + p → n + V_μ and, in particular, for many decay processes for mesons, $\mu^+ \to e^+ + V_e + V_\mu$, similar of pions into muon and kaon

into pions. All the following hyperon decays are examples of weak interactions (except the decay Σ^0 (em) $\rightarrow \Lambda^0 + \gamma$):

a) Decay mode of neutral kaon, as summarized in the following:

$k_1^0 \rightarrow \pi^+ + \pi^-, (69\%)$ or $\rightarrow \pi^0 + \pi^0, 30.7\%$

$$k_2^0 \longrightarrow \pi^0 + \pi^0 + \pi^0 \qquad 23.5\%$$

$$\longrightarrow \pi^+ + \pi^- + \pi^0 \qquad 11.5\%$$

$$\longrightarrow \pi + \mu + \nu \qquad 27.5\% \qquad \left. \begin{array}{l} \\ \\ \\ \end{array} \right\} \quad \tau = 5.77 \times 10^{-8} \text{ sec}$$

$$\longrightarrow \pi + e + \nu \qquad 37.4\%$$

$$\longrightarrow \pi^+ + \pi^- \qquad 0.15\%$$

a) Decay of (Λ^0) hyperon (heavier than nucleon, but lighter than deuteron)

b) $\Lambda^0 \longrightarrow p + \pi^- \quad\nearrow \quad 66.3\%$

$\searrow n + \pi^0 \quad 33.7\%$

It is important to note that the neutrinos and the antineutrinos are exclusively associated with the weak interactions, as the photons are only associated with the electromagnetic interactions. The other particles can respond to different interactions without any exclusiveness. For example, the pions are connected with the strong nuclear force, as the Yukawa theory tells us, but they are radiative capture affected by electromagnetic interaction, and they decay under the weak interaction.

5. Gravitational interaction. This is due to the gravitational force, which is the weakest force on the small scales and sizes compared with other forces. But it is so great on the large scales and sizes; it plays the

main roles in binding energy among the cosmological constituents such as our solar system. This force is a tractive force, as it is known, proportions directly with the product of the masses and inversely with the squared distance separating them. The proportionality constant expresses the strength of the force. This constant, known as the Newton constant, is given as $G = 6.67 \times 10^{-11}$ Newton .m^2. kgm^{-1}. It is possible to construct dimensionless coupling constant from this as $Gm^2/\hbar c$, considering (m) the nucleon mass, then, $Gm^2/\hbar c$ will be of value equal to about $2 \times 10^{-39} = g_g$. Comparing this with the nuclear force ($g_n = 1$), one finds that $g_g/g_n \cong 10^{-40}$, which is very small. And $g_{em}/g_n \cong 10^{-2}$, $g_{wi}/g_n \cong 10^{-13}$.

This data indicates that when the nuclear force is considered, the gravity force and weak interaction force are negligible, and the electromagnetic force could be considered as perturbation term in Hamiltonian of the system, since the decays of particles are governed by these forces, and such types of interactions are associated with the characteristic times of the occurring events. Therefore, the relative strengths of these interactions might be estimated on the bases of the lifetimes of the excited states. The strong interaction (nuclear) occurs in time equal to 10^{-23} sec (it is the time for a nucleon to transverse the diameter of the nucleus, about 10^{-13} cm. = 1fm for the light nuclei), corresponding to the range of this force, which is about 2–3 fm. The time for the electromagnetic interactions to occur is about 10^{-18} sec. And for the weak interactions, the time is 10^{-10}–10^{-8} sec, which is of the order of the lifetimes of most decay processes, which are considered to be quite long compared with the electromagnetic and nuclear interactions. Hence, the lifetime could be considered as good index to the type of interaction. This actually leads to the force nature that is acting on the particle, whether in the case of interacting or decaying. On such basis of force-strength range, it is expected that the main features of the nuclear particles structures can be defined by the strong interactions. Therefore, the electromagnetic and weak interactions might be considered to be small effects as perturbation terms in the Hamiltonian of the system under study. Therefore, the classification of the considered effect of the type of interaction is based on its strength and the order the problem treated under. In the zero-order description of the particle, the electromagnetic interaction, weak interaction, and presumably strange particles decays, i.e., all particles that would be stable are ignored, except those that can react by strong interactions (such as $\pi^- + p \rightarrow n$ + kinetic energy, which would still occur, but π^- would be stable in the absence of other baryons or mesons). The next order or the step of approximation will include the electromagnetic interactions. So the photon, which does not exist in the zero-order approximation, will clearly appear here and induce the electromagnetic decay to the neutral pion, π^0, gamma emission de-excitation of nuclear excited states and other more subtle effects, such as mass difference between the proton and the neutron, between $\pi^\pm$ and π^0 and between $\Sigma^\pm$ and Σ^0. In this order of approximation, the particles $\pi^\pm$, $\mu^\pm$ and strange particles are still stable. Introducing the weak interaction into the total interaction will lead to the remaining instabilities and also cause small additional shift in the mass of some particles.

VIII-2-2—Conservation Laws

The well-established laws of conservation in physics are

1. mass-energy conservation;

2. linear momentum conservation;

3. angular momentum conservation;

4. electric charge conservation.

If the decays' behaviors of some well-established particles are carefully noticed, an astonishing idea is cleared out. Some particles do not decay, although these conservation laws are satisfied in decay processes. In other words, why do some unstable particles decay in a particular mode of decaying? Also, why are some particles stable against any mode of decaying? The answer is clear: since there are different processes of decays corresponding to different kinds of fundamental interactions, the unstable particle chooses the suitable mode of decay that is allowed by certain laws of physics. These laws are the natural laws of conservation. The four laws mentioned before are the main features of physical interactions (interaction, scattering, collision, and reaction) on the level of the macro and micro physical systems in nature. By detailed studies of the behaviors of the discovered particles, it is found that additional laws of conservation must be satisfied so the decay takes place. This is why some unstable particles do not decay although the conservation of the four laws is satisfied in the processes of the decay. These additional conservation laws are necessary to answer the following physical questions:

1. Why is the free neutron unstable, and why does it decay under the effect of the weak interactions? Why it does not decay under the effect of the strong interaction to proton and π^- ? Or why does it decay under the effect of the electromagnetic to proton and photon? Or why does it decay to neutrino and antineutrino under the weak

interaction? It is known that the decay of neutron by the strong interaction is ruled out due to the law of mass conservation, where $m_p + m_{\pi^-} > m_n$. Also, its decay by the electromagnetic interaction is prohibited by the charge conservation law, where both the photon and the neutron are neutrals.

2. Why is the free proton stable? And why it does not decay under any of the four known forces (interactions)? Also, why it does not decay into positron and photon, where there is no violation to the main laws of conservation?

On these bases, the following might be asked:

If the four established laws of conservation in the processes of interactions and decays are satisfied, then why don't they occur? The answer is quite obvious: there must be some other types of conservation laws imposed by nature that have to be satisfied so the interactions or the decays take place. By deep investigations, it is found that according to the basic nature of these particles, there are, in fact, additional conservation laws that must be satisfied in order for their interactions or decays to occur. These particles are classified according to their masses: heavy or light, hadrons or leptons, respectively. The hadrons are also divided into baryons and mesons. Such detailed classifications will be seen later on. The additional conservation laws are the baryon number and the lepton number, so the laws of conservation became as follows:

1. Mass-energy, $\sum E_i = constant$
2. Linear momentum, $\sum P_i = constant$
3. Angular momentum, $\sum L_i = constant$

4. Electric charge, $\sum e_{i_=}$ constant
5. Baryon number, $\sum B_i$ = constant
6. Lepton number, $\sum L_e$ = constant-electron type
                            $\Big\}$ $\sum L_\mu$ =constant-muon type

These six laws of conservation, under the four types of interactions known in nature, are universally valid. Also, as it will appear later on, they are not the only conservation laws in nature. The classifications of the continuing discoveries of particles with new families, new laws of conservation rise. In addition to hadrons, photon, and leptons, there are now the hyperons, the strange and the so-called resonance particles. Such laws are the strangeness number and isotopic spin. According to the eightfold-way theory proposed by Gellman in 1963, it is required for conservation laws to have eight quantities. All these ideas and concepts will be discussed in the next sections consequently. The concept of baryon number is related to the charge, so B is (+1) for the positive charge, (−1) for the negative charge, and B = 0 for neutral particle (charge = 0); therefore, for the proton B = +1, and B = 0 for the neutron, and so on for other baryons.

An important remark to be noted is this: strongly interacting particles in a certain sense are dynamic structures, and their existence is owed to the same forces through which they mutually interact. On the contrary, the leptons have no such origin of dynamic structures. Even the necessary forces have never been found clearly. The bizarre spectrum of the states begins with one of zero energy, sharply terminating at the muon with energy $\approx$ 150 MeV, bearing no resemblance to any dynamic spectrum seen before. No well-established explanation of the leptons existence is found yet. Of course, this is rather the philosophical case, subjected to the future of the theory of the particles physics. At the international most advanced Large Hadron Collider in Geneva, Switzerland, great scientific efforts are going on to deeply investigate the nature foundations before and directly within 10^{-6} sec after the big bang events, including the way some of these particles get their masses and others with no mass. The first claimed discovery in 2012 is the existence of the Higgs fields with its agent to its effect, the Higgs boson, which presumably gives the other particles their masses through a kind of collision mechanism.

VIII-2-3—Bootstrap Dynamics in Creating Particles

As mentioned before, so many nuclear particles are divided among families. Each family group shares certain quantum mechanical quantum numbers and properties, such as spins, isotopic spins, and strangeness, etc. For example, this strangeness classifies those particles created by the strong interactions, but they decay by the weak interactions. This division among families is analogous to the periodic table of the atoms in atomic physics. The symmetries among these families emerge from bootstrap mechanism. So what is this mechanism? The bootstrap mechanism is based on the concept that is equivalent to the notion democracy governed by Yukawa forces. Each strongly interacting particle is conjectured to be a bound state of those channels with which it communicates, owing its existence entirely to forces associated with the exchange of particles that communicate with (grassed) channels. Each of these later particles, in turn, owes its existence to a set of forces to which the original particle makes a contribution. In other words, each particle helps to generate other particles, which, in turn, generate it. Or the strongly interacting particles are inducing the strong interaction. It is a kind of bootstrap functioning.

The theory of eightfold way (mentioned earlier), by its name, as stated before, requires conservation of eight quantities arranging the known strongly interacting particles into family groups of eight or tens. All particles of a definite family have the same intrinsic spin (J) and parity but with a different mass, charge, isotopic spin, strangeness, and other parameters. This theory can predict the existence of missing family members, but not the existence of new families. It does not provide any similar arrangement for the weakly interacting particles, which are lighter and interacting so slowly.

This theory also offers organizing scheme for the strongly interacting particles and predicting the behaviors of particles within the defined family. Some of the so-called nuclear particles are a kind of nuclear resonances, appearing as definite peak, due to the nuclear strong interaction, where the energy of the incident particle is in resonance with the binding energy of the target. Therefore, the resonance, in fact, means compound states of two particles. One way of feeling this resonance from a particle is comparing lifetimes. The resonance exists only for 10^{-23} sec (nuclear time), while the particle lives for at least 10^{-10} sec.

VIII-3—The Concept of Strangeness Theory

Due to the great range of the interaction strength, it is not expected that the electromagnetic or the weak interactions will have effects within the nuclear interactions (collisions). This fact can be noticed where no electrons, photons, and neutrinos are observed as direct products of nuclear collisions. Also, it is expected that baryons or mesons, which are unstable toward the decay into lighter baryons and/or mesons, decay rapidly via the strong interaction, but it was contrary to this expectation. They are decaying so slowly compared with the expected rate of decaying via the strong interaction, although for example, K-mesons decay into pions, and hyperons decay into pions and nucleons, but through very slow processes. This behavior raises a serious question about the mechanism by which they are decaying, although they are created through strong interactions. Their lifetime is 10^{-10} sec instead of 10^{-23} sec for nuclear force. The first thought was may be these particles possess a large angular momentum, about five to six units, which, together with the nuclear attractive potential, might yield unstable system, having a potential barrier that retard its decay. This proposal hypothesis failed experimentally. Many different suggested ideas tried to reconcile the rapid production of these particles with their quite slow decaying. The only one that stands out at present, for both its simplicity and ability to relate all known baryons and mesons to each other, is the so-called strangeness theory, proposed by Gell-Mann and Nishijima.

One of the main concepts of this theory is, the strange particles cannot be produced singly in nuclear collisions. Usually, they are produced in association with other strange particles. Each strange particle is carrying away some quantized quantity, which the other needs in order for it to decay. Thus, in nuclear interactions, the total amount of this quantity is conserved, but its conservation, similar to the conservation of the isotopic spin, is not quite rigorous. It might be violated in the processes of decays. This is due to the fact that the decay occurs under the weak interactions, so these particles are created by the strong force (strong interaction), but they decay under the effect of the weak interactions. Because of such behaviors, they are termed as strange particles, so that assumed quantum number, called the strangeness number, indexes these particles and plays a main role in their physical characteristics; it is labeled as S. Therefore, it looks more specific for the strangeness theory to be based upon a generalization of the concept of isotopic spin, which is based too on the assumption that the nuclear forces are charge independent. As it was mentioned previously, the idea of the isotopic spin (T) proposes that each nuclear particle (nucleon) possesses a certain total isotopic spin (T) and that each possible projection of (T) along a certain axis, say, $\xi - axis$, appears as a different charge state of the corresponding nuclear particle.

For analytical approach, let A be the mass number of the particle nucleus and (T) be its isospin quantum number. Its $\xi - component$ is T_ξ, and let Q be the charge in unit of electron charge (e). Then the relations between Q, T, T_ξ, and A are given by

$$Q = T_\xi + 1/2\,A \quad , \quad -T \le T_\xi \le + T \text{ (integer steps)}$$

Now, take the nucleons (p, n) and the pions (π^+, π^0, π^-), where for the nucleons $A = 1$ and for the pions, $A = 0$. Hence, $T_\xi = 1/2, -1/2$ for the proton and the neutron, respectively $= 1$. For the pions, $A = 0$,

1. π^+, $Q = +1$, $T_\xi = 1$

2. π^-, $Q = -1$, $T_\xi = -1$ $\qquad\qquad$ triplet

3. π^0, $Q = 0$, $T_\xi = 0$

For the nucleons,

1. The proton, $Q = +1$, $A = 1$, $T_\xi = 1/2$ $\qquad$ for high-energy doublet

2. The neutron, $Q = 0$, $A = 1$, $T_\xi = -1/2$

The usual theory of the isospin (charge independence of nuclear force) asserts that in all nuclear collisions, the total isospin is conserved as well as its various components. Thus, the cross section for a given nuclear reaction (collision or scattering) is the same for all T_ξ (T-projections). For example, take the following reactions:

a. $\pi^+ + p \rightarrow \pi^* + \pi^0 + p$

b. $\pi^- + n \rightarrow \pi^- + \pi^0 + n$

Both reactions must have the same cross section, because they belong to the same total isospin $T_{3/2}$. As pointed out before, the strangeness theory for the strange particles introduced the so-called strangeness quantum number S. Accordingly, Q takes the form [Q = T_ξ + 1/2 A + 1/2 S]; S is the integral quantum number. Therefore, the quantum numbers' *quantities*, which are conserved in all nuclear reactions, including those involving the production of strange particles, include the S too. This conclusion implies that all baryons and mesons must have definite values for T, A, and S. For the non-strange particles $S = 0$, as the nucleons, but S $\neq$ 0 for the strange particles. For a given A and S, the various projections T_ξ produce an isospin multiplet. Each component corresponds to a different charge state. Therefore, the various baryons and mesons are analogous to the fine-structure multiplet of the optical spectroscope. This concept is equivalent to the energy splitting of the eighteen distinct quantum states for the valence electron in the sodium atom in the ($n = 3$) state. This can be looked up in any atomic physics text. This atomic phenomenon is successfully used to describe many grouping of baryons and mesons within the hadron family; on this basis, it is possible to form particle types into groups within the hadrons family. Figure 2, page 282 shows the mass splitting of the baryons, first under the semi-strong interaction, which separates the baryons into four families, two nucleons. Their average masses are about 939 MeV and lambda (Λ) with mass 116 MeV, three sigma particles (Σ) with an average mass of 1193 MeV, and two Xi particles (Ξ) with an average mass of 1317 MeV. From this data, it is possible to estimate the splitting due to the semi-strong interaction that is in the order of 10^2 MeV.

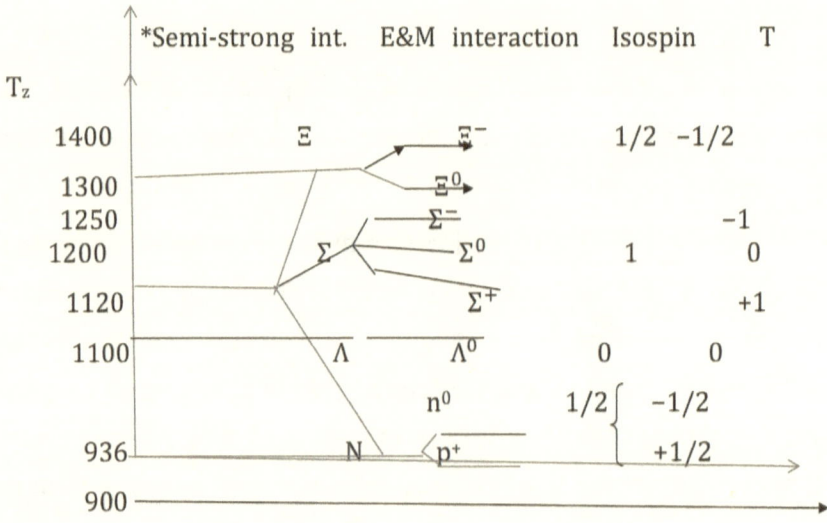

Figure 2: schematic diagram of mass splitting of spin 1/2 baryons by semi-strong and electromagnetic interactions (approximate scaling)

On this basis, the eight hadrons (n, p, Λ^0, Σ^+, Σ^0, Σ^-, Ξ^0, Ξ^+) can be considered as different states of a single primordial hadron. In this case, if, by any means, the semi-strong interaction is turned off, all the eight hadrons would have identical properties. Therefore, there will be (2T + 1) multiplet for hadron family. For example, for the nucleon, there are 2 (n + p); hence, 2 = 2T + 1, yielding T = 1/2, where T_ξ (p) = 1/2 and T_ξ (n) = −1/2, for high energy; for low energy $T_{\xi(p)}$ = −1/2, $T_{\xi(n)}$ = 1/2.

For pions, there are three different charge states, i.e., triplet, so $3 = 2T + 1$, which leads to $T = 1$, then the component $T_\xi = +1, 0, -1$ to $\pi, {}^+\pi^{0,} \pi^-$, respectively. It is also the same for the three sigmas, $3 = 2T + 1$, which implies that $T = 1$, and the T_ξ components are $+1, 0, -1$ to $\Sigma^+, \Sigma^0, \Sigma^-$, respectively. Using the relation $Q = T_\xi + (1/2) A$, $A = 0$ for the sigmas, then the values for T_ξ to the three sigmas are satisfied, as they are above. Let $A = 0$ and $Q = 0$, a neutral meson, which might be so difficult to detect experimentally. Taking $S = -1$, there will be three possibilities involving multiplet charges; they are

1. $A = 1$, $T = 1/2$ yields $Q = 0, -1$, which may be identified with antiparticles of kaons, k^-, k^{0-};

2. $A = 1$, $T = 0$, then from $Q = T_\xi\, 1/2\, A + 1/2\, S$, $Q = 0$, so it yields a neutral singlet baryon, which might be identified with the Λ^0 particle, since no charged baryon of mass near the mass of this neutral lambda has been observed;

3. $A = 1$, $T = 1$, where $2T + 1 = 3$, which yields baryon with triplet, as pointed above. It evidently corresponds to the three sigma particles if the neutral one is considered. Neutral sigma is more massive than the neutral lambda. But they have the same $S = -1$.

It decays presumably into neutral lambda and matter or by electromagnetic coupling into neutral lambda and gamma radiations if it is free. Finally taking $S = -2$, it is possible to obtain a doublet baryon with $A = 1$ and $T = 1/2$, then $Q = 0, -1$, or singlet or triple-double baryons with $A = 2, T = 0$ and $A = 2, T = 1$, respectively. Strange particles having $A = 2$ have not been observed, but the doublet $S = -2, A = 1, T = 1/2$ is considered as Xi hyperon. Hence, for convenience, the following diagram may illustrate in general the stated material:

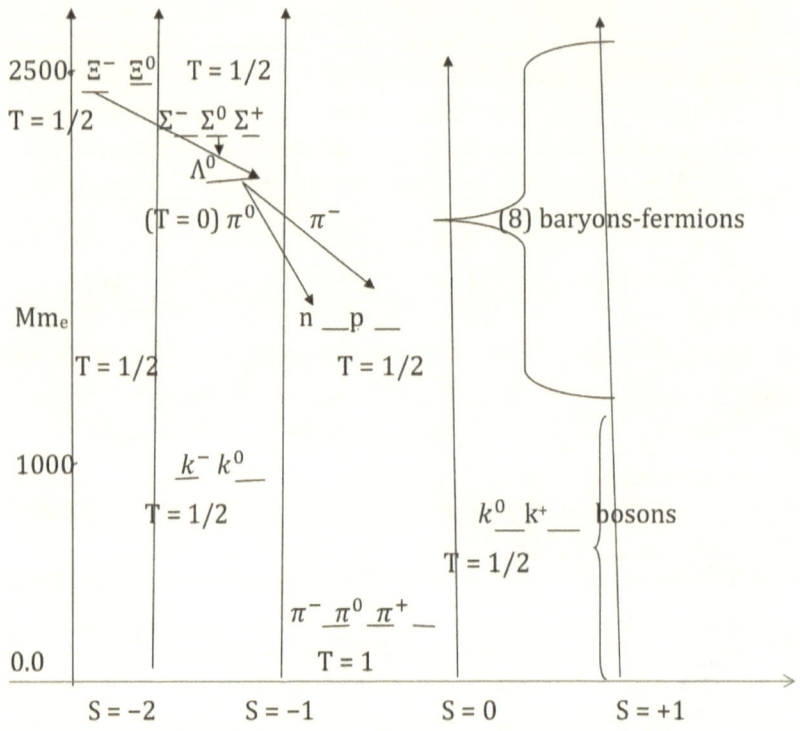

Figure 3: particle states according to the strangeness theory

It is important to notice that the change in isotopic spin changes the charge state. Also, the antibaryons are different from the baryons, but the anti-bosons do not differ from the bosons. This fact leads to the concept of the baryon number (B), which is the quantum number characterizing baryons particles; therefore, B = ±1 for the baryons and zero for the bosons. On this basis, the charge can be represented by the formula $q = T_\xi + \dfrac{S+B}{2} = T_\xi + a$, where $a = S + B/2$ = charge displacement or in general, $q = T_\xi + (1/2) A$, as written before, A is the mass number. Remember, A = 1 for the nucleons and $T_\xi = ±1/2$ for them. Their charges are 1, 0 for the proton and neutron, respectively, as it is. Now let Y be defined as the hypercharge given by the formula Y = B + S, so it is 1 for the proton (S = 0, B = 1) and 1 for the neutron too for the same reason. It is clear too that it is zero for sigma but −1 for Xi.

If q^ is considered as the charge average number for a group such as p and n, i.e., $q^ = (q_p + q_n)/2 = (1 + 0)/2 = 1/2$. Then $Y_{group} = 2q^ = 1$. Therefore, the hyperon charge for the nucleons is Y = +1. It is Y = 0 for sigma particles, and it is Y = −1 for Xi particles. So the charge number q of a particle is clearly related to the hypercharge and the isotopic spin component T_ξ. As has been shown for the baryons mass splitting under the effect of the semi-strong and electromagnetic interactions, it is possible to show that for the mass splitting for the zero-spin mesons under the effects of the same interactions. The following diagram shows that schematically:

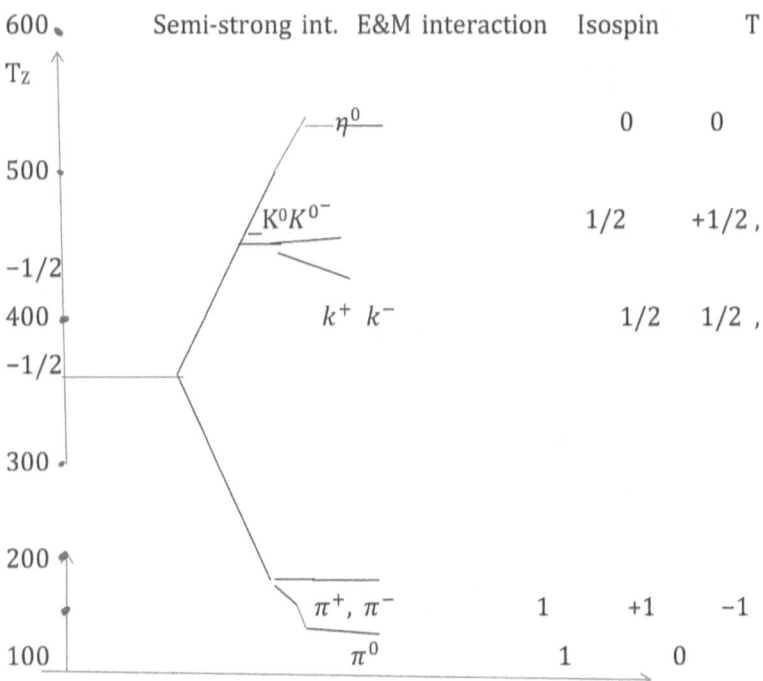

	Semi-strong int.	E&M interaction	Isospin	T
	η^0		0	0
	$K^0 K^{0-}$		1/2	+1/2,
	$k^+ \; k^-$		1/2	1/2,
	$\pi^+, \; \pi^-$		1	+1 −1
	π^0		1	0

600

Tz

500

−1/2

400

−1/2

300

200

100

Figure 4: mass splitting of the spin-zero mesons by the semi-strong and E&M interactions

Important remark: It is known that only the baryons and the mesons (hadrons family) have non-zero spin; photons, leptons are all assigned zero isospin. Also, as it is true for all quantum concepts, the isospin quantum number is also associated with a law of conservation. This law is of special character; it is stated as follows:

(i) For any strong interaction, the total isospin magnitude is conserved; the final magnitude is as the one before the interaction.

(ii) For any strong interaction or electromagnetic interaction, the total of T_ξ after the reaction is as it was before the reaction.

So it can be concluded that T and T_ξ are conserved under strong interactions. In addition to that, the T_ξ is also conserved under the electromagnetic interactions.

Some Examples Clarifying the Isospin Conservation Principle

a. Strong interaction (collision)

$$p^+ + p^+ \rightarrow \Lambda^0 + k^0 + p^+ + \pi^+$$

$$T = 1/2 + 1/2 = 1, 0 + (-1/2) + 1/2 + 1 = 1$$

$$T_\xi = 1/2 + 1/2 = 1 , 0 + (-1/2) + 1/2 + 1 = 1$$

T, T_ξ, conserved

Before   After

b. π^0-decay under electromagnetic interaction

$$\pi^0 \rightarrow \gamma + \gamma$$

$T = 1 \neq 0 + 0 = 0$, T is not conserved

$T_\xi = 0,\quad 0 + 0 = 0,\ T_\xi$ is conserved

c. Λ^0-decay by weak interaction

$\Lambda^0 \rightarrow \quad p + \pi^-$

$T = 0 \quad \neq \quad 1/2 + 1 = \quad 3/2$ $\Big\rvert$ Both T and T_ξ are not conserved

$T_\xi = 0 \quad \neq \quad 1/2 + (-1) = -1/2$

Now T is not conserved in lambda decay process, so according to (i), such decay, first, is not occurring via fast strong interaction; second, since its component is not conserved too, therefore, this decay is not happening via strong or electromagnetic interactions, which suggest that the lambda decay might take place via much more slow weak interaction. It can be found as good scientific relations between the hypercharge Y, the strangeness S, and the charge of the particle q and the component of the isotopic spin T_ξ as follows:

$Y = S + B$, $S = Y - B$, $q = T_\xi + 1/2(S + B)$, $\Rightarrow Y = 2(q - T_\xi)$,

$S = 2(q - T_\xi) - B$

As it was shown before, the total charge q is conserved under any interaction, and T_ξ is also conserved under the strong and the electromagnetic interactions. Consequently, the hypercharge is conserved under these interactions too. Similarly, since q, B, and T_ξ are conserved, then S is conserved. Table 1 lists the values of q, Y, T_ξ, and S for the nine baryons. Of course, there are the antibaryons, which have the same properties of the baryons but with opposite signs.

Table (1), Values of q, Y, $T\xi$ and S for hadrons, lifetime $\tau \approx 10^{-23}$

Family group	Particles	q	Y	T	S
pions(π)	$\pi^+ \pi^0 \pi^+$	1.0,-1	0	+1,0,-1	0
Mesons B=0,	Kaon (k^-) k	+1	1	1/2	1
	k^0	0	1	-1/2	1
	k^{0-}	0	1	-1/2	1
	k^{--}	-1	-1	+1/2	-1
Eta	η^0	0	0	0	0
Nucleons	p^+	1	1	1/2	0
	n^0	0	1	-1/2	0
Lambda	Λ^0	0	0	0	-1
Baryons B=1, Sigma	Σ^+	+1	0	+1	-1
	Σ^0	0	0	0	-1
	Σ^-	-1	0	-1	-1
Xi (Ξ)	Ξ^0	0	-1	1/2	-2
	Ξ^-	-1	-1	-1/2	-2
Omega	Ω^-	-1	-2	0	-3

Some Application Examples on Y and S Properties in Particles Decays

So far, it has been demonstrated that Y and S quantum numbers are conserved in both the strong and the electromagnetic interactions. Hence, it is quite useful to give certain examples to clarify this fact to the readers.

Example 1: Take the following interaction:

p + p (proton-proton collision) $\rightarrow \Lambda^0 + k^0 + p^+ + \pi^+$

Go back to the table. Y, in this interaction, is conserved, which is of usefulness to explain that some decays of excited states to particles have to proceed by the slow weak interactions.

Example 2: Sigma particles decays—they can decay by two modes of decays:

$1- \Sigma^+ \rightarrow P^+ + \pi^0$, or

$2- \Sigma^+ \rightarrow n^0 + \pi+$ $\tau = 0.8 \times 10^{-10}$ sec

Y = 0 Y = 1 + 0 = 1   Y is not conserved

Notice that the lifetime is within the scale of the decaying via weak interaction, as pointed out earlier. This indicates the decay of sigma is not allowed via the strong interaction, which also means the hypercharge is not necessary to be conserved under the weak interaction.

Example 3: conservation of strangeness

As stated previously, the strangeness is imposed as a characteristic of certain strange particles, which are created through strong interactions but decay via weak interactions. Therefore, it is represented by a certain quantum number of strangeness termed as S; it is shown in the table above, for number of particles. As shown before, S = Y−B, so for the nucleons (n) = 1−1 = 0, for pions, S (π) = 0 − 0 = 0

Now, take the reaction in example 1:

$P^+ + P^+ \rightarrow \Lambda^0 + k^0 + P^+ + \pi^+$

S = 0 + 0 = 0   ,   −1 + 1 + 0 + 0 = 0, →S is conserved

It is important to notice the fact that the strange particles are not produced singly; they are produced in association with others, via strong interaction, so this is an important requirement to keep the strangeness quantum number conserved. All this information about these particles are experimentally confirmed. This information is *a* clue of studying their properties via their tracking through bubble and cloud chambers. The assumption that T, A, q, and S should be conserved in all strong interaction is true and important. This leads to some important consequences, which are proved in the laboratory, that can be summarized as follows

Strange particles can only be produced in association with other particles, as stated before. In the collision type of intraction, where incoming particles are of S = 0, it is expected to find strange particles within the product only, such as

a) $\Lambda^0 + K^{0,+}$

b) $\Sigma^{+,0,-} + K^{0,+}$

c) $k^{-,0} + k^{0,+}$ (a)

d) $\Xi^- + 2\,k^{0,+}$

These predictions with the experiment are excellent.

Other associated productions that have been observed in the cloud and the bubble chambers are $\Lambda^0 + \Theta^0$, $k^- + k^{+,0}$, $\Sigma^{\pm} + \Theta^0$, $k^0 + k^0$,and $\Xi^- + 2\,k^0$. No cases such as $\Lambda^0 + \Lambda^0$, $\Lambda^0 + k^-$, $\Sigma^{\pm} + k^-$ are found.

In the case of colliding strange particles with ordinary particles, the only interaction that will occur is the one that conserves the strangeness quantum number. This includes the simple scattering interaction, where there is a certain scattering exchange charge between a colliding particle, or the incident particle, and the target, such as $k^0 \leftrightarrow k^+$, $\Sigma^+ \leftrightarrow \Sigma^-$ and other interactions as follows:

1) $K^- + P \rightarrow \Sigma^- + \pi^+$

2) $\Sigma^+ + n \rightarrow \Lambda^0 +$ P (S is conserved)   (b)

3) $\Xi^- + P \rightarrow \Lambda^0 + \Lambda^0$

It can be noticed that the strangeness theory predicts unsymmetrical properties for k+ and k− interacting with ordinary matter. For example, kaon k+ can interact only through simple scattering or charge-exchange scattering, while kaon k− can interact via both of these interactions, in addition to the reaction via which lambda and sigma group are produced. This phenomenon creates a symmetrical picture that is born by observations. It constitutes strong evidence that the theory of the strangeness has to be corrected.

Also, it is observed that if Σ^- is absorbed at rest by a nucleus, the only reactions that conserve S will rapidly take place. These include the reactions (a, b) and the following:

1. $\Sigma^- + p \rightarrow \Lambda^0 + n$

2. $\Xi^- + p \rightarrow \Lambda^0 + \Lambda^0$ S—conserved --------------(c)

As clarified before, free decay of strange particles is restricted because, first, their creation is associated with other particles production, as stated previously; second, the strangeness quantum number S is conserved. Hence, it is possible to introduce here that the assumption $\Delta S = 0$ holds rigorously only for the strong and the electromagnetic interactions, but this is not always true, because weak interactions also exist in nature too, which allow transition to occur slowly with $\Delta S = \pm 1$. In such a way, the decays of all strange particles are placed on the same basis. Then this will account in a most satisfactory way for the approximate equality of most known strange particles lifetimes. The most astonishing example of the operation of this selection rules ($\Delta S = \pm 1$) is noticed in the hyperon Ξ^-. It decays into two steps: Λ^0 then to the proton, also under 2.85 GeV energy of p-p collision, which is experimentally observed in the following, as shown in the diagram below:

$P + P \rightarrow K^0 + \Lambda^0$ (1) on the diagram, then,

$k^0 \rightarrow \pi^+ + \pi^- (2), \Delta S = -1$

$\Lambda^0 \rightarrow P^+ + \pi^- (3), \Delta S = +1$

The conservation law of S here is obviously violated.

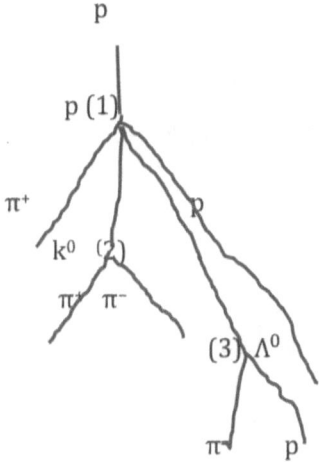

Figure 4: p-p collision products

The observed lifetimes of these two decays are about 10^{-19} sec.

This indicates that they occur via weak interaction. Since such decays are practically observed in the bubble chamber, as illustrated in the above schematic diagram for proton-proton collision, it indicates that the strangeness quantum number, S, is not conserved under weak interaction. From the inspection of the unstable hadrons' lifetimes, it shows that all the decays happen by the weak interactions, where S is not conserved, except π^0, Σ^0, and η^0, decay is not violating S; hence, they decay via the rapidly electromagnetic interaction. But even though there is a selection rule for this decay under the effect of weak interaction, it states (for weak interactions) that the strangeness quantum number S should change by unity or zero, i.e., $\Delta S = \pm 1$ or 0.

As an example, take the decay of the neutron (beta decay):

$n^0 \rightarrow p^+ + e^- + \bar{\nu}_e$

$S = 0 \quad 0 \quad 0 \quad 0 \Rightarrow \Delta S = 0$

It is important to notice that some weak interactions do not involve leptons. This is quite obvious in k^0 and Λ^0 decays, as shown before, where $\Delta S = \pm 1$.

VIII-4—Parity and Particles

Parity is a very important physical concept related to the symmetry of a physical system in its microstate. It plays an important role in quantum mechanics and quantum field theory. It is treated fairly well in chapter 6 of this book. It has to do mathematically with the inversion, reflection, and reversing of the coordinates of the system through the origin of this coordinates, such as $\vec{r} = -\vec{r}$, or $x \to -x$, $y \to -y$ and $z \to -z$. It is from right to left through the origin, changing a right-handed coordinate system into a left-handed coordinate system. This is a general or classical description to the concept of parity. In quantum mechanics and quantum field theory, as stated above, it is an important operator playing an intrinsic role in building the state of the atomic, nuclear, and subnuclear systems, such as leptons, hadrons, hyperons, and quarks. So this physical dynamic operator plays a main role in studying the structures of these very mini microscopic particles. As it is well established, such microscopic systems are described by a wave function or, more mathematically in formalism, eigenfunction, denoted by $\psi(r)$. This function contains the whole information about the state of the system. It is a statistical type of function; therefore, $\psi(r)^2$ represents the probability density of the particle described by this eigenfunction, which is the same for $\psi(r)$ or $\psi(-r)$, so there are just two possibilities for the eigenfunction: either $\psi(r) = \psi(-r)$, and its parity is even with value +1, or $\psi(r) = -\psi(-r)$, and its parity is odd with value (-1). As it was shown in chapter 6, these values ± 1 are the eigenvalues of the operator (π_0) or P^; hence, P^$\psi = +1\psi$, or P^ $\psi = -1\psi$.

If the parity is conserved, there is no preference for the right or the left, which means that if a certain process takes place in nature, it will be the same if it is looked through a mirror, and the probability density will be on equal scale. Before 1956, this was a universal law, but then, it was found that it is not in the weak interactions like beta decay (see chapter 7). It was found that particles with 1000 m_e have two different modes of decay. They are τ *and* θ particles (mesons); their decays are

$$\tau^0 \rightarrow \pi^+ + \pi^0 + \pi^-$$

$$\theta^0 \rightarrow \pi^+ + \pi^-$$

The pions spin is (0), and its parity is odd (-1), i.e., $0^{-1} \equiv I^P$. The angular distribution of the τ- decay products indicates that the orbital angular momentum is $\ell = 0$. It leads to $I^P = 0^{-1}$, for τ-meson; for θ -meson, $I^P = 0^{+1}$. Therefore, (τ) has odd parity, and (θ) has even parity. As pointed out in chapter 6 (beta decay), the experimental results showed that the mass of these two thought mesons was the same $(1000\ m_e)$ and the same lifetime. This experimental result indicates that these particles are identical but differ in their decay mode, in which the parity is not conserved. This has led T. D. Lee and Yang to prove it theoretically and suggest that the parity might also not be conserved in beta decay too; hence, $\tau = \theta$ = K-meson, having $I^P = 0^-$, just like the π-meson (pion). In fact, the charged K-meson decays in more than one mode as follows:

$K^\pm \rightarrow \pi^0 + \mu^\pm + \nu + 233$MeV (3.4%), $K^\pm \rightarrow \pi + e + \nu + 358$ MeV (4.8%).

$K^\pm \rightarrow \mu^\pm + \nu + 388$ MeV (63%), or $K^\pm \rightarrow \pi^\pm + \pi^0 + 219$ MeV (21.5%), this is (θ) mode of decay.

$K^\pm \rightarrow \pi^\pm + \pi^+ + \pi^- + 75$ MeV (5.5%), $\quad$ τ- mode of decay

$K^\pm \rightarrow \pi^\pm + \pi^0 + \pi^0 + 84$ MeV (1.7%)

As it is known, K^+ and K^- are antiparticles. K^+ and K^0 constitute a doublet of 1/2 isospin, just like the neutron and the proton. Also, K^{0^-} is the antiparticle of K^0, similar to the neutron. K^- and K^{0^-} are antikaons. Their strangeness has to be S = −1, opposite to that of the kaon if it is assumed baryon conservation in associated production. Also, they are neutrals; they differ only by S, because they have no spin and no charge. K^0 has two components: k_1^0 and k_2^0. They differ in the mode of the decay as such:

$k_1^0 \rightarrow \pi + \pi = 2\pi$, lifetime $\tau = 0.9 \times 10^{-10}$ sec

$k_2^0 \rightarrow 3\pi$, lifetime $\tau = 5.8 \times 10^{-8}$ sec

These experimental results with the experiment done by Wu et al., Garwin et al., in 1957, proved that the parity is not conserved in the weak interactions such as the beta decay confirming the suggestion of T. D. Lee in 1956. But it is conserved in the strong interactions and the electromagnetic interaction. Take the decay of the neutral pion π^0: $\pi^0 \rightarrow \gamma + \gamma$, which occurs via the electromagnetic interaction. Let the neutral pion be at rest, then the produced gammas must leave in the opposite direction with equal momentum P to keep the total linear momentum conserved. Now the spin of this pion is zero, but each gamma (photon) has a spin of unit of $I_\gamma = 1$. To conserve the angular momentum in this decay, the senses of the spin J of the photon should be opposite. For such physical phenomenon, there are two possible orientations of J and P: first, J is aligned with P (momentum) for each photon (right-handed); second, J is not aligned with P for each photon (left-handed). These two possibilities are shown in the diagrams a and b, respectively, below:

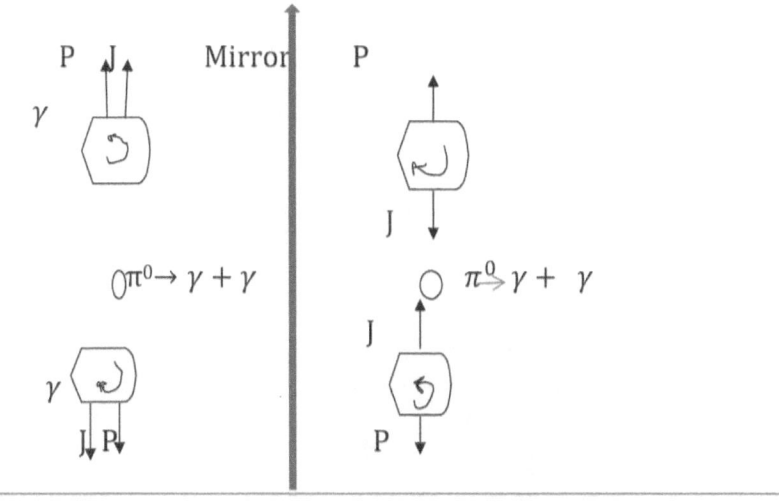

(a) Right-handed particle   (b) Left-handed particle

Figure 5: two a priori possibilities of π^0 decay into two gammas

Diagram b is the mirror image of a, relative to the vertical line dividing the two decays.

From this diagram, it is noticed that a, which is right-handed decay to the right-handed photons, must equal the number of left-handed photons. This conclusion is experimentally approved. Take the decay of $\pi+$, $\pi^+ \rightarrow \mu^{+^-} + v_\mu$. Again, the linear momentum conservation demands that $P_{\mu^-} = -P_{\nu_\mu}$ and the angular momentum $J_{\mu^-} = -J_{\nu_\mu}$. This is shown in figure 6 below. It is analogous to the neutral pion decay (page 298). The decay of π^+ into left-handed particles (b) is the mirror image of its decay into right-handed particle (a).

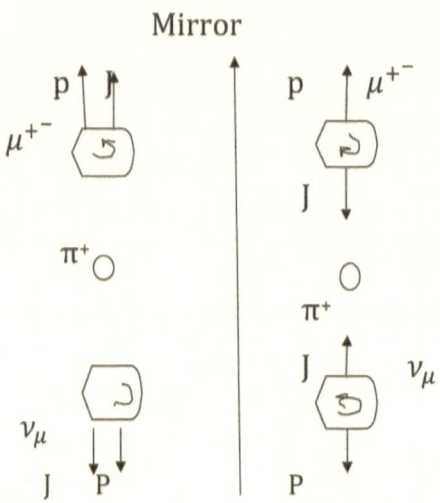

(a) Right-handed   (b) Left-handed image of a

Particle   particle

Figure 6: two a priori possibilities for the decay of π^+

(b) is mirror image to (a).

Experimentally, it is shown that if π^+ is observed in the initial system and it is at rest, it decays to μ^{+^-} in a left-handed state and left-handed neutrinos V_μ, only. For the case of π^-, it is found that it decays into the same muon and the neutrino but in the right-handed picture. On this basis, the neutrinos and the antineutrinos have different intrinsic handedness (refer to chapter 6). The neutrinos are exclusively left-handed; the antineutrinos are exclusively right-handed. The helicity for the antineutrino is +1, and it is −1 for the neutrino; therefore, they are distinguishable particles. It is important to notice that although only left-handed, (μ^{+^-}) are emitted in the decay of (π⁺), when it is at rest in reference system. So it is claimed that any particle of finite rest mass, such as the muon particle, would not have just one intrinsic handedness for all inertial observers. To clarify this situation, suppose that an anti-muon traveling upward, with its spin pointing upward too, constituting a right-handed particle, is viewed by an observer at rest relative to the meson, as shown in a. If the particle is viewed from a reference frame traveling upward at a higher speed than the speed of (μ^{+^-}), then it looks like it's moving downward In this case, the momentum p of the particle is reversed, but its angular momentum J is not. So the particle is seen by the observer as a left-handed particle.

Since μ^{+^-}, with measured rest mass, may be of both right-handedness and left-handedness. The left-handed only appears in the decay of π⁺ at rest, which implies that the neutrino must always be left-handed. This is in accord with the constant speed (c) of the neutrinos of about zero mass ($m_v \cong (5 \times 10^{-4} m_e) \cong 2.5 \times 10^{-4}$ MeV. Recent studies claim that the neutrino is oscillating, which means it has mass. This also indicates no frame of reference for which the velocity vector of the neutrinos can be reversed to. The nonconservation of parity initiates the theoretical and the experimental physicists to reinvestigate the other two conservation laws, which somehow are related to the parity concept conservation, in weak interaction like beta decay. These two laws are charge conjugation, which claims that if a particle is replaced by its antiparticle, the processes of the interaction are unchanged, so charge conjugation is conserved. The second law is the time reversal, which states that if the processes of the reaction are going forward or backward, these processes are the same; it is time invariance (T).

This law might be stated as follows: If a motion picture portrays a possible process, then the motion picture seen is unfolded as the film is running backward, portraying a possible process. Hence, there are three separate important physical dynamic operators: P, C, and T. They are separately conserved, in the strong and the electromagnetic interactions, as experiments confirmed that are conserved in weak interactions, except for infrequent violation in the decay of the neutral mesons (kaon). Also, the parity (P) and charge conjugation (C) are separately not conserved in weak interactions, but their combined operator (CP), i.e., their combined operation in the weak interactions, is conserved. Later in 1977, a theorem was enunciated about (CPT) that requires an invariance under the combined operation of these three operators. This indicates that if this situation is physically real, then the combined operations of C, P, and T must lead to another realized physical situation in nature. Consequently, the CPT theorem is the following: There has to be certain relations by the physical parameters of the particles and their antiparticles. As an example for clarification, take the neutron and the proton, where they represent two different charge states to a single particle called nucleon. Therefore, the Pauli exclusion principle allows two nucleons to occupy the same state with the same usual quantum numbers, provided that one is a neutron and the other is a proton. This is analogous to the electron states that one is spinning up (1/2) and the other is spinning down (−1/2), but their spins are the same magnitude. Therefore, this idea is taken, with some modifications, over to be used in particles physics.

To avoid any possibility of confusion, the isospin term is used, which is an intrinsic property related to the nuclear particles of different families. As mentioned earlier, it is denoted by T, and its component along $\xi-axis$ is given by T_ξ, in isospin space or charge space, where

$T_\xi = q - Y/2 = q - (B + S)/2$ (refer to the previous sections concerning Y, B, and S). Hence, the conservation laws valid for all particles interactions, as stated before, are (1) mass-energy, (2) linear momentum, (3) angular momentum, (4) electric charge, (5) baryon number, (6) lepton number for lepton particles, (7) CPT. These conservation laws are not quite universal for all interactions, as shown in table 2 below:

Table 2: Particles Conservation Laws

Quantity	Strong int.	E&M int.	Weak int.	Remarks
Isospin	Yes	No	No	None
Isospin component,T, Yes	No	No	None	
Equivalently, B, S	Yes	No	No	None
Parity , P ,	Yes	Yes	No	None
Charge conjugation C, Yes	Yes	Yes	None	
Time reversal, T,	Yes	Yes	Yes	↓
(except infrequent neutral kaon decay)				
CP	Yes	Yes	Yes	↓
(except infrequent neutral kaon decay)				

VIII-5—Observed Nuclear Particles, Entity, and Properties

VIII-5-1—Introduction

So far, the theoretical background of the particles is explored; the conservation laws govern their interaction, the selection rules that control their decays, and their interaction behaviors under different strengths of different forces in nature. Also, their quantum numbers, which play the main rules in their classifications into certain families of equivalent physical properties, were discussed. In the beginning, in the thirties, it was thought that these particles are the electron, which was discovered in 1897, the proton, discovered in 1916, and the neutron, discovered in 1932, where the proton and the neutron constitute the nucleus of the atom. These three particles (e, p, n) each has a rest mass, as shown in chapter 1, with energy $E = mc^2$, where c is the velocity of light. In 1905, the light was interpreted as a stream of photons by Einstein. The photon is quanta of energy equals to $(h\nu)$; (h) is the well-known Planck constant, and (ν) is the frequency of the photon wave. The photon represents its own family with zero rest mass; it is emitted as absorbed radiations outside or inside the nucleus. The photon, denoted by (γ), was discovered by Maxwell through his famous equations, which unified the electric field and the magnetic field into the electromagnetic field, in 1865, as an electromagnetic wave moving with a constant velocity, which is the light velocity, constituting an important postulation in the theory of special relativity by Einstein (1905). In virtue of its motion, it has an energy given by, as mentioned above, $h\nu$, which leads to mass in motion equals to $E/c^2 = h\nu/c^2 = h\nu/cc = h/c$. $\nu/c = h/c. 1/\lambda = h/\lambda .1/c = P$ (momentum)$/c = p/c = mv/c$; if v = c, then mv = mc = p $\rightarrow$ mc/c = m gm.

These particles, as it will appear later on, form different families of hundreds of numbers, started by light ones and going up to heavier ones. The mass depends actually on the energies used in the collision between particles and the type of the accelerators or the colliders used in the experiments designed for studying the particles or other purposes. Today, the greatest collider in the world is the Large Hadron

Collider (LHC) at the boundary between Geneva, Switzerland, and France. It is the European Institute for Nuclear Research. The energy used is 14 terawatt, to study many physical phenomena, such as the universe after 10^{-6} seconds from the big bang event, where the plasma of the quarks and the gluons are the beginning of the creation of the universe matter, and also to study the many proposed theories about the number of the dimensions according to string theory, the supersymmetry particles, the source of the particles masses, and the Higgs boson or bosons!

If the particles are classified according to their masses from the lower mass to the higher mass in terms of the electron mass (0.511 MeV), it starts with the light muon (207 m_e), followed by kaon (K), lambda (Λ), sigma (Σ), Xi (Ξ), and omega (Ω). Each species exhibits a number of neutral and antiparticles. These particles, as mentioned earlier, are produced, and their physical properties are very well studied in the laboratories with the aid of the high-energy accelerators (200–2000 GeV), through the eighties and the nineties of the twentieth century.

As it is fairly well-known, these particles open the door of understanding of the universe and the matter constituting it. That is because, except for very rare occasions, the observed matter is made of these particles and described in terms of them. Going back to the nucleus, the atom, the molecule, the compound and finally, the matter, where the nucleons (proton and neutron) are strongly bounded by the strong nuclear forces via the virtue pions ($\pi^+ \pi^0 \pi^-$) as messengers to this short range force to form the nucleus of the atom, where the electrons, which are too much lighter than the proton ($m_p = 1845\ m_e$), are tied to the nucleus by the virtual photons (γ), moving around under the effect of the coulomb force, between the nucleus with positive charge and the electrons with negative charge. These virtual photons are considered as messenger particles carrying the effects of the electromagnetic fields between interacting charged particles. The long-lived naturally radioactive nuclides remaining in the earth usually decay into stable and unstable daughters, resulting in an emission of radiations of high energy (~MeV) such as $\alpha, \beta, and\ \gamma$. The abundance of real photons in the earth's environment is too large. They are coming from the stars' activities and the unlimited inelastic scattering (collisions) between the earth atoms and molecules. The existence of the positron (e^+) is quite limited, due to the fact that its creation requires relatively high energy (~1.02 MeV), because it is created in association with the electron through the pair production process as a result of light incident on matter with energy not less than 1.02 MeV. Also, the real pions are detectable in fewer quantity than the electrons and positrons. That is due to their large rest masses, which are about 140 MeV. Therefore, for the particles of limited kinetic energy, 1000 MeV or less, the detectable particles on the earth are only fifteen. They are two nucleons, three pions, muon-anti-muon pair, electron-positron, two neutrino pairs (electron neutrino-antineutrino and muon neutrino-antineutrino), photon, and graviton. The stable particles of these in free space include the proton, photon, electron, positron, the electron and the muon neutrinos, their antiparticles, and the graviton. The unstable particles in free space are the neutron, the three pions, and the muon and its anti muon. From the previous sections, it was shown that the neutral pion decays by the electromagnetic interaction, but the neutron, the charged pions, the muon, and the anti-muon decays by the weak interaction. Later on, due to the use of advanced accelerators and colliders with higher energies, as mentioned earlier, tens of new particles were discovered, with such huge numbers of particles. How are the elementary or the fundamental particles defined? The fundamental particle is defined as a particle with no dimension and is structureless. So according to this definition, the only particles that might be considered fundamental or elementary now (2013) are the following particles: e^+, e^-, V_e, V_e^- , V_μ, V_μ^- , and the quarks. The following table ( ) lists the already established particles directly or indirectly, which actually exist for a time greater than 10^{-21} seconds. The table excludes those with existing time shorter than this time, probably within 10^{-23} sec. These particles are called resonance particles, which decay by strong interactions, as will be discussed later. The table (3) lists these particles according to their rest masses (or energy = mc^2) increasingly. Also, they are classified into groups according to family of common properties as follows:

1.   The photon of spin 1 represents the category by itself.

2.   The leptons (meaning "light" in Greek) is family with spin 1/2, so they are statistically fermions; they are e⁻, e^+, V_e, V_e^- , μ^+, μ^-, V_μ , V_μ^- .

3.   The mesons (meaning "intermediate" in Greek), a family with spin (I = 0), include the pions (π^+, π^0, π^-), the kaons (k^+, k^0, k^-, k_1^0 , k_2^0), eta (η), and the charmed D-particles. They are statistically bosons.

4.   The baryons (meaning "heavy" in Greek) are a family with spin of half-integer (1/2, 3/2, 5/2, etc.). They are statistically fermions. This family includes the nucleons (p, n) and the more massive particles called the hyperons (meaning "overweight" relative to other particles). They are lambda (Λ), sigma (Σ), Xi (Ξ), and omega (Ω). Since baryons and mesons are created by a strong force and also interact by strong force, they are called hadrons (meaning "strong" in Greek). These hadrons have integral charge

in unit of proton charge: +1, 0, −1. Also, to each particle, there is a corresponding antiparticle marked with a bar over it. These antiparticles have the same properties of its counter particles, except a reversal sign in these properties, as it is well-known.

The neutral particles, such as n, $\gamma, k^0, \pi^0, and \eta^0, are$ their own antiparticles themselves. Also, μ^+ and μ^- are each an antiparticle to the other. Each particle of these particles has a distinctive mean lifetime, which is the average lifetime from its creation to its decay into many identical particles. From the following table it can be noticed that only nine of the listed particles and the antiparticles are stable against spontaneous decay. They are (1) the photon, (2) the electron, (3) the two neutrinos, (4) their associated antiparticles, (5) the photon.

Recently, some attempts to measure the average lifetimes of the proton and the electron had been tried to cope with some theoretical predictions or claims. The predicted average lifetime is 10^{31}–10^{33} years for the proton, and it is greater than 10^{21} years for the electron.

This raises a question. If these mean lifetimes are possible and the estimated lifetime of the universe is about fourteen billion years = 14 $\times 10^9$, which is too small compared to the mean lifetimes of both the proton and the electron, then the proton and the electron are living for more than three times of the age of the present universe, which is predicted to continue for another fourteen billion years! This raises a philosophical problem that should be thought about, but not here. Also, each particle has an intrinsic spin that classically means the particle is spinning about its internal axis, but this spin is quantum mechanical concept. Its value is given by $[J(J + 1)]^{1/2}\hbar$; J is the spin quantum number, and $\hbar$ *is Plank constant* $/2\pi$. The value of J is either integer or half-integer. Those particles with integer spin are bosons, as mentioned before, and those with half-integer are fermions that obey the Pauli exclusion principle. So no two identical fermion particles can occupy the same quantum state. The Pauli exclusion principle plays a repulsive role between the two particles. This action is applied on all particles with spin of half-integers like the leptons and the baryons. Those so-called field particles, such as the photon and the mesons, which have integer spins that are called bosons, do not obey the Pauli exclusion principle. Therefore, they can occupy the same quantum state regardless of their number. As stated before, each particle has a hypercharge (Y) and isospin. Its component along a certain axis is related to its (Y), as pointed out previously. In table 3, the mode of the decay to each unstable particle is given, but the decay modes of particles of a fraction less than 5%, although observed, are not listed in the table. It is important also to note that there is a complementary decay, in which it is possible to replace the particle by its antiparticle, such as

$n^0 \rightarrow p^+ + e^- + \bar{V_e}$, which can be replaced by

$^-n^0 \rightarrow p^- + e^+ + V_e$ with the same lifetime of about 10^3 sec

The lifetimes in this table are 10^{-23} sec for strong interactions, 10^{-19} -10^{-15} sec for fast electromagnetic interactions, and 10^{-10} -10^3 sec for slow weak interactions. The resonance particles are of very short average lifetimes ranging from 10^{-24} to 10^{-21} seconds, which will be dealt with in the following sections.

ALI A. ABDULLA

Table 3: Particles with Average Lifetime τ>10⁻²¹ Seconds

Table 3: Particles with Average Lifetime $\tau > 10^{-21}$ Seconds

<u>Family Name Symbol Anti.P. Rest E MeV τsec. S Y T Principal</u>
<u>decay mode</u>

Non photon γ same   0 ∞ 1   0   0,   stable

Hadron

  Lepton, neutrino $\nu_e \nu_\mu$ $\bar{\nu_e}, \bar{\nu_\mu}$ 0 ∞ 1/2   0 0   stable

L_e no.=+1for e⁻,e⁺, e⁻ e⁺   0.511 ∞ 1/2   0 0   stable

L_μ no.=+1, muon μ^- μ^+ 105.6569 2.197x10⁻⁶ 1/2 0 0 ,$\mu^- \to e^-$
$+\bar{\nu_e}+\nu_\mu$

For μ^+, μ^-

HADRONS

Mesons   Pions $\pi^+ \pi$ $\bar{\pi^-}\pi^+$ 139.59 , 2.602x10⁻⁸ , 0   0   1
,$\pi^+ \to \mu^+ + \nu_\mu$ (100%)

                π^0 same   134.964 , 0.83x10⁻¹⁶ ·0   0   1,
$\pi^0 \to \gamma + \gamma$(98.9%)

            Kaons   K⁺ K^{--} 493.70 ,1.237X10⁻⁸ , 0 +1 1/2 ,

                        $K^+ \to \mu^+ + \nu_\mu$(63.9%)

                        $\to \pi^+ + \pi^0$ (21%)

                        $\to \pi^+ + \pi^0 + \pi^-$(5.6%)

            $K^0 K^{0-}$ 497.7

            $K^0 = 1/2\ [K_S^0 + K_L^0]$

K_S^0 $\qquad$ 0.893×10^{-10} $\qquad$ $K_S^0 \to \pi^+ + \pi^-$ (68.7%)

$\qquad\qquad\qquad\qquad\qquad\qquad\qquad$ $\to \pi^0 + \pi^0$ (31.3%)

K_L^0 $\qquad$ 5.18×10^{-8} $\qquad$ $K_L^0 \to \pi^0 + \pi^0 + \pi^0$ (21.%)

$\qquad\qquad\qquad\qquad\qquad\qquad\qquad$ $\to \pi^+ + \pi^- + \pi^0$ (12.2%)

$\qquad\qquad\qquad\qquad\qquad\qquad\qquad$ $\to \mu^\pm + \pi^\pm + \nu_\mu$ (27.1%)

$\qquad\qquad\qquad\qquad\qquad\qquad\qquad$ $\to \pi^\pm + e^\pm + \nu_e$ (39.1%)

Eta $\quad$ η^0 same 548.8 $\quad$ 7×10^{-19} 0 0 0 $\quad$ $\eta^0 \to \gamma + \gamma$ (38%)

Charmed D $\quad$ D^0 $\quad$ D^{0-} 1863 $\quad$ $\sim 10^{-13}$ 0 0 1/2 $\quad$ $D^0 \to K^+ + \pi^-$

D^+ $\quad$ D^- 1868 $\quad$ $\sim 10^{-13}$ 0 0 1/2 $\quad$ $D^+ \to K^- + \pi^+ + \pi^+$

BARYONS

Proton $\quad$ P^+ $\quad$ P^- $\quad$ 938.280 $\quad$ ∞ $\quad$ 1/2 $\quad$ 1 $\quad$ 1/2 $\qquad$ Stable

Neutron $\quad$ n^0 $\quad$ n^- $\quad$ 939.57 $\quad$ 0.9×10^3 $\quad$ 1/2 $\quad$ 1 $\quad$ 1/2

$\qquad\qquad\qquad\qquad\qquad\qquad\qquad$ $n \to p + e + \nu_e$ (100 %)

Lambda $\quad$ Λ^0 Λ^{0-} 1115..6 $\quad$ 2.5×10^{-10} $\quad$ 1/2 $\quad$ 0 $\quad$ 0 ,

$\qquad\qquad\qquad\qquad\qquad\qquad\qquad$ $\Lambda^0 \to p + \pi^-$ (64%)

$\qquad\qquad\qquad\qquad\qquad\qquad\qquad$ $\to n + \pi^0$ (36%)

Baryon no.=1

Sigma $\quad$ Σ^+ $\quad$ Σ^- $\quad$ 1189.4 $\quad$ 0.8×10^{-10} 1/2 $\quad$ 0 $\quad$ 1 ,

$\qquad\qquad\qquad\qquad\qquad\qquad\qquad$ $\Sigma^+ \to p + \pi^0$ (53%)

$$\rightarrow\ n + \pi^{+}\ (47\%)$$

	Σ^{0}	Σ^{0-}	1192.5	$<10^{-14}$	1/2	0	1. $\Sigma^{0} \rightarrow \Lambda^{0}+\gamma$
	Σ^{-}	Σ^{+}	1197.4	1.49×10^{-10}	1/2	0	1, $\Sigma^{-} \rightarrow n + \pi^{-}$
Xi	Ξ^{0}	Ξ^{0-}	1314.9	3×10^{-10}	1/2	-1	1/2, $\Xi^{0} \rightarrow \Lambda^{0} + \pi^{0}$
	Ξ^{-}	Ξ^{+}	1321.3	1.65×10^{-10}	1/2	-1	1/2, $\Xi^{-} \rightarrow \Lambda^{0} +$
Omega	Ω^{-}	Ω^{+}	1672.2	1.3×10^{-10}	3/2	-2	0 , $\Omega^{-} \rightarrow \Xi^{0} + \pi^{-}$?

$$\rightarrow \Xi^{-} + \pi^{0}\ ?$$

$$\rightarrow \Lambda^{0} + k^{-}$$

Note: Σ^{+} decays twice faster than the decay of Σ^{-}. It has two dominant decay modes; Σ^{-} has only one mode of decay.

VIII-5-2—Some Historical Remarks on These Particles

It might be of interest to review the historical background of these important particles, which play the main role in building the matter origin and virtually in the being of the universe and our being. So they will be looked over briefly in the following sections.

1—The Electron and the Positron

As it is well-known, the first particle discovered by J. J. Thomson in 1897 was the electron. It has played the direct role almost in all physical phenomena, except those involving the nuclei and the gravitation, principally because its (e/m) is relatively large, where (e) is its charge in coulomb unit, and (m) is its mass in gm. The electron is characterized by the following data:

Mass, Me = 9.108×10^{-31} kgm = 0.5487×10^{-3} amu = 0.511 MeV/C^2

Charge, q_e = -1.6020×10^{-19}C, = -4.802×10^{-10} e.s.u = -1 electron

Spin= 1/2 in $\hbar$ unit.

Magnetic moment (μ_S) = -0.92837×10^{-23} Am.

Mean lifetime—stable—proposed >10^{21} sec.

2—The Positron

It is an abbreviation to "positive electron" ("posit-ron"), discovered in the cosmic rays by Anderson in 1932, which has proved the Dirac theory of electron, postulating the existence of a positive particle as antielectron. It differs only by the charge sign; it is +1, and it is stable too. But in case it meets the electron, both will be annihilated into gamma radiations. It is possible that some times before the annihilation, the electron and the positron form a positronium. It is like an atom; its ground state configuration could be either $^1S_0$ or $^3S_0$, and the annihilation processes are indifferent for these two possible states, due to an operation of certain selection rules. The singlet state decays with lifetime about 8 $\times 10^{-9}$ seconds into two gammas with equal energy leaving in opposite direction. The triplet state decays within lifetime of 7×10^{-6} seconds into three gammas with total energy of about $2(m_e c^2)$ = 1.022 MeV. Therefore, the creation of e⁻, e⁺ pair requires at least 1.022 MeV. The electron-positron annihilation in matter is of very important use in studying and investigating the structure and the properties of different materials.

3—Photon

As it is known, the light is energy in very fast and unreached motion by other particles in vacuum. Its speed is measured to be about 2.997 × 10^8m/section. In 1865, Maxwell had shown it as an electromagnetic wave. In 1905, Einstein proposed its speed as a universal constant, as one of his special theory of relativity assumptions. During his trying to interpret the photoelectric effect, he proposed that the light is composed of quanta. Each one has an energy equal to $h\nu$, following Planck work on blackbody radiations in 1900, so he called this quanta a photon. The photon, in fact, is the electromagnetic field messenger particle carrying the mutual interacting effects between the charged particles. The following are the main properties of the photon:

Rest mass = 0

Magnetic moment = 0

Charge = 0

Spin = 1 in unit of $\hbar$

Mean lifetime = ∞

Due to the fact that the light is of electromagnetic nature, its interaction with other particles is via the charge, electric moment, and magnetic moment of the particles.

4—Proton

This particle is quite important in the life of the universe in general. It is one of the main constituents of the atom nucleus; it was discovered probably in 1916. It is the nucleus of the hydrogen; therefore, it is playing the great role in building the universal constituents. It constitutes almost half the mass of the nuclei of the elements in the universe; it is a good subject to study for the possibility of unifying the nature forces, but that needs a great accelerator of huge size, probably as large as our sun. The main properties of the proton are the following

Mass = 1.67×10^{-27} kgm = 1.007593 amu = 938.211 MeV = 1836.12 m_e.

Charge = 1.60206×10^{-19} C = +1 electron

Spin = 1/2 in $\hbar$ unit

Isospin = 1/2, isospin projection = 1/2

Magnetic moment = 1.41044×10^{-26} Am. $\equiv 2.79275$ μ_n (nuclear magneton)

Mean lifetime—stable—estimated to be about 10^{31}–10^{33} years

As the case of the electron, there is an antiproton too (p⁻), with the same properties, except its charge is (−1e); therefore, if it meets the proton, they will be annihilated into radiations. The proton usually interacts with matter, and due to its charge, it loses energy as it is moving via matter. In case it has enough energy, it creates nuclear reaction, as already shown in the nuclear reaction previously. For high energy (~3 GeV), the proton's reactions lead to new particles, which will be shown in the next sections, also in the Large Hadron Collider experiment in CERN, in Geneva, Switzerland (2008–2013).

5—Neutron

When the proton was discovered, still the mass of the nucleus, which constitutes about 99.7% of the atomic mass, is bigger than the mass of the proton, about more than two times; therefore, a serious research is carried on to find out the mysterious point about this scientific problem. In 1932, Chadwick had discovered a particle with zero charge (neutral) and mass that is a bit bigger than the mass of the proton, so accordingly, the nucleus constitutes the protons and the neutrons. Their sum represents the mass number of the nuclei, and both the proton and the neutron are different charge states to a particle known as the nucleon. These two different charge states give to the nucleon two degrees of freedom, in addition to its special coordinates and spin space. These two charge states provide the nucleon particle a new space called isospin space or charge space, characterized by the so-called isospin. As shown in chapter 6, the neutron and the proton, under certain physical conditions, can be transferred from one into the other as in beta decay. The main properties of the neutron are as follows

Mass = 1.008982 amu = 939.505 MeV = 1838.65 me

Charge is zero, neutral

Spin = 1/2 in $\hbar$ unit

Magnetic moment = −1.91315 μ_n

Isospin = −1/2, isospin projection = −1/2

Mean lifetime = 0.9×10^3, unstable

Decay, $n \rightarrow p + e^- + \nu^- + 0.783$ MeV

Notice that this is exoergic decay, $m_n > m_p$

Also, for the neutron, there is antineutron. If they meet, they will be annihilated into the pi-mesons ($\pi^+ \pi^0 \pi^-$).

Neutron Interactions with Matter

The neutron is unstable particle in free space; its mean lifetime is about 0.9×10^3 sec. Beta decay is one way of its interaction inside the nucleus. It is a kind of weak interaction, and this was treated in chapter 6. Inside the nucleus, it strongly interacts with the proton under the strong nuclear force through mutually exchanging the pion mesons of the nuclear force field, so they transform one into another. This can be represented by

1-p + p $\rightarrow$ p + n + π^+, or 2-p + n $\rightarrow$ p + p + π^-

From (1), it can be found that the energy of the left side is 1876.422 MeV, while the energy of the right side is 2017.716 MeV, so this reaction is endoergic type; therefore, the colliding protons must be supplied with energy bigger than 2017.716 in order for the reaction to occur. Also, the reaction 2 is endoergic for the same reason, since the energy of the right side is 2016.422 MeV, which is bigger than that of the left side of the reaction, which is 1877.716. Hence, both p and n have to be accelerated with energy bigger than 2016.422 MeV so this reaction can occur. These two reactions can be graphically represented as

n   p   n   p

    π   π

n   p   p   p

The other possible interaction between the nucleons is the so-called charge-exchange scattering, as mentioned earlier in chapter 1. In this type of scattering, the proton collides with the neutron at rest, exchanges charge with it, leaving the proton almost stationary, and the neutron is moving fast, or vice versa. This is illustrated as follows:

1) $p^{\rightarrow} + n(\text{at rest}) \rightarrow n^- + p^-$

2) $n^{\rightarrow} + p(\text{at rest}) \rightarrow p^- + n^-$

It is well-known that since the neutron is neutral, it does not lose energy by ionization when it passes through matter. Also, it is fact that the neutron is not affected by coulomb barrier, because it has no charge; therefore, it strongly interacts with nuclei. But this is not valid with the proton, which is a charged particle greatly affected by coulomb barrier. Another important property to the neutron is, at low energies, it strongly reacts with certain nuclei such as those that lack a neutron to complete the shell. But it weakly interacts with those nuclei whose neutron shells are closed shells. This is probably due to the Pauli exclusion principle because the neutron is a fermion. Moreover, if it passes through matter composed of nuclides that are totally inert toward the neutrons, the interaction will be elastic nuclear collision, which will consequently be slowed down, due to the fact that the neutron lab system kinetic energy is diminished in each collision with an amount equal to that carried away by the struck nucleus; this fact is quite important in preparing thermal neutrons, which is to be used in thermal nuclear reactors, where the fuel is uranium 235. Of course, such mechanism is more effective if the mass of the struck nucleus is quite small. Hence, the hydrogenous materials, such as the paraffin, are the most useful to be used for preparing thermal neutrons.

6—Neutrino

The neutrino has a scientific story related to beta decay, when conservation laws of energy and angular momentum seemed violated if it is taken as p $\rightarrow$ n + β^+ or n $\rightarrow$ p + . To avoid this scientific problem, the neutrino particle was proposed by Pauli, then it was the answer, so beta decay should take the form

1-p $\rightarrow$ n + β^+ +v , 2-n $\rightarrow$ p + β^- + v^-

By investigation of these two decay processes, it is clear that the conservation of energy and angular momentum are valid. So this important physical particle, by assumption, exists and is experimentally approved, which is now within the lepton family. It has the following properties:

M_v ~<5 × 10^{-4} ≈ 0, it vibrates; charge = 0, spin = 1/2, τ = ∞, stable

Magnetic moment <10^{-8} μ_B, probability is zero

Like other physical particles, it has an antiparticle called antineutrino, denoted by v^-. In general, for each particle, there is an antiparticle with the same properties, except for the sign of the charge, but is this true too for the neutrino? As shown in chapter 6 (beta decay), the neutrino and the antineutrino are not identical particles because they have different helicity! This fact has been confirmed experimentally by Wu et al. in 1957, based on T. D. Lee and Yang's theory about the nonconservation of parity in beta decay in 1956 (refer to chapter 6 for the details). Hence, under such properties, the neutrino does not interact with matter; therefore, it is too difficult to be detected. The only possible interaction is the so-called inverse beta decay. This interaction is extremely weak. The cross section of this interaction is $\sigma \approx 10^{-48}$ m^2 = 10^{-24}barns, which is very small.

7—μ-Meson (Muon)

In his trying to interpret the feature of the nuclear force, which combines the nucleons inside the nucleus, Yukawa proposed, in 1935, theoretically a particle of mass about two hundred times the electron mass, which carries on the effect of this strong force between the nucleons. In 1937, a particle with mass close to this suggested mass and (±) charge was discovered in cosmic rays by Noddermyer and Anderson by analyzing tracks of cosmic rays in the cloud chamber. The mass of this particle was found to be about 240 times of the electron mass. They named it first as mesotron, but the established scientific name now is mu-meson, which is abbreviated to muon (μ-meson). Its mass is established to be about 206.93 times the electron mass. But the mass of the anti-muon (μ^-) is 206.76 times the electron mass. But it was discovered that this muon is not the particle suggested by Yukawa but that the pions ($\pi^+ \pi^0 \pi^-$) are what he was looking for. Muon Particle is found in nature in both μ^+ and μ^-, where the second muon is the anti-muon. It is classified as lepton but not as an elementary particle and unstable. So the properties of the muon are summarized as follows:

Mass = 206.93 m_e = 105.7 MeV

Magnetic moment = 1.0026 ($e\hbar / 2\, m_u$)

Mean lifetime = 2–22 × 10^{-6} sec.

Charge = +1, −1, spin = 1/2

The decay mode processes are

$$\mu^+ \rightarrow e^+ + V_e + V_{\mu}^- + 106 \text{ MeV}$$

$$\mu^- \rightarrow e^- + V_e^- + V_{\mu} + 106 \text{ MeV}$$

The muon interacts with matter via its charge, its magnetic moment and through beta decay weak interaction. Sometimes, the muon μ^- passes through matter. when it comes to rest, it might be captured in the Bohr orbit of an atom, forming mu-mesic atom in heavier atoms. The ground state orbit of the mesic atom is so small the muon moves within the nucleus itself. Like the positronium atom, a positive muon can form a neutral semi-atomic system with an electron called muonium, as in the following diagrams:

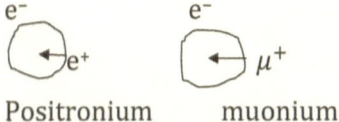

Positronium   muonium

No annihilation between the positive muon and the electron with emission of two gammas has been observed. Theoretically, annihilation between μ^+ and μ^- could occur but practically has not been observed. However, the muon looks as a bit bigger brother to the electron, but the problem of their difference in mass is not yet solved.

8—Pions ($\pi^+ \pi^0 \pi^-$)

The pions' particles are also found in cosmic rays. They are the Yukawa-predicted particles, which play the role of transmitting the field of nuclear force effect between the nucleons inside the nucleus. Their mass is about 270 times the electron mass, which equals to about 139 MeV. Also, they are produced as a result of the nucleons' collisions or when nucleons are bombarded by photons with energy that the rest energy of the pion (139.6 MeV) is supplied in the center of mass system. These reactions are given by

$p + p \rightarrow p + n + \pi^+ -140$ MeV $-(139.6 \sim 140) \sim M_{\pi^+}$

$\rightarrow P + P + \pi^0 -135$ MeV$— M_{\pi^0} = 135.1$ MeV

$\rightarrow p + n + \pi^0 -135$ MeV

The threshold energy in the laboratory system is given by

$E_{th} = Q[1 + m_1/m_2 + Q/2m_2c^2]$—(refer to chapter 7)

where m_1 is the mass of the bombarding particle,

m_2 is the mass of the target particle,

Q is the energy balance.

The threshold energy to produce the pion is about 290 MeV; therefore, increasing the bombarding energy of two or more pions can be emitted. The charged pi mesons can be detected by any of the visual chambers so that the charge and the mass can be obtained by the amount of the ionization and the orbit radius in a magnetic field.

Yukawa pointed out that the binding between nucleons is a kind of exchanging force due to the exchange of pions just analogous to the photon exchange between interacting charged particles in electromagnetic field. Therefore, the photons and the pions are called messengers of the field interactions. The nucleon-nucleon interactions via the exchange of pions can be represented by Feynman diagrams as follows:

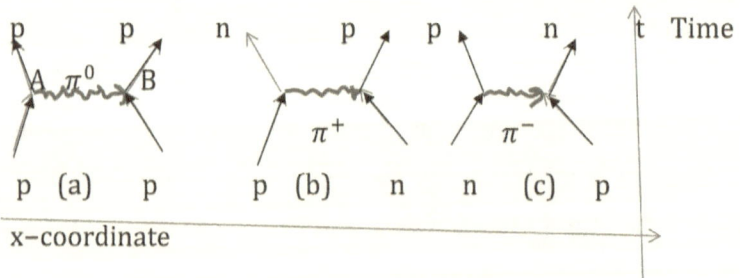

In figure 7, the proton creates and emits a virtual π^0, at the vertex A. After a short time Δt, a second proton absorbs this neutral pion at vertex B. The energy conservation law, during this short time of the existence of the pion, might be violated, but this violation is consistent with the uncertainty principle, $\Delta E \, \Delta t \approx \hbar$, where ΔE is the "borrowed" energy through the exchange processes, which is restored again when the pion is absorbed by the second proton at vertex B. Figures b and c show the pion exchange in other possible nucleons collisions. The nucleon-nucleon interaction is of a short range; it is about 1.4 fm. This range limits the traveling distance of the pion before it gets absorbed by the other proton at vertex B. If its speed is considered the speed of light, then the time of its existence can be calculated to be $\Delta t \approx 1.4 \times 10^{-13}/3 \times 10^{10}$ cm/ sec $\approx 0.5 \times 10^{-23}$ seconds. This time is, in fact, the range of nuclear strong interaction. It is called nuclear time. Accordingly, the borrowed energy, as mentioned above, can be calculated using uncertainty principle to be $\sim 2 \times 10^{-11}$ joules = 130 MeV. Now, if this energy is considered as the rest mass of the pion, then using $E = mc^2 = 130$ MeV to find that $m_\pi = 200 \, m_e$, as expected by Yukawa in 1935. Also, this leads, because of the short-range character, to the expectation that the fields particles associated with strong interactions have finite masses in the order of few hundreds of the mass of the electron. These pions that were suggested by Yukawa are confirmed experimentally via (1) high-energy nucleons collisions, between the nucleons in cosmic rays and those in the earth atmosphere, (2) using the laboratories under certain controlled scientific conditions, with the aid of accelerators and colliders of high accelerating energies (200 MeV to 2000 GeV) or today, the energy in the level of tera (CERN). The following are the typical collisions to produce the pions:

$$p + n \rightarrow p + n + \pi^0$$

$$\rightarrow p + n + \pi^- + \pi^+$$

It can be noticed from these two possible reactions that some of the kinetic energy, indeed, is transformed into a rest mass of the produced pions. Therefore, there is no doubt that the higher energy used the heavier particles expected to produced, as the equation E = m c^2 is telling. So the energy is the origin of everything, and the mass is energy in content. It is important to recognize that the pions are the nuclear field's particles and that the photons are electromagnetic field particles; therefore, they differ in certain points such as the following;

1. The photon has integral spin, while the pions have zero spin.

2. The photon is neutral, has no charge, has only one zero-charge state, while the pions have three charge states: +1, 0, −1.

3. The rest mass of the photon is zero due to the infinite range of the electromagnetic interaction, while the rest mass of the pions is finite, which accounts for the short range of the nuclear field.

4. The photon in free space is stable, but the free pions are unstable.

Although the pion's properties are summarized before, it might be useful to summarize them in table 4 .

Table 4: Pion Properties

Item	$\pi+$	$\pi0$	$\pi-$
Rest mass	139.569 MeV	134.964 MeV	139.569 MeV
Charge	+1	0	−1
Spin	0	0	0
Magnetic moment	0	0	0
Mean lifetime	2.603×10^{-8} sec	8.3×10^{-17} sec	2.603×10^{-8} sec
Decay mode	$\pi^+ \rightarrow \mu^{+-} + \mu_\nu,$ 99.9990% $\rightarrow e^+ + \nu_e$	$\pi^0 \rightarrow \gamma + \gamma,$ 98.85% $\rightarrow e^+ + e^-(1..15\%$	$\pi^{--} \rightarrow \mu^- + \nu_\mu^-$ $\rightarrow e^- + \nu_e^-$

It is important to notice the following:

1. Both charged pions decay by weak interaction resemblance to beta decay, as can be seen from their mean lifetimes. They are within the range of the weak interactions.

2. Neutral pion decays to two photons with about 99%. Its mean lifetime is about 8×10^{-17} sec, which indicates that its decay is under the electromagnetic interaction.

The weak interaction is of a short range about 100 fm = 10^{-15} cm. Therefore, the virtual messenger particles of this weak interaction field are expected to be of great masses. Abdus Salam and Weinberg postulated such particles in their theory of unifying the weak interaction field with the electromagnetic field in the sixties. In the eighties, the postulation was experimentally confirmed by discovering these particles. They were denoted as w⁺, w⁻. Their masses are about eighty-five times the proton mass (80 GeV), and z⁰ is heavier, about ninety-five times the mass of the proton (9o GeV); it is neutral (called eighties, the postulation was experimentally confirmed by discovering these particles. They were denoted as w⁺, w⁻. Their masses are about eighty-five times the proton mass (80 GeV), and z^0 is heavier, about ninety-five times the mass of the proton (90 GeV); it is neutral (called heavy photon). This weak interaction (electroweak interaction) can be represented by Feynman diagrams as follows:

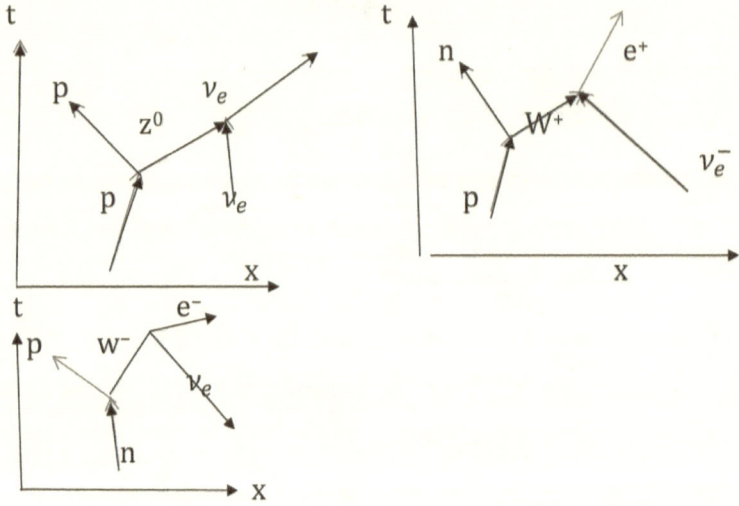

Figure 8: Feynman diagrams for weak interaction via the exchange of virtual particles w +, w⁻, and z^0

Before concluding, it is necessary to regard the gravitational force. Its virtual particle is the graviton, which is not detected yet. Its spin is supposed (2), and its mass is zero. The force range is infinity. From the previous information, one might conclude

a. Mean lifetime of particle-decaying particles by strong interactions ranges from 10^{-25} to 10^{-19} seconds;

b. Mean lifetime of particles decays by E&M interaction ranging from 10^{-19} to 10^{-16} seconds;

c. Mean lifetime of particles decays by weak interaction ranging from 10^{-15} to 10^3 (neutron) seconds. In table 5 is a summary of the characteristics of the four fundamental interactions as shown in the following table:

Table 5: Characteristics of the Four Fundamental Interactions

Force Type	Re.to.str.Int	Time sec.	F. particles	M_0MeV	range	
Strong	1	10^{-23}	Mesons	140	1fm	π^+ π^0 π^-
E&M	10^{-2}	10^{-17}	Photon	0	∞	0
Weak	10^{-13}	10^{-10}	w^+, w^-, z^0	80–90 GeV	$>10^{-2}$fm	$w^\pm$
z^0 1						
Gravitation	10^{-40}	10^{17}	G	0	∞	G 2

These forces are naturally created through natural processes. Physicists think they belong to one great force; therefore, they are trying to work out a great unified theory (GUT). The differences between these forces are due to the nature of series processes in which the universe went through from the starting ignition of creating the universe until the present. The present status of physicists toward these forces is to find out the theory of everything using the great Large Hadron Collider at CERN. They think the universe is created from quantum vacuum, out of breaking symmetry for 10^{-43} sec (Planck time) and borrowing energy for non-sensed time, based on the uncertainty principle. Therefore, there must be one force that is the source of other forces. So it might be possible to find a theory that explains explicitly how these forces came out.

9—K-Meson (Theton θ and Taun τ)

As mentioned before, these particles are developed out of magnetic cloud chamber, where pictures were produced by cosmic rays, studied by Rochester and Butler. K-mesons were classified within the strange particles, just like the hyperon particles, because of their completely superfluous existence in the theory of matter and their comparatively long lifetimes. From the high-energy nucleon-nucleon collisions, it was shown that these hyperons interact quite strongly with nuclear field, but they decay through the weak interaction. Their mean lifetimes are within the range 10^{-10}–10^{-8} seconds. The discovery of the puzzling particles θ and τ of mass about 10^3 m_e leads to $\theta \approx \tau =$ K-mesons, which indicates that there are different types of K-mesons, which are classified as

1. K-mesons of type taun (τ)—characterized by its decay into three charged pions: $\tau^{\pm} \rightarrow \pi^+ + \pi^+ + \pi^- + 75$ MeV

2. K-meson (θ-meson or theton)—it is neutral, decays into two charged pions: $\theta \rightarrow \pi^+ + \pi^- + 214.6$ MeV

3. θ^+, counterpart of θ^0-meson—decays as

$\theta^+ \rightarrow \pi^+ + \pi^0 + 219.1$ MeV

Other properties of K-mesons were mentioned in the previous sections that can be returned to it.

10—Hyperons

These particles are described in table (3). They are classified as strange particles, closely related to the K-mesons. Their masses are greater than the neutron mass but less than the mass of the deuteron. According to their masses' increase, they are as follows:

1. Λ^0

It is the first and the best known hyperon, neutral, unstable, decays into nucleons and pion with mean life time of about 2.3×10^{-10} sec such as $\Lambda^0 \rightarrow p + \pi^- + 40$ MeV(exoergic) 66.3%

$\rightarrow n + \pi^0$, 33.7%, mass = 115.4 MeV, one charge space, isospin T = 0, spin = 1/2, parity = +1. Due to its similarity to the neutron, it was speculated that it might be an excited state to the neutron.

2. *Sigma* (Σ) hyperon particle

This is the second type of hyperon, was first observed and explained in the cosmic rays, but later became a product of bombarding the nucleons by the beam of π^- with an energy above the threshold 880 MeV such as $\pi^- + n \rightarrow \Sigma^- + k^0$. Since k^0 is produced in this interaction, it indicates that the strangeness of sigma must be S = −1, because k^0 has S = 1, and n, p have S = 0.

Other general properties are shown in table 6.

3. Xi(Ξ)

This is a well-established hyperon, was observed as a cosmic ray induced as negative particle, where the track in the chamber suddenly changed direction and a V-like track produced nearby. The following diagram illustrates that

$\Xi^- \rightarrow \pi^- + \Lambda^0 \rightarrow p + \pi^-$

$\tau = 1.75 \times 10^{-10}$ sec

Its counterpart decays to

$\Xi^0 \rightarrow \Lambda^0 + \pi^0 + 70$ MeV
$\tau = 3.05 \times 10^{-10}$ sec

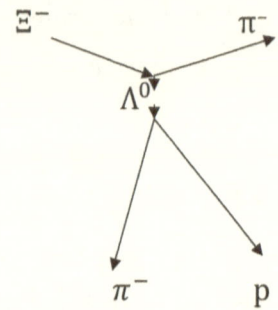

Also, it can be produced by bombarding the nucleons with pion beam with energy above the threshold energy of 2.23 GeV, as follows:

$\pi^- + p \rightarrow \Xi^- + k^+ + k^0 + (>2.23$ GeV)

This type of decay requires that the strangeness S of the Xi must be −2. Why? It has two charge state + and 0. Its spin is ½; parity = 1. Its isospin is 1/2; the mass for Xi plus is 1320.8 MeV, and for the neutral is 1314.3 MeV.

11—Resonance Particles

The resonance concept was pointed out earlier. It is a physical phenomenon that has totally changed the concept of the particles physics. These resonance particles occur in the processes of the strong interactions. The resonance concept says that all these particles come about from resonance cases, so what does the resonance mean physically? To answer, take as an example an electron of controllable kinetic energy bombarding hydrogen atom. Let the atom be in its ground state, where from quantum theory, the internal energy of this atom is quantized. Each state of the quantization has specific properties. Hence, if the kinetic energy of the incident electron is less than 10.2 ev, the energy required to raise the atom to its first excited state, then the collision is elastic (no loss of energy), and the atom remains in its ground state. But if the electron kinetic energy equals or exceeds 10.2 ev, inelastic collision takes place, and there is a sudden decrease in the kinetic energy and the atom of (H) making transition to its first excited state, then within a certain time, it goes back to its ground state, emitting gamma radiations (photons). This shows the excitation, by bombarding particle, to a quantized state leading to a resonance behavior, which is the probability for an inelastic excitation process to be maximum, when the kinetic energy of the incident particle (electron) equals the internal excitation energy of the atom. Or it falls off if the kinetic energy is less or larger than the excitation energy of the atom. This phenomenon is well-known in atomic physics. Here, the atom in its first excited state, considered as created particle, is distinct from its status in the ground state. It has different mass $[m_0 + m (10.2)]$ and different angular momentum (excited state). This so-called created particle is usually unstable; therefore, quickly ($\sim 10^{-8}$ seconds), it decays to (H + γ) via the electromagnetic interaction that will be stalled in its ground state and (γ) emitted. This decay time is actually too long in the electromagnetic interaction; it is because the electron and the proton are separated by large distance compared to the strong and the weak interactions. It is atomic distance (10^{-8} cm), where the electromagnetic force $\propto 1/r^2$.

In high-energy physics, where the energy in the range (GeV-tera), the resonance phenomenon often occurs, creating resonance particles is quite easy, by bombarding nucleons by nucleons, or by pions with kinetic energy of the range mentioned above. In 1952, the proton is bombarded by pions with energy 40–220 MeV. The first resonance particle produced of the lowest mass in the baryon group is the so-called delta (Δ) with mass ($m_\Delta \cong 1230$ MeV = 1.230 GeV). But in the beginning, it was not defined as such.

The following figure shows this example: the resonance, peak of the number of pion incidents scattered by the proton as target, where the kinetic energy of the pions is above 200 MeV (total center of mass energy is 1.230 GeV), which indicates that an inelastic scattering had taken place, in addition to elastic scattering too:

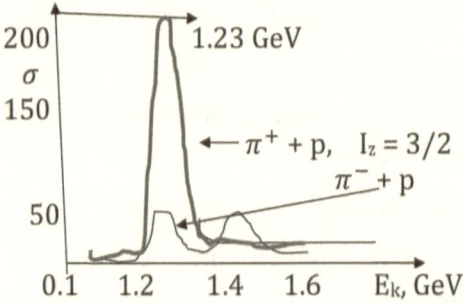

This can be explained as follows: the proton is excited to a higher state of energy, where the baryon particle is created with energy higher than the proton initial energy, which is the delta particle. For certain times, it decays to p and π^+. This can be represented as

$$\pi^+ + p^+ \rightarrow \Delta^{++} \rightarrow \pi^+ + p^+;$$ the decay time can be estimated.

Using the uncertainty principle $\Delta E \, \Delta t \approx \hbar$, where ΔE = 100 MeV, to be $\tau \approx 10^{-23}$ seconds. This means delta particle (Δ) decays under the strong interaction. From this example, it can be claimed that the resonance phenomenon, with the uncertainty principle, can be used to find the mean lifetimes for unstable particles within the range of over 10^{-25} to 10^{-19} seconds, for shorter times than 10^{-25} seconds, which is equivalent to $\Delta E \cong 6$ MeV. This is a spread energy comparable to the rest energy of the intermediate particle, so it is not sure whether the resonance particle does exist within the energy distribution curve or if it can be distinguished. Also, for $\tau > 10^{-19}$ sec, it is found for $\Delta E = 6$ MeV, and pions with kinetic energy in the range of the GeV; increasing difficulties rise in resolving energy differences. Many resonances, mesons, and baryons are discovered with rest masses ranging from 0.5 GeV to >10 GeV. Table 6 shows some established resonance particles, within the meson family. They are hadrons with integral angular momentum. Table 6 lists some resonance particles in the baryon family. They are hadrons with half-integer angular momentum. Their fundamental properties are clearly mentioned. The universally valid conservation laws, as mentioned before, and the partial conservation fixed before are found to hold for all interactions, even those involving resonance particles.

Table 6: Some Established Resonance Particles in Meson Family

Meson	$E_{0, MeV}$	$\tau, 10^{-23} sec$	qxe	JPxℏ	T	St,	Charm, C	Most probable decay
ρ	773	43.3	0.±1	1⁻	1	0	0	π π
ω	782	6.6	0	1⁻	0	0	0	π⁺ π⁰ π⁻
K*	892	1.3	0,+1	1⁻	1/2	1	0	K π
K*⁻	892	1.3	0,−1	1⁻	1/2	-1	0	K⁻ π
η⁻	958	>60.0 0	0⁻ 0	0	0			η π π
φ	1020	16.0	0 1⁻	0	0	0	0	K⁺ K⁻
f	1271	0.37	0	2⁺	0	0	0	π π
Λ₂	1310	0.65	0,±1	2⁺	1	0	0	ρ π
K⊕	1420	0.6	0,+1	2⁺	1/2	1	0	K π
K⊕⁻	1420	0.6	0,−1	2⁺	1/2	−1	0	K⁻ π
f ⁻	1516	1.6	0	2⁺	0	0	0	K K⁻
ω⁻	1667	0.45	0	3⁻	1	0	0	ρ⁰ π⁰
g	1690	0.37	0,±1	3⁻	1	0	0	π π
ψ	3098	984	0	1⁻	0	0	0	π π
PC }	3267							
	3504	? 0		2⁻ 1	0	0		ψ, γ
ψ⁻					?			π⁺ π⁰

ALI A. ABDULLA

Table 7: Some Established Resonance Particles in Baryon Family

Baryon	E_0,MeV	τ,10^{-23}sec.	J^Px$\hbar$	qxe	T	St.no.	Charm,C
Most probable decay							
Nucleons							
N_1	1470	0.3	$1/2^+$	0,+1	1/2	0	0
$N\pi$							
N_2	1520	0.5	$3/2^+$	0,+1	1/2	0	0
$N\,\pi$							
N_3	1535	0.5	1/2	0,+1	1/2	0	0
= =							
N_4	1670	0.4	$5/2^-$	0,+1	1/2	0	0
= =							
N_5	1688	0.5	$5/2^+$	0,+1	1/2	0	0
=							
N_6	1700	0.4	$1/2^-$	0,+1	1/2	0	0
=							
N_7	2220	0.2	9/2	0,+1	1/2	0	0
=							
N_8	2650	0.2	$13/2^{-?}$	0,+1	1/2	0	0
=							
N_9	3030	0.2	17/2?	0,+1	1/2	0	0
= =							
Δ_1	1232	0.6	$3/2^+$	-1,0,+1,+2	3/2	0	0
= =							
Δ_2	1650	0.4	$1/2^-$	= = = =	3/2	0	0
= =							
Δ_3	1670	0.3	$3/2^-$	= = = =	3/2	0	0
= =							
Δ_4	1950	0.3	$7/2^+$	= = = =	3/2	0	0 =
=							
Δ_5	2420	0.2	$11/2^+$	= = = =	3/2	0	0
= =							
Δ_6	2850	0.2	?	= = = =	3/2	0	0
= =							
Δ_7	3230	0.1	?	= = = =	3/2	0	0
= =							
Lambda (Λ) ---				0 0	o	-1	0
N K⁻							
Λ_1	1520	4.4	$3/2^{-O}$		0	-1	0
= =							
Λ_2	1815	0.8	$5/2^+$		0	0	-1

0	=	=					
Λ_3	2100	0.3	$7/2^-$	0		0	-1
0	=	=					
Λ_4	2350	0.5	?	0		0	-1
0	=	=					
Λ_5	2585	0.3	?	0		0	-1
0	=	=					
Sigma(Σ)				-1,0,+1	I	-1	0
N K$^-$							
Σ_1	1385	1.9	$3/2^+$	-1,0,+1	1	-1	?
Λ π							
Σ_2	1670	1.3	$3/2^-$	-1,0,+1	1	-1	?
N K$^-$							
Σ_3	1765	0.6	$5/2^-$	-1,0,+1	1	-1	?
= =							
Σ_4	1915	0.7	$5/2^+$	-1,0,+1	1	-1	?
N K$^-$							
Σ_5	2030	0.4	$7/2^+$	-1,O +1	1	-1	?
= =							
Σ_6	2250	0.7	$7/2^+$	-1,0 ,+1	1	-1	?
= =							
Xi(Ξ)				-1,0	1/2	-2	0
Ξ π							
Ξ_1	1530	7.2	$3/2^+$	-	-	-	-
- -							
Ξ_2	1820	1.3	--				
Λ K$^-$							
Ξ_3	1940	1.0	--				
-Ξ π							

VIII-5-3—Classification of These Physical Particles

As already seen, there is a huge number of the so-called elementary particles, with big quantity of data concerning their properties and characteristics. This situation might be confusing because they have no systematic behaviors. Therefore, many suggestions were proposed to bring this huge data to an organization and regularity for these particles to be of fruitful studies. The most promising system for that is the so called unitary symmetry. It has found a striking way to arrange the particles according their similar spins and their parities, but with different hypercharge and isospin. As an example, take the spin and parity as O^- and $1/2^+$ of the mesons and the baryons, as in table 8 below:

Table 8

Spin and Parity		$I^P = O^-$					$I^P = 1/2^+$			
T		-1	-1/2	0	+1/2	1	-1	-1/2	0	+1/2
Y	1		k^0		k^+			n		p
	0	π^-		π^0		π^+	Σ^-		Σ^0 Λ	Σ^+
	-1		k^-	η	k^{0-}		Ξ^-		Ξ^0	

It is well-known in physics that the real basic results of any symmetry classification are the multiplet. An example is the multiplet $(2J + 1)$ in atomic physics, where this is the possible values of $m_{j,}$ the projection of J along the direction of applied magnetic field on the system. Remember the Zeeman effect of splitting the atomic spectral lines to different energies. Therefore, as a physical phenomenon on the microsystem scales, it is not strange that it is applicable to the high-energy physics problems. So there is considerable evidence. The multiplet structures eight and ten (octet and decuples) are naturally recognized with many different values of isotopic (isospin) and parity, which characterizes several occurrences. The following figures show this possible multiplet, which is firmly confirmed and strongly supports the symmetry principle concept. It is real evidence to its occurrence.

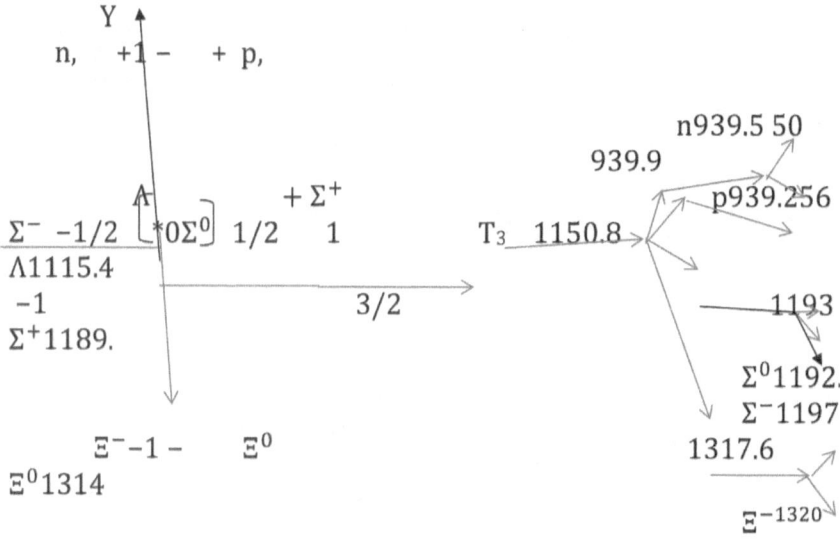

Figure 9: multiplet structure (8)

Also, the resonance particles can be classified diagrammatically according to the mass and the Z-component of the isospin, T_z, as follows, where the multiplet is given by $(2T + 1)$:

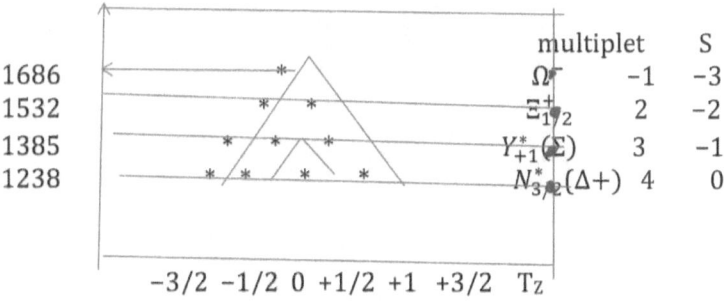

Figure 10: mass versus T_z for established baryon resonance particles

Ω^--particle discovered by bombarding the proton with the K^--meson at Brook Haven, about one hundred thousand pictures were taken, counting the track length of the meson to be about 10^6 feet. These pictures were partially analyzed, searching for the more characteristic modes of decay of omega (Ω^-) particle. These events can be interpreted as follows:

$$K^- + p \rightarrow \Omega^- \pm K^+_{s=1} + K^0_{s=1} \Rightarrow S(\Omega) = -3$$

$$\Xi^0 + \pi^-$$

$$\hookrightarrow \Lambda^0 + \pi^0 \dashrightarrow \gamma_1 + \gamma_2$$

$$\hookrightarrow P + \pi^- \quad \hookrightarrow e^+ + e^- \quad \hookrightarrow e^+ + e^-$$

These interactions are identified from the analysis of the momentum and gab-length. It is found that Ω^- as hyperon particle with mass 1686 ± 12 MeV/C^2, charge (−1 e), B = 1, T = 0, and S = −3.

The following diagram shows the multiplet of the structure (10), for some resonance particles, which is somehow of high mass:

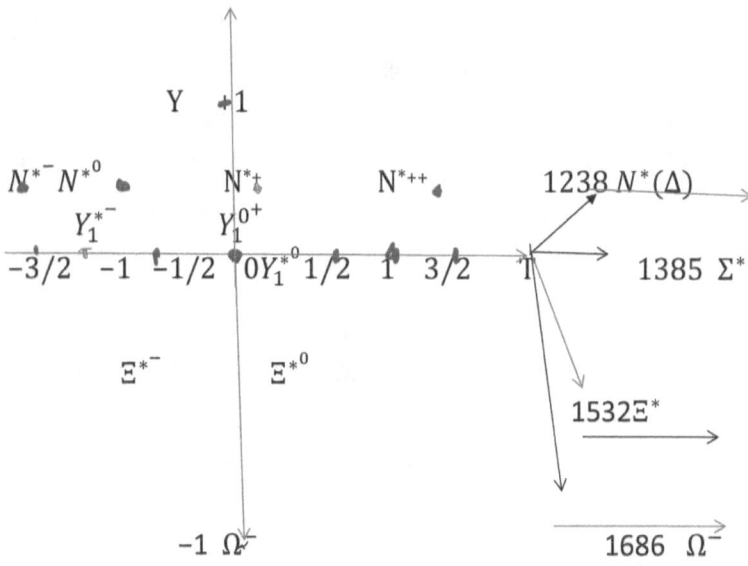

*indicates resonance particle; Y is hypercharge

Figure 11: resonance particles

From this diagram, it is clear that N^* represents Δ particles, Y^* represents sigma (Σ) particles, and Ξ^* is Xi particles. As stated earlier, the resonance particles are an indication that as the reaction's energy goes higher, it is possible to discover new particles, due to the fact that the origin of everything is energy.

VIII-5-4—Regge Theory for Particles

The theory is looking for the dynamic effect relating those particles with the same hypercharge (Y) and isospin (T). It shows that the mass of the particle is dependent of the total angular momentum and the parity (I^P). This relation is represented by smooth mathematical curves, the mass versus the total angular momentum, or the energy versus the angular momentum. These smooth curves are called Regge trajectories. The energy is the eigenvalue of the the radial solution of the Schrodinger equation to find the scattering amplitude of high-energy hadrons.

It is based on the rule that states that only states separated by two full integers can lie on the same trajectory and have equal spaced energies, with identical slopes for each case. As the theory claims further, giving more to the high-energy behaviors of various hadron reaction amplitudes make definite predictions available to direct experimental tests. But although it is so attractive, it has met with huge difficulties theoretically.

Hence, in spite of its ambition, it remains as a theoretical hypothesis. For the baryons and the mesons, recently (2012), it is found that the trajectories are not quite parallel and so smooth. Depending on new data, it is found that the theory of Regge trajectories is not quite general that the trajectories are truly parallel and smooth. For illustration, the figures 12a and 12 b can show examples for these new findings (The Regge trajectories are physically dealing with the analyticity of the scattering amplitude in principle and they are well known to physicits.

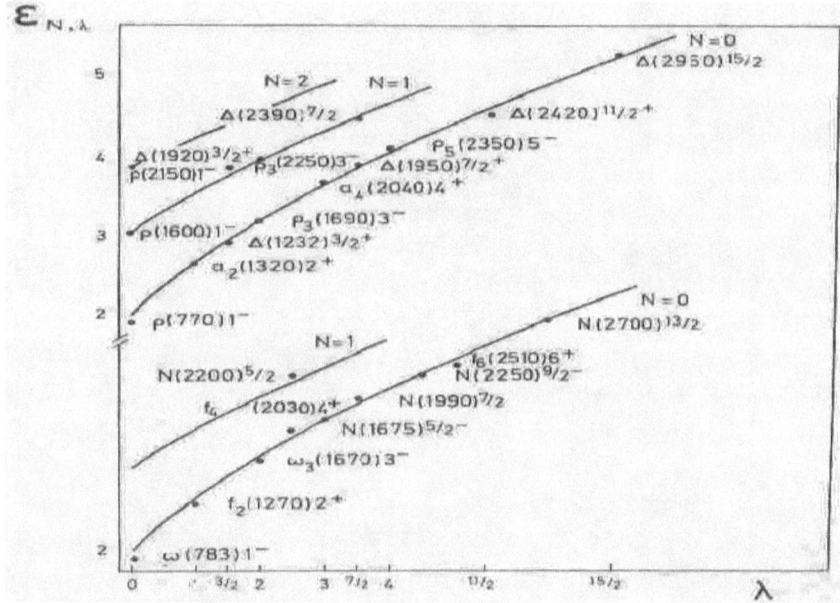

Figure 12a: $\varepsilon_{N\lambda}$ versus λ

where $\varepsilon_{N\lambda}$ is the eigenvalue, N is the principal quantum number, and λ is the orbital angular momentum.

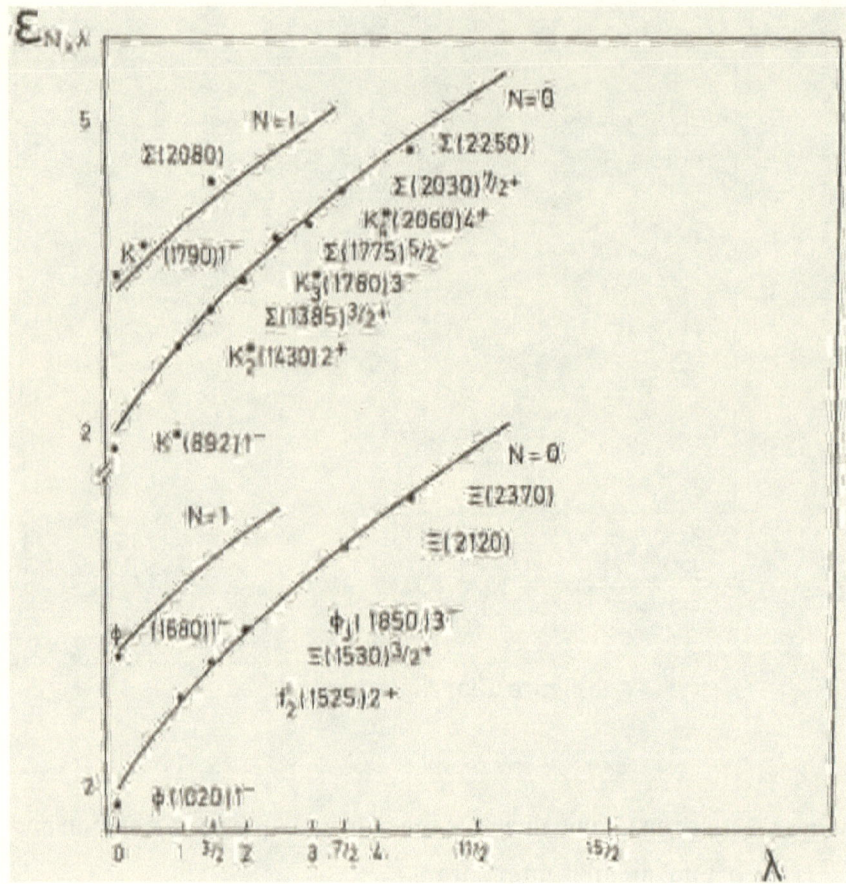

Figure 12b: $\varepsilon_{N\lambda}$ versus λ

Previously (in 1961), if one looks at the baryons N(939, 1/2$^+$), Λ

(1115, 1/2$^+$), $\Xi(1318, \frac{1}{2}+)$, and Λ (1405, 1/2$^+$), one notices that they

are aligned on the 1/2$^+$ spin line (Y-coordinate), each corresponding
to its mass (X-coordinate). The baryons Δ (1238, 3/2$^+$), N(1688, 5/2$^+$),
and Δ (1920, 7/2$^+$) are located ascendingly on a trajectory connecting
the first and the baryon (1920, 7/2$^+$). The baryon N(939, 1/2$^+$) is
located on a trajectory connecting it to the baryon Δ (1815, 5/2$^+$). The

other three baryons were proposed to lie on three different trajectories crossing the 5/2⁺ spin-x-coordinate line with three different masses of baryons to be found! The numbers inside the parentheses are the mass and the spin, respectively. The case of the mesons is quite unclear. The mesons π (137, 0⁻), K(496, 0⁻), and η (548, 0⁻) lie on the mass axes (x-coordinate) with zero spins. The mesons P (759, 1⁻), ω (782, 1⁻), K(886, 1⁻), and φ (1029, 1⁻) lie on spin 1-x-axes. The meson f(1250, 2⁺), the spin 2-x-axes line. If trajectories are drawn, they will be not quite parallel and smooth. Figures 13a and b show that. Under these circumstances, it looks like the theory is not quite workable. This theory, in fact, concerns with the analytical properties of the scattering, which is developed under many models, which takes the logarithmic form of the trajectories, leading to the pomeron theory. This theory is still in process of being well verified by more experimental tests.

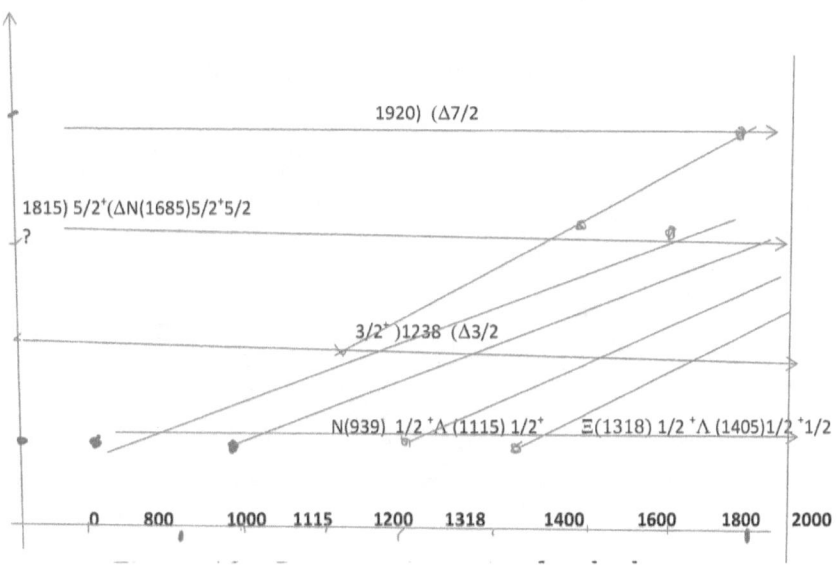

Figure 13a: Regge trajectories for the baryons

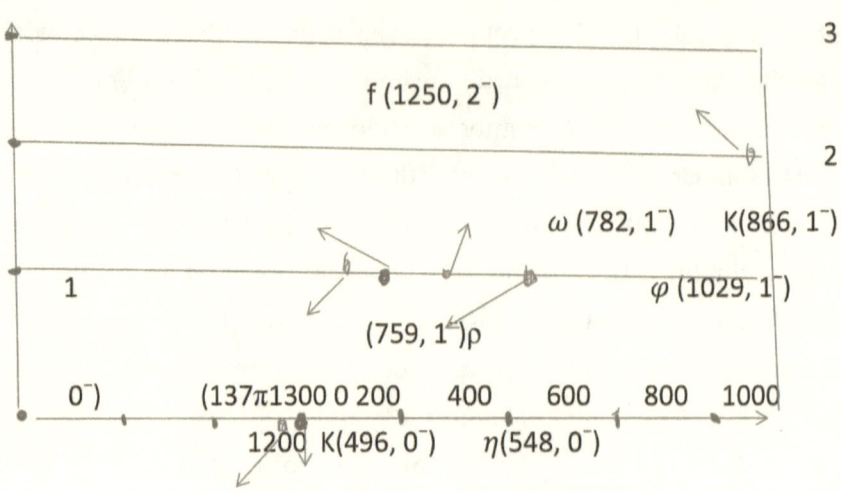

Figure 13b: Regge trajectories for mesons

VIII-6—The Quark Model for Hadron Structures

Table 9: Quantum Properties of Leptons

	Lepton symbol	Charge q	Lepton no.		Spin
		q	L_e	L_u	
Leptons	e^-	−1	1	0	1/2
	ν_e	0	1	0	1/2
	μ^-	−1	0	1	1/2
	ν_μ	0	0	1	1/2
Anti leptons	e^+	+1	−1	0	1/2
	$\bar{\nu}_e$	0	−1	0	1/2
	μ^+	1	0	−1	1/2
	$\bar{\nu}_\mu$	0	0	−1	1/2

It can be noticed from this table that each lepton of these eight particles is distinguished by at least one quantum number. The quark particles, which are hypothetically introduced as eight quarks-antiquarks, also differ by certain quantum properties, combine as groups to form the known hadrons (baryons and mesons). It is found that the nucleons are formed of three different quarks, and the mesons are formed of quark and antiquark (as pair). Physically, the nucleon has two possible charge space states, proton state and neutron state, which form two-dimensional spinor, such as

$$\xi = \begin{vmatrix} p \\ n \end{vmatrix} \equiv \begin{vmatrix} \xi_1 \\ \xi_2 \end{vmatrix} \quad \text{—for nucleon}$$

$$\xi^- = \begin{vmatrix} p^- \\ n^- \end{vmatrix} \equiv \begin{vmatrix} \xi_1^- \\ \xi_2^- \end{vmatrix} \quad \text{—for antinucleon}$$

These spinors form the basic isospin doublets of the system. Higher isospin multiplet can be constructed by forming direct products of ξ, ξ^-, or both. These states are specifically written as

$$\left. \begin{aligned} \tfrac{1}{\sqrt{2}}(\xi_1^- \xi_1 + \xi_2^- \xi_2) &= \tfrac{1}{\sqrt{2}}(p^- p + n^- n) \\ \xi_1 \xi_2^- &= p\,n^- \end{aligned} \right] \quad \begin{aligned} &\text{For singlet I = 0} \\ &2I + 1 = 1 \end{aligned}$$

and

$$\left. \begin{aligned} \tfrac{1}{\sqrt{2}}(\xi_1^- \xi_1 - \xi_2^- \xi_2) &= \tfrac{1}{\sqrt{2}}(p^- p - n^- p) \\ &\text{For triplet I = 1} \\ \\ \xi_1^- \xi_2^- &= p^- n^- \end{aligned} \right] \quad 2I + 1 = 3$$

This can be stated in another way, where the direct product of the two isospin doublets breaks down into isospin singlet and isospin triplet.

Symbolically, this might be written as $2 \times 2 = (1) + (3)$. It is known that the pion is isospin triplet. Also, the nucleon-antinucleon combinations have the isotopic spin properties resemblance to the pion. The nucleons and their antinucleons, as physical particles, are considered to be the non-strange member of the quark triplet. The third member is required so that the strangeness can be introduced. This member is the proposed (λ) quark with strangeness, S = -1. Hence, the two components of the spinor, of course, will be three components, including (λ) the quark; therefore, the spinor takes the form

$$\xi = \begin{bmatrix} p \\ n \\ \lambda \end{bmatrix} = \begin{bmatrix} \xi_1 \\ \xi_2 \\ \xi_3 \end{bmatrix}$$

And the corresponding antiparticle system is represented by

$$\xi^- = (p^- \ n^- \ \lambda^-) = (\xi_1^-, \xi_2^-, \xi_3^-)$$

which transform in such way that $\xi\xi^-$ is invariant.

The p, n quarks form isodoublet (T = 1/2, 2T + 1 = 2) of strangeness, S = 0 , and λ quark is isoscalar of strangeness, S = -1. The baryon number assigned to each quark is 1/3, due to the fact that the nucleon is composed of three quarks, and since the hypercharge is given by Y = S + B = 1/3 + 0 = 1/3 for p and n, and 2/3 for λ, also q = T_ξ + 1/2 Y, which leads to -q = 2/3e, 1/3e for p, n quarks, respectively, and 1/3e for quark λ. This is summarized in table 10 below:

(Table 10)

Quark	B	I	T	Y	S	Q
P	1/3	1/2	1/2	1/3	0	2e/3
n	1/3	1/2	1/2	1/3	0	1e/3
λ	1/3	0	0	2/3	-1	1e/3

The quarks form two basic triplets represented by the following graphs:

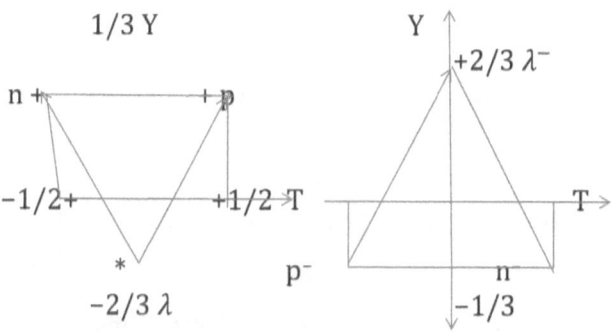

For particles   For antiparticles

It is possible to form higher multiplet by forming all possible pairs of quark-antiquark, nij $= \xi_i^- \xi_j^-$ (i, j = 1, 2, 3). This yields nine possible states, which can be written as $3 \times 3 = 1$ singlet + 8 octet. The octet of quark-antiquark can be represented in the following graph:

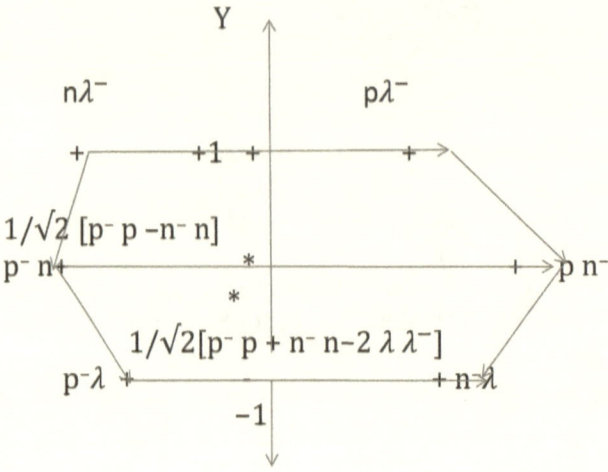

Figure 12: quark-antiquark states

Similarly, the baryon states are supposed to be three quark-bound states (uud ≡ *proton*, udd ≡ n) with angular momentum ordered in singlets, octets, and decuplets. The lowest states are those whose angular momentum is zero. There are ten states with spin 3/2 forming decuplet of baryons with $I^p = 3/2^+$. This can be illustrated as follows:

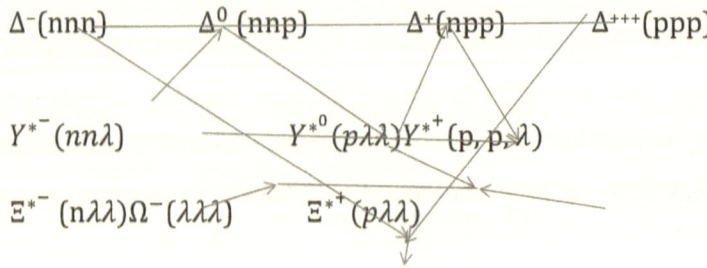

Ω^- was the missing tenth particle predicted by Gell-Mann, discovered in 1962, with negative charge and hypercharge = −2. It was shown earlier that it is a product of the interaction $k^- + p$, producing $k^+ + k^0$ and Ω^-.

So its strangeness no. must be S = −3. On these facts, the proposed quark model for the hadrons, baryons, and mesons is well established. Table 11 shows the discovered quarks with their antiquarks so far; they represent three generations according to the time of their discoveries.

Table 11

	Quark symbol	Charge	B.no	St,	Charm	Spin	Mass MeV	Generation
	u	2/3	1/3	0	0	1/2	360	First
Quarks	d	−1/3	1/3	0	0	1/2	360	=
	c	2/3	1/3	0	1	1/2	1500	second
	s	−1/3	1/3	−1	0	1/2	450	=
	t	-	-	-	-	-	174 GeV	third
	b	-	-	-	-	-	5000?	=
	u⁻	−2/3	−1/3	0	0	1/2		
	d⁻	1/3	−1/3	0	0	1/2		
Anti-	c⁻	−2/3	−1/3	0	−1	1/2		
quarks	s⁻	1/3	−1/3	1	0	1/2		
t⁻ --?								
b⁻ --?								

These classified quarks are, in general, called flavors. The size of the quark is about 10^{-15} m, which is so small, can be considered dimensionless (point) and is, therefore, considered a fundamental particle. Their masses are based on approximate estimation from the scattering experiment measurements. That is due to the fact that they still practically cannot be isolated to be subjected individually to experimental measurements. This feature is a very inherent property in the quarks' binding force. If they are stretched, a very strong force will oppose that, but within their usual content size, they are relaxed in their motions. Also, each quark possesses three kinds of color with allowed values of red, green, and blue. These color values were introduced to preserve the Pauli exclusion principle in these particles. In addition to those properties, the quark forces are attractive type forces to construct the nucleons, proton or neutron, such as

$$\text{uud} \rightarrow p_{1/2}^{+} \rightarrow p \rightarrow n + e^{+} + V$$

where n is composed of udd. This indicates one of quark u is converted to quark d, i.e., $u \rightarrow d + e^{+} + V$,

also, $d \rightarrow u + e^{-} + V_{e}^{-}$ in the decay of the neutron to proton.

As it is known, the mesons are also constructed of one quark and antiquark, which are colorless.

Chapter Nine

Weak Interaction

IX-1—Introduction

Some brief ideas were given about the main features of the weak interaction in general, previously. This field is somehow complicated in nature, so its physics is not easy to understand, but it is quite important to be learned with concern. Thus, the concepts and the restrictions in this field are so useful to be covered from the very beginning. In this chapter, the most possible to be learned will be discussed, getting the benefits of the previous subjects in consideration, especially chapter 8, to make the discussion quite clear to the readers in general and specially to the students. Also, a review of at least a part of the field of weak interaction is necessary to be given so that the students can easily be oriented to the original literature. In some cases, like some semi-leptonic hyperons decays, it is necessary for the review be fairly complete. Some emphasis will be put on the techniques and the tricks, which are necessary in the actual computations of lifetimes or spectrum in weak processes. Hence, the fields of weak interactions are proven to be fruitful for the so-called particles physics.

IX-2—The Weak Interaction Process

For any type of reactions, as was mentioned in chapter 8, there are certain characters such as the lifetimes and the cross sections. By determining these physical parameters, the type of the reaction will be known. In chapter 8, it was pointed out that the lifetime of the particle decay via the weak interaction is in the range 10^{-10} to 10^3 seconds, and the interaction cross section is in the order of 10^{-38} cm² or less. In spite of the possible large number of final states, experiments can be divided into two cases, according to their initial state, either with one particle in the initial state characterized with lifetime such as the neutron decay, n $\rightarrow$ τ = 10^3sec $\rightarrow$ p + β^- + V or

$$\mu^+ \rightarrow \tau = 2.2 \times 10^{-6}\text{sec} \rightarrow e^+ + V_e + V_{\mu}^-$$

or with two particles in the initial state such as

$$V_{\mu}^0 + n^0 \rightarrow p^+ + \mu^-$$

Accordingly, either a neutrino is involved, as in the above examples, or it may involve hypercharge (Y), as in the following interaction: $\Lambda \rightarrow$ P$^+$ + π^- (refer to chapter 8).

IX-3—Classes of Weak Interaction

Phenomenologically, it can be distinguished into five classes of weak processes; they are

1. leptonic processes such as the decay of μ^+;

2. semi-leptonic processes with $\Delta Y = \Delta S = 0$ such as V_{μ} + n $\rightarrow$ p + μ^-;

3. semi-leptonic processes with lΔYl = 1;

4. hadronic processes such as $\Lambda \rightarrow$ p + π^-;

5. CP violating processes.

The fact that no process with l∆Yl = 2 occurs in nature is corroborated very well by experiments. Table 1 gives limits on the branching ratios for l∆Yl = 2 processes (Particle Data Group, Physics Letter 398/1, April, 1972).

Table 1

Process Experimental Limit

$\Xi^0 \rightarrow p_i V_e^- <1.3 \times 10^{-3}$ $l = e, \mu$

$\Xi^0 \rightarrow p\,\pi^- <0.9 \times 10^{-3}$

$\Xi^- \rightarrow n_i V_i^- <10^{-2}$

$\Xi^- \rightarrow n\,\pi^- <1.1 \times 10^{-3}$

IX-4—Construction of Weak Interaction Theory

To construct a theory dealing with a weak interaction, it might start with the assumption that mesons can be thought of as bound states of fermion-anti-fermion systems such as pp⁻, nn⁻, etc. Hence, on this basis, it can be started with the underlying proposal that each weak interaction process involves exactly four fermions. Thus, the weak interaction Hamiltonian H_w is dependent of four fermion field operators, which is usually written as

$$\psi(r) = \frac{1}{3\sqrt{2\pi}} \int d^3 P \sqrt{\frac{m}{P_0}} \times$$

$$\sum_{r=1}^{2} [a_r(p) u_r(p) e^{-ipx} + b_r(p) V_r(p)\, e^{ipx} \tag{1}$$

Here, it is convenient to go into some brief details using Dirac equation and Dirac operators. The following form for Dirac equation will be used:

$$(i\eth - m)\,\Psi\,(x) = 0\,, \qquad \eth = \partial_\mu \gamma^\mu \tag{2}$$

with the commutation relations

$$\gamma_\mu \gamma_\nu + \gamma_\nu \gamma_\mu = 2\,g_{\mu\nu} \tag{3}$$

The matric $g_{\mu\nu}$ has negative space-like and positive time-like components. Therefore, the hermiticity properties of γ^- matrices are

$$\gamma_0 \gamma_\mu \gamma_0 = \gamma_\mu^{+} \tag{4}$$

As it is well-known, the anti-matric tensor is defined as

$$\sigma_{\mu\nu} = 1/2\,[\gamma_\mu \gamma_\nu - \gamma_\nu \gamma_\mu] \tag{5}$$

and the pseudoscalar is defined as

$$\gamma_5 = -i\gamma^0 \gamma^1 \gamma^2 \gamma^3 = \begin{pmatrix} 0 & -1 \\ -1 & 0 \end{pmatrix} \tag{6}$$

where each entry in the matrix of equation (6) is, of course, 2 × 2 matrix. The normalization of the spinor is such that

$$u^-u = 1 = u^+u = E/m \tag{7}$$

A typical weak interaction process is shown in figure 1:

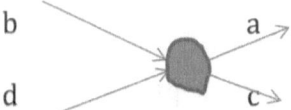

Figure 1: general four-fermions process

Now, what is the kind of space-time dependence of the four field operators? To answer such question, it might be assumed that the currents should be local; this means that the pairs of the operators ought to be taken at the same space-time point. In addition, without going into details, the currents assumed with vector nature are types of vector-axial vector mixtures.

Here, however, it is not yet assumed that the equal mixtures, as in the case of vector-axial (V-A) theory or the derivative coupling, is excluded. Under these assumptions, the weak interaction Hamiltonian given as

$$H_\omega = \frac{G}{r^2}\int d^4x d^4y \psi_a^-(x)\Gamma_\mu \psi_b(x)k^{\mu\lambda}(x-y)\psi_c^-(y)\Gamma_\lambda \psi_d(y) \tag{8}$$

Γ_λ, for the lepton current, where the V-A is always assumed, it takes the form

$$\Gamma_\lambda = \gamma_\lambda \, (1 + \gamma_5) \tag{9}$$

The simplest structure tensor $k^{\mu\lambda} \, (x)$ is given by

$$k^{\mu\lambda} \, (x) = g_{\mu\lambda} \partial^{(4)} (x) \tag{10}$$

This is "the four-fermions model." But in this case, there is still ambiguity concerning the order of the particles a, b, c, and d. It stems from the Fierz transformations:

$$\overline{\Psi}_a \, (x) \, \gamma_\lambda (1 - \gamma_5) \Psi_b \, (x) \, \overline{\Psi}_c \, (x) \, \gamma^\lambda (1 + \gamma_5) \, \Psi_d \, (x) + \overline{\Psi}_a \, (x) \, \gamma_\lambda (1 + \gamma_5) \, \Psi_d \, (x) \, \overline{\Psi}_d \, (x) \gamma^\lambda (1 + \gamma_5) \, \Psi_{b(x)} \tag{11}$$

Current is connected to a scalar-pseudoscalar current with a different ordering of the particles such as

$$\overline{\Psi}_a (x) \, \gamma_\lambda (1 + \gamma_5) \Psi_b \, (x) \, \overline{\Psi}_c \, (x) \, \gamma^\lambda (1 + \gamma_5) \, \Psi_d \, (x) = -2 \, \overline{\Psi}_a (x) \, (1 - \gamma_5$$
$$)C \, \Psi_C^{-T} (x) \Psi_d^T (x) \, C^{-1} (1 + \gamma_5) \, \Psi_b \, (x) \tag{12}$$

where C is 4 × 4 matrices with the following properties:

$$\left. \begin{array}{l} C^{-1} \gamma_\mu \, C = - \gamma_\mu^T \\[8pt] C^{-1} \gamma_5 \, C = \gamma_5^T \\[8pt] U^- C^T = V^T \\[8pt] C^T = -C \end{array} \right\} \tag{13}$$

As long as the weak interactions are regarded as genuine four-fermions interactions, equations (11) and (12) are pure mathematical identities. However, a physical difference jumps up, as soon as the weak interaction is mediated by a boson. It might be recognized, in passing, that the use of the Fierz-transformed interactions of equation (12), together with an intermediate boson, lead to a scalar-pseudoscalar coupling, which is renormalizable. But the paid price for this advantage is, the currents lose their simple properties with respect to baryon number and lepton number. In such a case, the intermediate boson would be a particle carrying both the baryon and the lepton numbers different from zero. Strictly speaking, there would be all kinds of scalar intermediate bosons, with muon, plus electron number, with baryon electron number, and so on. But such assumptions need to be checked against experiment.

The weak interaction Hamiltonian of equation (8), together with the definition of the structure tensor given in equation (10) and the vertex operator of equation (9), allow for a complete calculation of the decay of polarizing muon. The process of such decay is given by

$$\mu^+ \rightarrow e^+ + V_e + V_\mu^-.$$

If ρ is the degree of polarization of the initial muon, (E) and (q) are the energy and the momentum of the emitted electron, where θ is the angle between the polarization axis of the muon and the direction of the emitted electron. The spectrum is given as

$$N(E,\omega) = \frac{G^2}{12\pi^4} Q \, E \, E_0 [3 - 3x + (2/3)\rho(4x - 3) + 6\eta \Sigma(1/x - 1)$$

$$- \rho(q/E)\cos\theta][1 - x + 2\delta(4/3x - 1)] + 0(\epsilon^2)$$

$$(14)$$

where $\epsilon = \dfrac{m_e}{m_\mu}$, $E_0 = \dfrac{m_\mu + m_e}{m_\mu}$, $x = E/E_0$

As it is seen from equation (8), H_ω does not contain free parameters except (G); therefore, the spectrum can be calculated also without further parameters. The four parameters included in equation (14) are to serve the test of the theory. Table 2 gives the predicted theoretical values with their best experimental fits (S. E. Derenzo et al. *Physics Letter*, 288, 401, 1968).

Table 2: Parameters of Muon Decay

Parameter	V-A Prediction	Experimental Value
Michel ρ	3/4	0.7518 ± 0.0026
Shape δ	3/4	0.7540 ± 0.00085
Asymmetry ζ	−1	−0.973 ± 0.014
Eta η_0		−0.12 ±0.21

Hence, the theoretical Hamiltonian [equation (8)] is in good agreement with the experimental fits, as table 2 clarifies that. So equation (8), which is built on the assumption introduced by Herbert Pitchman, can be considered as the foundation on which the theory of weak interaction is based, with careful checking of each new prediction against the experiment results. After equation (11), it was stated that an ambiguity was still there, concerning the order of particles in the interaction Hamiltonian. If it is considered, each current should contain either leptons or baryons; such ambiguity is immediately removed. In other words, in semi-leptonic process, the currents should have definite properties with respect to baryon number and lepton number. Once this situation is taken for granted and the order in muon decay is specified, an empirical rule can be stated as follows:there are no neutral lepton currents; many experiments had confirmed this empirical rule quite well. Table 3 gives the present upper limits (Particle Data Group, *Physics Letter*, 39B / 1 April, 1972).

Table 3: Absence of Neutral Lepton Currents

Decay Mode Branching Ratio Limit

$$K^{\pm} \rightarrow \pi^{\pm} \ e^{+} e^{-} < 4.1 \times 10^{-7}$$

$$\rightarrow \pi^{\pm} \mu^{+} \mu^{-} \ < 2.6 \times 10^{-6}$$

$$\rightarrow \pi^{\pm} \nu \nu^{-} \ < 1.2 \times 10^{-6}$$

$$K_{l}^{0} \rightarrow \mu^{+} \mu^{-} \qquad \left. \begin{array}{l} 1.2 \quad +0.8 \\ = 1.2 \quad -0.4 \times 10^{-8} \end{array} \right]$$

Equation (10) represents the simplest choice for the structure tensor $K_{\mu\lambda}$ (x), where the case of the intermediate boson model is the first nontrivial possibility, where this structure tensor describes the propagation of the intermediate vector boson. It is given by

$$K_{\lambda\mu}(p) = \frac{\sqrt{2g^{2}}}{G} [g_{\lambda\mu} - \frac{P_{\lambda}P_{\mu}}{M_{\omega}^{3}}] \frac{1}{P^{2} - M_{\omega}^{2}} \qquad (15)$$

Equation (15) represents the Fourier transformation of the structure matrix $K_{\lambda\mu}$. Here, it should be noted, in the intermediate boson model, that the basic Hamiltonian is for the simple boson-two-fermion vertex.

Therefore, the Hamiltonian in equation (11) for this case is an effective one, for four Fermi processes mediated by intermediate vector boson. So it is possible to observe, in the intermediate vector boson model, that the decay of the muon can no longer be parameterized in the given by equation(14). In this case, the best parameterization is given by

$$N(x) = \frac{G^2 m_\mu^5}{96\pi^3} x^2 [3 - 2x + \zeta x(2 - x)] \tag{16}$$

where $\zeta = m_\mu^2 / m_\omega$ \hfill (17)

So far, the possible weak interaction Hamiltonian is discussed in general. Now the actual computations have to be considered, comparing that with the available experimental data. It is important here to remember that so far, no non-locality has been actually determined. Hence, the so simpler case will be taken, and the weak interaction Hamiltonian is given as

$$H_\omega = \frac{G}{\sqrt{2}} \int d^4x J_\lambda^+(x) J^\lambda(x) \tag{18}$$

where $J_\lambda(x)$ is the total weak current given by

$$J_\lambda(x) = l_{\lambda(x)} + \alpha j_\lambda^\pi(x) + \beta j_\lambda^k(x) \tag{19}$$

where l_λ is the lepton current given by

$$l_{\lambda(x)} = \sum_l \Psi_l^-(x) \gamma_\lambda (1 + \gamma_5) \Psi_{v_l}(x), \quad l = e, \mu \tag{20}$$

j^{π}_{λ} and j^{k}_{λ} are the two hadronic currents that carry the quantum numbers of charged pions and kaons, respectively, multiplied by α and β as an extra strength factor. These factors should satisfy the condition

$$\alpha^2 + \beta^2 = 1 \tag{21}$$

From (21), it can be concluded that $\alpha = \cos \theta$, $\beta = \sin \theta$.

The angle (θ) is the so-called Cabibbo angle, first introduced by Gell-Mann and Levy (*Nuovo cimento*, 16, 705, 1958).

IX-5—The Selection Rule lΔYl $\leq$ 1

ΔY is the change in the hypercharge (chapter 8). Hamiltonian given by equation (18) is consistent with this selection rule. But its support from the semi-leptonic processes is not too overwhelming; this is illustrated in table 4. In weak hadronic processes, there is a very convincing one, the $k^0_1 - k^0_2$ mass difference. It is $\Delta Y = 2$ phenomenon, and quantitatively, it is of second order in the weak coupling constant, proving that $\Delta Y = 2$ processes are absent to first order in the coupling constant (G).

Table 4: Limits for lΔYl = 2 Transition

Transition	Limit
$\Xi^0 \rightarrow p_\iota \nu_e$	1.3×10^{-3} for $\iota = e, \mu$
$\rightarrow p\,\pi^- <0.9 \times 10^{-3}$	
$\Xi^- \rightarrow n\,e^- \bar{v_e} <10^{-2}$	
$\rightarrow n\,\pi^-$	$<1.1 \times 10^{-3}$

Assuming the selection rule $|\Delta Y| \leq 1$, there is a total number of thirty-four semi-leptonic baryon decays compatible with kinematic constraints. In this number, muonic and electronic modes of the otherwise same decay were not counted separately. Tables 5, 6, and 7 show these thirty-four decays processes.

Table 5: Hypercharge Conserving Semi-Leptonic Baryon Decays

Observed Decays Not Yet (1978) Observed

$$n \rightarrow p + e^- + \bar{v_e} \quad \Sigma^0 \rightarrow \Sigma^+ + e^- + \bar{v_e}$$

$$\Sigma^- \rightarrow \Lambda + e^- + \bar{v_e} \quad \Sigma^- \rightarrow \Sigma^0 + e^- + \bar{v_e}$$

$$\Sigma^+ \rightarrow \Lambda + e^+ + v_e \quad \Xi^- \rightarrow \Xi^0 + e^- + \bar{v_e}$$

Table 6: I ΔYI = 1, Decays of Baryons with Three Final Particles

$\Delta Y = \Delta Q$	Notes	$\Delta Y = -\Delta Q$	Notes
$\Lambda \rightarrow p + e^- + \bar{v_e}$	Observed	$\Sigma^+ \rightarrow n + e^+ + v_e$	observed?
$\Sigma^0 \rightarrow p + e^- + \bar{v_e}$		$\Xi^0 \rightarrow \Sigma^- + e^+ + v_e$	
$\Sigma^- \rightarrow n + e + \bar{v_e}$	observed		
$\Xi^0 \rightarrow \Sigma^+ + e^- + \bar{v_e}$			
$\Xi^- \rightarrow \Lambda + e^- + \bar{v_e}$	observed		
$\Xi^- \rightarrow \Sigma^0 + e^- + \bar{v_e}$			
$\Omega^- \rightarrow \Xi^0 + e^- + \bar{v_e}$			

Table 7: $|\Delta Y| = 1$, Decays of Baryons with
More than Three Final Particles

$\Delta Y = \Delta Q$ $\Delta Y = -\Delta Q$

$\Lambda \rightarrow P + \pi^0 + e^- + \overline{v}_e$ $\Lambda \rightarrow n + \pi^- + e^+ + v_e$

$\rightarrow n + \pi^+ + e^- + \overline{v}_e$ $\Sigma^0 \rightarrow n + \pi^- + e^+ + v_e$

$\Sigma^0 \rightarrow p + \pi^0 + e^- + \overline{v}_e \; \Sigma^+ \rightarrow p + \pi^- + e^+ + v_e \rightarrow n + \pi^0 + e^+ + v_e$

$\rightarrow n + \pi^+ + e^- + \overline{v}_e \; \Omega^- \rightarrow \Xi^- + \pi^- + e^+ + v_e$

$\Sigma^- \rightarrow n + \pi^0 + e^- + \overline{v}_e \rightarrow \Xi^- + \pi^- + \pi^0 + e^+ + v_e$

$\rightarrow p + \pi^- + e^- + \overline{v}_e \rightarrow \Xi^0 + \pi^- + \pi^- + e^+ + v_e$

$\Xi^= \rightarrow \Lambda + \pi^0 + e^- + \overline{v}_e$

$\Omega^- \rightarrow \Xi^0 + \pi^0 + e^- + \overline{v}_e$

$\Omega^- \rightarrow \Xi^- + \pi^+ + e^- + \overline{v}_e$

$\rightarrow \Xi^0 + \pi^0 + \pi^0 + e^- + \overline{v}_e$

$\rightarrow \Xi^0 + \pi^+ + \pi^- + e^- + \overline{v}_e$

$\rightarrow \Xi^- + \pi^+ + \pi^0 + e^- + \overline{v}_e$

It can be noticed that there are no muonic decay processes with ΔY = 0. If a "hadron charge change" is defined by $\Delta Q \equiv (Q_f - Q_i)$, where Q_f and Q_i are the final and the initial charges of the hadrons, respectively. Then only one of the six decays with ΔY = 0 has $\Delta Q = -1$; the other five decays have $\Delta Q = +1$. Also, the hypercharge change in tables 6 and 7 is defined as $\Delta Y = Y_f - Y_i$.

There is an inherent asymmetry in the quantum number of the baryons (no positive strangeness occurs among baryons); hence, only two of the nine $|\Delta Y| = 1$ decays with three final particles have $\Delta Y = -\Delta Q$. But there is no kinematic reason why the decay Σ^+-decay, for example, would not occur with similar frequency as for the $\Sigma\Sigma --$ decay. Yet it takes two to three years before another possible Σ^+-decay is found, and only few possible candidates have been accumulated so far. On the other hand, several hundreds of Σ — decays are collected so that it can be set up for another empirical selection rule,

$$\Delta Y = \Delta Q \qquad\qquad (22)$$

Keeping in mind that it might not be a strict rule, but it certainly holds to a good accuracy (see table 8 next).

Table 8: The Selection Rules for $\Delta Y = \Delta Q$

$$\frac{\Sigma^+ \to e^+ n v_e^-}{\Sigma^- \to e^- n v_e} < 0.018 \qquad\qquad \frac{\Sigma^+ \to \mu^+ n v_\mu^-}{\Sigma^- \to \mu^- n v_\mu} < 0.0995$$

$$\frac{K_{L_4}^+ (\Delta Y = -\Delta Q)}{K_{L_4}^+ (\Delta Y - \Delta Q)} < 0.006$$

The Gell-Mann-Nishijima formula $Q = I_3 + Y/2$ (refer to chapter 8), where (I) is isotopic spin, allows to build up selection rules relations with the isotopic spin. From this formula, it can be written as

$$\Delta Q = \Delta I_3 + 1/2 \; \Delta Y \tag{23}$$

Consequently,

$\Delta Q = \Delta Y \Rightarrow \Delta I_3 = +1/2$, $\Delta Q = -\Delta Y \Rightarrow \Delta I_3 = 3/2$; therefore, $\Delta I \geq 1/2, \; \geq 3/2$, respectively. ΔI is defined as $|I_f - I_i|$.

Thus, $\Delta Q = -\Delta Y$ processes require at least $\Delta I = 3/2$; therefore, they are more complex structures than $\Delta Q = \Delta Y$ processes, which are compatible with $\Delta I = 1/2$. Similar relations can be derived for hadronic processes. Since ΔQ is always zero in this case, equation (23) can be modified to become

$$\Delta I_3 = 1/2 \; \Delta Y \tag{24}$$

(hadronic decay), so it can be obtained that

$$|\Delta Y| = 1 , \Delta I_3 = \pm 1/2, \Delta I \geq 1/2 \tag{25}$$

$$|\Delta Y| = 2 , \Delta I_3 = \pm 1, \Delta I \geq 1 \tag{26}$$

So the one-half isotopic spin $I = 1/2$ rule, as a hadronic process, restricts the hypercharge change to unity.

Now, for strange particles, the charge Q is given by

$$Q = I_3 + 1/2(S + B) \tag{27}$$

where $\Delta Q = \Delta I_3 + 1/2(\Delta S + \Delta B)$ (28)

S is the strangeness quantum number;

B is the baryon quantum number.

Equation (27) is applied to the baryons and the bosons involved in the decay. Therefore, there are two cases:

Case 1: $\Delta Q = 0$, $\Delta S = \pm 1$, $\Delta B = 0$; hence, if $\Delta Q = 0$, that indicates that the decay is non-leptonic.

Case 2: $\Delta Q = \pm 1$, $\Delta S = \pm 1$, $\Delta B = 0$; this implies that

$\Delta I_3 = \pm 1/2$ or $|\Delta I_3| = 1/2$

Suppose the more restrictive rule $|\Delta I| = 1/2$ is adopted, or $\{I = 0\} \rightarrow$

$\{\text{mixture}(I = 1/2)$ and $(I = \dfrac{3}{2})\}$, in general $\Lambda \rightarrow \Big\rceil \; p + \pi^-$

$\rightarrow n + \pi^0$

However, under the $|\Delta I| = 1/2$ rule, $\{I = 0\} \qquad \rightarrow \Big\rfloor \; \{I = 1/2\}$

Using the usual rules for combining angular momenta, with the fact that π has $I = 1$ and n has $I = 1/2$, the pure isotopic spin state

$$(\chi_{1/2}^{-1/2}) \text{ is given as } \chi_{1/2}^{-1/2} = \sqrt{\frac{2}{3}} \; p\pi^- - \sqrt{\frac{1}{3}} \; n\pi^0 \qquad (29)$$

With total isospin 1/2 and z-component (−1/2), furthermore,

$$P \pi^- = \sqrt{\frac{1}{3}} x_{1/2}^{-1/2} + \sqrt{\frac{2}{3}} x_{1/2}^{-1/2}$$

$$n\pi^0 = \sqrt{\frac{2}{3}} x_{1/2}^{-1/2} - \sqrt{\frac{1}{3}} x_{1/2}^{-1/2}$$

(30)

Now let (M) be the operator connecting χ_0^0 and $\chi_{1/2}^{I_3}$ states, then the

amplitude (Amp)$[\Lambda \rightarrow p + \pi^-] = <\chi_0^0 |M|p \pi^-> = \sqrt{\frac{2}{3}}$ Amp (a)

And Amp $[\Lambda \rightarrow n + \pi^0] = <\chi_0^0 |M|n\pi^0> = -\sqrt{\frac{1}{3}}$ Amp—(b)

Hence, $= \frac{Amp(a)}{Amp(b)} = |\sqrt{\frac{\frac{2}{3}}{\frac{-1}{3}}}|^2 = (-\sqrt{2})^2 = 2$ (31)

Still for case 2, if $\Delta Q = \Delta I_3 + (1/2) \Delta S$

Now, assume $\Delta I_3 = \pm 1/2$. Then, $\pm 1 = \pm 1/2 \pm 1/2$, so for $\Delta Q = 1 = 1/2 + 1/2 \rightarrow \Delta Q = \Delta S$. For $\Delta Q = -1 = -1/2 -1/2 = -1 \rightarrow \Delta Q = \Delta S$.

Assume $\Delta I_3 = \pm 3/2$, so

$\Delta Q = 1 = 3/2 - 1/2$

$\Delta Q = -1 = -3/2 + 1/2$

Therefore, $| \vec{I} | = 1/2$ implies $\Delta Q = \Delta S$, and $\Delta S = -\Delta Q$ implies that

$| \Delta \vec{I} | \geq 3/2$. So $K^0 \rightarrow \pi^+ + e^- + v$, $\Delta Q/\Delta S = +1/-1 = -1$

$k^0 \rightarrow \pi^- + e^- + v$, $\Delta Q/\Delta S = -1/-1 = 1$

where $\Delta Q = Q_K - Q_\pi$, $\Delta S = S_K - S_\pi$

The decay $K^0 \to \pi^+ + e^- + v$ has been observed; hence, the $|\Delta I| = 1/2$ rule is not rigorously valid for leptonic decay.

IX-6— T - Θ Puzzle

In 1956, it was discovered that there were two kinds of k–meson with the same mass, charge, and lifetime. They only differ in their decays. They decay as follows:

$$\left.\begin{array}{l} \Theta \to \pi^+ + \pi^0 \\ T \to \pi^+ + \pi^+ + \pi^- \\ \quad \to \pi^0 + \pi^0 + \pi^+ \end{array}\right\} \tag{32}$$

Dalite (*Report Progress in Physics*, 20, 163, 1957) has showed that these particles are distinct particles, under the assumption of parity conservation and identical particles—that is to say,

$$P(\Theta) = (-1)^2(-1)^S \Theta \tag{33}$$

where $P(\Theta)$ is the parity of the particle (Θ) and S is its spin.

The final state of the particle T can be pictured as

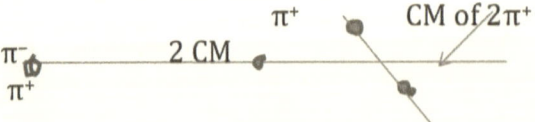

The motion can be divided into motion of two identical pions about their center of mass and the motions of the di-pion and the odd pion about the center of mass of the entire system. Let ℓ be the angular momentum of the di-pion system and $\mathcal{L}$ be the angular momentum of the di-pion and odd pion about the center of mass of the entire system. Let S_T be $\mathcal{T}$ spin, then,

$$\overrightarrow{S} = \overrightarrow{\ell} + \overrightarrow{\mathcal{L}} \quad \text{or} \quad 1 \, \mathcal{L} - \ell 1 \leq S \leq 1 \, \mathcal{L} + \ell 1$$

$$\text{So } P(\mathcal{T}) = (-1)^3 (-1)^{\ell} (-1)^{\mathcal{L}} \tag{34}$$

Since the di-pion consists of two identical baryons, $\ell = 0, 2, —$

$$\text{Hence, } P(\mathcal{T}) = (-1)^3 (-1)^{\mathcal{L}} \tag{35}$$

Suppose $S_\Theta = S_T = 0$, then $P(\Theta) = 1 = -P(\mathcal{T})$

Let these results be discussed further as follows:

a. Suppose π^- has maximum momentum, then $\ell = 0$, because the two π^+ travel parallel to each other. In this case, $\mathcal{L} = s_T$; therefore,

$$P(\mathcal{T}) = (-1)^3 (-1)^{S_T} = -P(\Theta) \tag{36}$$

b. Suppose π^- is at rest, then $\mathcal{L} = 0$, and $\ell = J$; therefore,

$P(\mathcal{T}) = -1 = -P(\Theta)$, since $\mathcal{L} = 0, 2, —$, and $S_T = S_\Theta$.

c. Suppose all the pions are collinear, then the wave function can depend only on one angle, so $T \sim Y_J^M (\Theta)$; hence, either $\mathcal{L}$ or $\ell = 0$. Thus, (T) $= (-1)^3 (-1)^J = P(\Theta)$. Note (a) is subcase to(c).

The experiment indicated that particle Θ is of 0^+ and T is of $0^-, 2^-$.

Now if T is of 2^-, $T \rightarrow \pi^+ \gamma$ is allowed.

But if T is of 0^-, $T \rightarrow \pi^+ \gamma$ is forbidden because the pion has no spin to balance the photon (γ) helicity. Therefore, such transition has not been seen yet. The simplest solution that can be assumed is, the parity is not conserved. However, in k-decay, there are no polarizations to use for testing parity nonconservation, so the decay of lambda might be used:

$\Lambda \rightarrow P + \pi^-$. This might be represented graphically as

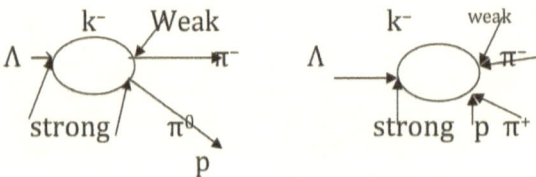

The above is the possible diagram, which, if mixed, could account for nonconservation of parity. The ($\sigma_\Lambda . p\pi$) term in Λ decay can be looked over. Apparently, $\mathcal{L} = s_\Lambda \pm 1/2$, where $\mathcal{L}$ is the p, π orbital angular momentum. Consider $\pi^+ p \rightarrow \Lambda + k_0$, a strong interaction:

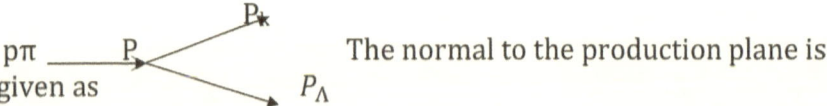

The normal to the production plane is given as

$$n^\wedge = \frac{\vec{P}\pi \times \vec{P}_\Lambda}{\vec{P} \times \vec{P}_\Lambda},$$ $\vec{S}_\Lambda \cdot \vec{n}$, is scalar, therefore, Λ emerge from this strong interaction polarization, even with parity conservation.

Consider $\Lambda \rightarrow P + \pi$, a weak interaction. $\vec{S}_\Lambda \cdot \vec{P}\pi$ is a pseudoscalar; therefore, the correlation between $\vec{S}_\Lambda$ and $\vec{P}\pi$ implies that parity is not conserved, as it will be clarified next.

Define a phase space as $\dfrac{1}{\psi} \sim (1 + a\vec{S}_\Lambda \cdot \vec{P}\pi)$, it is the asymmetry parameter. However, counting up and down asymmetry, it really measures a_p, where p is lambda polarization. Oversheth et al. (*Physical Review*, 129, 1795, 1963) found from this that three conclusions can be drawn:

a. Parity is not conserved.

b. By using the CPT theorem, it can be concluded that the decay is not invariant under charge conjugation. The following physical argument is given by Crawford (UCRL-10540). Consider the up-down asymmetry, neglecting the final state interactions; the final state is a plane wave. In this case, the spins are parallel; no z-component for the spins, represented by (A) graph:

(A) P ↑↑S_P P ↓ S_P

    Λ ●   Apply Time Reversal   → Λ ●

      π↑$|S_\Lambda$ π ↓$|S_\Lambda$

This process corresponds to some matrix elements squared l<ilMlf>l².
Complex conjugating interchanges the initial and the final states.

Complex conjugate (B) reflects the system in a mirror perpendicular

S_Λ (parity operation) as follows:

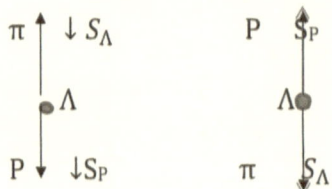

Due to CPT theorem, C = TP, if C is invariance, holds. Thus, the
probability for case A should be the same as that for (B), excluding
final state interactions. In addition to this, there should be no up-down
asymmetry. Final state interaction allows a = 0.2 without violating C
invariance. The a = 0.62 value implies failure of C invariance.

c. Lee and Yang (*Physical Review*, 109, 1755, 1958) have proved that

1. $\left| < \zeta > \right| \le \dfrac{1}{2S_A + 2}$

where ξ is the cosine of the angle between the decaying pion
momentum and the Z-axis. The $<\xi>$ is the average of ξ, with respect
to $I(\xi)$, its distribution.

2. If $I(\xi) = 1/2(1 + a\xi)$ and $< \zeta > \le \dfrac{1}{6S_A}$

$<1\ \xi\ 1> = 1/2 \int_{-1}^{+1} \xi(1 + a\xi)\, d\xi = [1/2\ \xi^2 + a\ 1/3\ \xi^3] = 2/3$

Therefore,

$$|a| \leq \frac{1}{2S_A} \left. \begin{cases} \text{if } S_A = 1/2, \text{ then } |a| \leq 1 \\ \text{if } S_A = 3/2, \text{ then } |a| \leq 1/3 \end{cases} \right.$$

Hence, $S_\Lambda = 1/2$.

IX-7—Non-Relativistic Analysis $\Lambda \rightarrow P + \pi^-$

If $S_\Lambda = 1/2$, then the πp state must be S-wave or p-wave. So the connecting operator between the initial and the final states M is given by

M= S + P $\vec{\sigma} . \hat{q}$, $\hat{q} = \vec{P}\pi/|P \pi|$.

S is scalar due to the S-wave; P is a pseudoscalar due to the P-wave. The factor $\vec{\sigma} . \hat{q}$ is pseudoscalar too. In the first form, there is one vector $\vec{P}\pi$, $\vec{p}\rho$ or the relative momentum. There is one pseudovector $\vec{J}$. In the initial state, $\vec{J} = (1/2) \vec{\sigma}$; in the final state, $\vec{J} = \mathcal{L} + (1/2)\vec{\sigma}$.

Then, $\psi_i (1/2)\vec{\sigma} . \hat{q} \psi_i = \psi_\rho (\mathcal{L} + (1/2) \vec{\sigma}) . \hat{q} \psi_\rho$ and $\vec{\mathcal{L}} . \hat{q} =$

$(\vec{r} \times p_r) . \hat{q} = 0$. r is the relative vector $\vec{p}$; r is the relative momentum parallel to q^. The final state is assumed as plane wave (no interaction). Hence,

$$(\psi_i \sigma . q^\wedge \psi_i) = (\psi_f \vec{\sigma_p} . \hat{q} \psi_f) \tag{36}$$

Thus, the expectation of $\sigma^{\rightarrow} . q\hat{}$ is equal to the expectation of the proton helicity [this is the basis of Cronin and Overseth's determination of (a)].

The matrix element is $(\psi_p^+ M \psi_i)$, where both states, of course, are spinor. The final state is a combination of scalar and proton spinor. The projection operators are

$$P_\Lambda = (1/2) (1 + n_\Lambda\hat{} \cdot \sigma^{\rightarrow}) \text{ for Lambda}$$

$$P_p = (1/2)(1 + n_p^{\rightarrow} \cdot \sigma^{\rightarrow}), \text{ for proton, where } n_\Lambda\hat{} \text{ is the lambda}$$

polarization and $n_p\hat{}$ is the proton polarization.

So $|M|^2 = (\psi_p^+ M, \psi_i)(\psi_i^+ M^* \psi_f) = \text{Tr}(M P_\Lambda M^* P_p)$ (37).

To see how these traces can be used, to evaluate the square matrix elements, it is necessary that the methods of the relativistic quantum mechanics be reviewed to see how easy for it to be extended.

By factoring the brackets, using the following vectors products,

$$(\sigma^{\rightarrow} \cdot q\hat{})(\sigma^{\rightarrow} \cdot n_\Lambda\hat{}) = n_\Lambda\hat{} \cdot q\hat{} + i\sigma^{\rightarrow} \cdot (q\hat{} x n_\Lambda\hat{})$$

$$(\sigma^{\rightarrow} \cdot q\hat{})(\sigma^{\rightarrow} \cdot n_p\hat{}) = n_p\hat{} \cdot q\hat{} + i\sigma^{\rightarrow} \cdot (q\hat{} x n_p\hat{})$$

$|M|^2$ will be

$$|M|^2 = 1/4\text{Tr}[\{(S + P n_\Lambda\hat{} \cdot q\hat{}) + \sigma^{\rightarrow} \cdot (Pq\hat{} + S\ n_\Lambda\hat{} + P_i \cdot (q\hat{} x n_\Lambda\hat{})\}\{(S^* + P^*$$

$$n_p\hat{} q\hat{})$$

$$+ \sigma^{\rightarrow} (P^{\rightarrow} q\hat{} + S^* n_p\hat{} + P)(q\hat{} x n_p\hat{})$$ (38)

It is possible to continue with the aid of the vector identity

$$(\hat{n}_\Lambda \cdot \hat{n}_p) = (\hat{q} \times \hat{n}_\Lambda)(\hat{q} \times \hat{n}_p) + (\hat{q} \cdot \hat{n}_p)(\hat{q} \cdot \hat{n}_\Lambda)$$

to obtain

$$\left|M\right|^2 = [1 + a\hat{n}_\Lambda \cdot \hat{q}] + \hat{n}_p \cdot [(1 + \hat{n}_p \cdot \hat{q})\hat{q} + G\hat{n} \times \hat{q} + \gamma(\hat{q} \times \hat{n}_\Lambda)\hat{q}] \quad (39)$$

where $a = 2\text{Re}(SP^*/(1 Sl^2 + lPl^2))$,

$G = 2\text{ Im}SP^*/(1 S l^2 + lPl^2)$, $\gamma = lSl^2 - lP^*l^2/l Sl^2 + lPl^2$

The last terms in equation (39) are only detectable if the proton is examined from the decay (Lee and Yang, *Physical Review*, 108, 1645, 1957, and R. Gatto, *Nuclear Physics* 5, 183, 1958).

To this, we end this chapter; the other details are left to the quantum field theory, to be included in our next book, *Quantum Mechanics*.

Appendix (A)

Some Solved Problems

Problem 1

For a moving electron about a nucleus, the total energy is the Hamiltonian function action itself.

The Solution

$$H = \frac{P_r^2}{2m} + \frac{P_\theta^2}{2mr^2} + V(r) = E$$

$^r e^-, m_e$

$$P_r = \pm \sqrt{\frac{-K\hbar^2}{r^2} + \frac{2me^2}{4\pi\varepsilon_o r} + 2mE} = \pm\sqrt{\frac{R}{r^2}} \qquad (1)$$

where $p_r^2 = k^2 \hbar^2$

$$R = a + br + cr^2$$

$$\left.\begin{array}{l} \\ \\ \\ \end{array}\right\} \qquad (2)$$

$$a = -K^2\hbar^2 \qquad , b = \frac{2me^2}{4\pi\varepsilon_o} \qquad , c = 2mE$$

Now,

$$\oint P_r dr = n_r h$$

The limits of the radius (r) at which (P_r) becomes zero are

$$r_1 r_2 = \frac{-b \pm \sqrt{b^2 - 4ac}}{2c} \tag{3}$$

We can write

$$\oint P_r dr = 2 \int_{r_1}^{r_2} P_r dr \tag{4}$$

Using the standard integral tables, you find

$$\int \frac{\sqrt{R}}{r} dr = \sqrt{R} + \frac{b}{2\sqrt{-c}} sin^{-1} \frac{-2cr - b}{\sqrt{b^2 - 4ac}} - \sqrt{-a} \, sin^{-1} \frac{br - 2a}{r\sqrt{b^2 - 4ac}}$$

At the upper limit and the lower limit, respectively, the quantities whose inverse sine we are taking is ±1 so that the difference of the inverse sine at the two limits is π, and the quantity $\sqrt{R}$ vanishes at both limits.

Thus, we have

$$\oint P_r dr = 2\pi \left(\frac{b}{2\sqrt{-c}} - \sqrt{-a} \right) = 2\pi \left(\frac{me^2}{4\pi\varepsilon_o \sqrt{-2mE}} - K\hbar \right)$$

$$= n_r h$$

$$2\pi\left(\frac{me^2}{4\pi\varepsilon_o}\frac{1}{\sqrt{-2mE}}\right) = h(n_r + K)$$

Then

$$\frac{4\pi^2(m^2e^4)}{(4\pi\varepsilon_o)^2(-2mE)} = h^2(n_r + K)^2 = n^2h^2$$

Because $n = n_r + K$ = total principal quantum number, where n_r is the radial Q number and K is the azimuthal Q number. Now let $(4\pi\varepsilon_o)$ be a unit.

$$\frac{-2\pi^2me^4}{E} = n^2h^2 \qquad \Rightarrow \qquad E = -\frac{1}{2}\frac{me^4}{n^2\hbar^2} \qquad (5)$$

Thus,

$$H = E = -\frac{1}{2}\frac{me^4}{n^2\hbar^2} \qquad (6)$$

Now to find the radius (r_n),

$$2\pi r_n = n\lambda \qquad\qquad , n = 1, 2, 3, \dots$$

$$r_n = n\lambdabar, \qquad \lambdabar = \frac{\lambda}{2\pi},$$

or $\qquad\qquad$ and since

$$\lambda = \frac{h}{P}$$

then

$$r_n = \frac{n\hbar}{m\upsilon_n},$$

(7)

Now the centrifugal force = the coulomb force, which yields

$$\frac{Ze^2}{r_n^2} = \frac{mv_n^2}{r_n} \Rightarrow mv_n = \frac{Ze^2}{r_n} \Rightarrow r_n = \frac{Ze^2}{mv_n^2}$$

(8)

From (7) and (8), we find

$$\upsilon_n \frac{Ze^2}{n\hbar}$$

(9)

Now substitute (9) in (7); you find

$$r_n = \frac{n^2\hbar^2}{me^2} \cdot \frac{1}{Z}$$

(10)

But $Z = 1$ for H, then

$$r_n = \left[\frac{n^2\hbar^2}{me^2} \right]$$

(11)

Using the data $\hbar = 1.054 \times 10^{-27}$ erg. sec,

$$m_e = 9.1 \times 10^{-31} \text{ gm}$$

$$e = 1.6 \times 10^{-19} \text{ coulomb}$$

So the smallest radius (ground state) when $n = 1$ is $r_1 = 0.5 \overset{\circ}{A}$ (Bohr-radius) substitution for m, ℓ and $\hbar$ into (6); you find

$$E_1 = -13.6 eV \hspace{4cm} \text{QED}$$

Problem 2

Consider the potential as $V(r) = -V_o e^{-ar}$. Use the wave function equation for this V(r) and apply the condition requiring only one bound state; see what you get.

The Solution

Write down the Schrodinger equation

$$\frac{d^2\psi}{dr^2} + \frac{M}{\hbar^2}[E - V(r)]\psi(r) = 0 \tag{1}$$

Now let $u = e^{-ar} \Rightarrow V(r) = -V_o u$ $\hspace{2cm}$ (2)

So that $u = 1$ for $r = 0$, $\quad u = \dfrac{1}{e}$ for $\quad r = \dfrac{1}{a}$, $u = 0$ for $r = \infty$

Now $\dfrac{d}{dr} = -au\left(\dfrac{d}{du}\right)$

Hence, equation (1) becomes

$$\frac{d^2\psi}{du^2} + \frac{1}{u}\frac{d\psi}{du} + \left(\frac{MV_o}{a^2\hbar^2}\frac{1}{u} + \frac{ME}{a^2\hbar^2}\frac{1}{u^2} \right)\psi = 0$$

(3)

For bound state $E < 0$. Let $E = -w$ and $a = \dfrac{1}{a_o}$

where a_o is the breadth of the potential well.

Equation (3) becomes

$$\frac{d^2\psi}{du^2} + \frac{1}{u}\frac{d\psi}{du} + \left(\frac{MV_o a_o^2}{\hbar^2}\frac{1}{u} + \frac{Mw}{\hbar^2}\frac{1}{u^2} \right)\psi = 0$$

(4)

Let

$$\frac{MV_o a_o^2}{\hbar^2} = \alpha^2, \quad \frac{Mw}{\hbar^2} = \beta^2$$

Then equation (4) becomes

$$\frac{d^2\psi}{du^2} + \frac{1}{u}\frac{d\psi}{du} + \left(\frac{\alpha^2}{u} + \frac{\beta^2}{u^2} \right)\psi = 0 \qquad (5)$$

Equation (5) is the differential equation for Bessel function. Therefore, its solution is given by

$$\psi = BJ_n\left(2\alpha e^{-r/2a_o} \right)$$

(6)

where $n = 2a_o(Mw)^{1/2}/\hbar$, which is the order of the Bessel function. The argument of Bessel function becomes small as (r) gets large; therefore, the first term of the ordinary expansion of (J) in a power series is sufficient.

Then we have

$$\psi = \frac{B}{n!}(2\alpha)^n e^{\left[\frac{-(Mw)^{1/2}}{\hbar}r\right]}$$ (7)

$$r \gg a_o$$

Equation (7) is apart from the constant factor, identical with

$$\left.\begin{array}{ll} \psi = C\ \sin[M^{1/2}(V_o - w)^{1/2}r/\hbar & r < a_o \\ \psi = A\ \exp[-(Mw)^{1/2}(r - a_o)/\hbar & r > a_o \end{array}\right\}$$ (8)

Now w is the eigenvalue, which can determined by the condition that $\psi(r = 0)$ must vanish.

So we have to find (n) for a given (V_o) from the condition

$$J_n(2\alpha) = 0$$ (9)

Then (w) can be calculated from the relation

$$n = \left[2a_o (Mw)^{1/2} / \hbar \right]$$

where

$$n^2 = \frac{4a_o^2 (Mw)}{\hbar^2} \Rightarrow w = \frac{n^2 \hbar^2}{4Ma_o^2}.$$

(10)

Now V_o should be greater than a certain limit so that equation (9) has a solution (n) at all.

This limit follows readily from the fact that the first zero (u_n) of $(J_n(u))$ moves toward smaller values of u when n decreases. Therefore, certainly $u_n > u_o$. By looking at page 237 of Jahnke–Emde, we find that the first zero of the Bessel function of order zero is $u_o = 2.4048$.

Hence, equation (9) has a solution only if

$$2\alpha > 2.4048$$

Substitute for

$$\alpha^2 = \frac{MV_o a_o^2}{\hbar^2}.$$

Then,

$$V_o > (\hbar^2 / Ma_o^2) \times 1.4457$$

If $V_o = 1.4457\hbar^2 / Ma_o^2 \Rightarrow$ solution of equation (9) will be n = 0, which implies that $w = 0 \Rightarrow$ no binding energy (BE). If the binding BE remains small compared to V_o, i.e., BE<V_o, then V_o must be only slightly larger than $(\hbar^2 / Ma_o^2) \times 1.4457$.

For different ranges of (a_0), Bethe and Bacher gave same values of $\left[MV_0 a_0^2 \hbar^{-2}\right]$. Some of them are listed below:

$$
\begin{array}{llll}
a_0 = 0 & 0.5 & 1 & 1.5 \\
n = 0 & 0.228 & 0.456 & 0.684 \\
\\
MV_0 a_0^2 \hbar^{-2} = 1.446 & 1.888 & 2.37 & 2.890 \\
V_0 = 59.5 a_0^{-2} & 310 & 97 & 53
\end{array}
$$

QED

Note:

$$
a_0 = \frac{1}{a}
$$

a is the parameter in $V = -V_0 e^{-ar}$.

Problem 3

Show that

1) $\vec{J} \times \vec{J} = i\vec{J}$

2) $[J^2, J_z] = 0$

The Solution

(1) $[J_x, J_y] = iJ_z$

(2) $[J_y, J_z] = iJ_x$ $\hbar = 1$

(3) $[J_z, J_x] = iJ_y$

Then multiply (1) by $\vec{e}_z$ and (2) by $\vec{e}_x$ and (3) by $\vec{e}_y$, where $\vec{e}_x$, $\vec{e}_y$, $\vec{e}_z$ are unit vectors, then you find

$$\vec{e}_x[J_yJ_z - J_zJ_y] + \vec{e}_y[J_zJ_x - J_xJ_z] + \vec{e}_z[J_xJ_y - J_yJ_x]$$

$$= i\{\vec{e}_x J_x + \vec{e}_y J_y + \vec{e}_z J_z\} \qquad \text{or}$$

$$\begin{pmatrix} \vec{e}_x & \vec{e}_y & \vec{e}_z \\ J_x & J_y & J_z \\ J_x & J_y & J_z \end{pmatrix} = i(\underbrace{\vec{e}_x J_x + \vec{e}_y J_y + \vec{e}_z J_z}_{\vec{J}}) = i\vec{J}$$

$\therefore \vec{J} \times \vec{J} = i\vec{J}$. This is the required result for (1).

Now find (2), i.e., $\left[J^2, J_z\right] = 0$

We know that

$$\therefore \left[J^2, J_z\right] = \left[J_x^2, J_z\right] + \left[J_y^2, J_z\right] + \left[J_z^2, J_z\right]$$

$$= 2J_x\left[J_x, J_z\right] + 2J_y\left[J_y, J_z\right] + 2J_z\overset{0}{\left[\cancel{J_z, J_z}\right]}$$

$$= 2J_x\left(-iJ_y\right) + 2J_y\left(iJ_x\right) = -2iJ_xJ_y + 2iJ_xJ_y = 0$$

$$\therefore [J^2, J_z] = 0 \qquad\qquad \text{(as it is required)}$$

Problem 4

Compute the matrix representation of J_x, J_y and J_z in |jm> representation, for j=3/2,2 and 5/2.

The Solution

To compute the matrix elements of J_x, J_y and J_z, the following relations have to be used:

(1) $J_+ \left| jm \right\rangle = \hbar \left[(j - m)(j + m + 1) \right]^{1/2} \left| j, m + 1 \right\rangle$ ...(raising operator)

(2) $J_- \left| jm \right\rangle = \hbar \left[(j + m)(j - m + 1) \right]^{1/2} \left| j, m - 1 \right\rangle$ ...(lower operator)

(3) $J_+ = J_x + iJ_y$ ... (raising operator)

(4) $J_- = J_x - iJ_y$ ... (lower operator)

(5) $J_x = \dfrac{1}{2}[J_+ + J_-]$, $J_y = \dfrac{1}{2i}[J_+ - J_-]$

(6) $[J_+, J_-] = 2\hbar J_z \Rightarrow$

(7) $J_z = \dfrac{1}{2\hbar}[J_+, J_-]$

Now

$$\left\langle j, m + 1 \left| J_x \right| jm \right\rangle = \frac{1}{2}\left[\left\langle j, m + 1 \left| J_+ \right| jm \right\rangle + \left\langle j, m + 1 \left| J_- \right| jm \right\rangle \right]$$

$$= \frac{1}{2}\hbar \left[(j - m)(j + m + 1) \right]^{\frac{1}{2}} \tag{8}$$

Remember, $(\langle j,m+1/jm+1\rangle)=1, \qquad \langle j,m+1/jm-1\rangle=0$

Now

$$\langle j,m-1|J_x|jm\rangle = \frac{\hbar}{2}[(j+m)(j-m+1)]^{\frac{1}{2}}$$

(9)

Verify this.

Now for J_y, following the same procedures, one finds

$$\langle j,m+1|J_y|jm\rangle = \frac{-i\hbar}{2}[(j-m)(j+m+1)]^{\frac{1}{2}}$$

and $\langle j,m-1|J_y|jm\rangle = \frac{+i\hbar}{2}[(j+m)(j-m+1)]^{\frac{1}{2}}$

for J_z, which is given by (7) and also from (5).

$$J_+J_- = J^2 - J_z(J_z-\hbar)$$

$$J_-J_+ = J^2 - J_z(J_z+\hbar)$$

So accordingly, we get

$$\langle jm|J_z|jm\rangle = \frac{\hbar^2}{2\hbar}[\langle jm|jm\rangle[j(j+1)-m(m-1)]]^{\frac{1}{2}}$$

$$\left[\langle jm|jm\rangle[j(j+1)-m(m+1)]\right]^{\frac{1}{2}}$$

$$\frac{\hbar}{2}\left[[(j+m)(j-m+1)]^{\frac{1}{2}} - [(j-m)(j+m+1)]^{\frac{1}{2}}\right]$$

We know $J_z \mid jm \rangle = \hbar m = \hbar m \Rightarrow$

$$\langle jm+1 \mid J_z \mid jm \rangle = \hbar m \langle jm+1 \mid jm \rangle = 0$$

$$\langle jm-1 \mid J_z \mid jm \rangle = \hbar m \langle jm-1 \mid jm \rangle = 0$$

$$\langle jm \mid J_z \mid jm \rangle = \hbar m \langle jm \mid jm \rangle = \hbar m$$

Therefore, $\langle jm \mid J_z \mid jm \rangle = \hbar m$ (12)

So far, we have found the formulas (8, 9, 10, 11, 12) of the matrix representation of J_x and J_z. Now we calculate these for the j values 3/2, 2, 5/2.

Now take

$$j = \frac{3}{2} \Rightarrow m = \frac{3}{2}, \frac{1}{2}, \frac{-1}{2}, \frac{-3}{2} .$$

From (8)

$$\left\langle \frac{3}{2} \frac{5}{2} \middle| J_x \middle| \frac{3}{2} \frac{3}{2} \right\rangle = \frac{\hbar}{2}[0] = 0 \qquad .m = \frac{3}{2}$$

$$\left\langle \frac{3}{2} \frac{3}{2} \middle| J_x \middle| \frac{3}{2} \frac{1}{2} \right\rangle = \frac{\hbar}{2} \left[\left(\frac{3}{2}, \frac{-1}{2} \right) \left(\frac{3}{2} + \frac{1}{2} + 1 \right) \right]^{1/2} = \frac{\hbar}{2} \left[\frac{b}{2} \right]^{1/2}$$

$$= \hbar \frac{\sqrt{3}}{2}$$

$$\therefore \left\langle \frac{3}{2}\frac{3}{2}\left|J_x\right|\frac{3}{2}\frac{1}{2}\right\rangle = \frac{\sqrt{3}}{2}\hbar \qquad, m = \frac{1}{2}$$

$$\left\langle \frac{3}{2}\frac{1}{2}\left|J_x\right|\frac{3}{2}\frac{-1}{2}\right\rangle = \hbar \qquad, m = -\frac{1}{2}$$

$$\left\langle \frac{3}{2}\frac{-3}{2}\left|J_x\right|\frac{3}{2}\frac{-3}{2}\right\rangle = \hbar\frac{\sqrt{3}}{2} \qquad, m = -\frac{3}{2}$$

From (9)

$$\left\langle \frac{3}{2}\frac{1}{2}\left|J_x\right|\frac{3}{2}\frac{3}{2}\right\rangle = \frac{\sqrt{3}}{2}\hbar \qquad, m = \frac{3}{2}$$

$$\left\langle \frac{3}{2}\frac{-3}{2}\left|J_x\right|\frac{3}{2}\frac{-1}{2}\right\rangle = \frac{\hbar}{\sqrt{2}} \qquad, m = -\frac{1}{2}$$

$$\left\langle \frac{3}{2}\frac{-1}{2}\left|J_x\right|\frac{3}{2}\frac{1}{2}\right\rangle = \hbar \qquad, m = \frac{1}{2}$$

$$\left\langle \frac{3}{2}\frac{-5}{2}\left|J_x\right|\frac{3}{2}\frac{-1}{2}\right\rangle = 0 \qquad, m = \frac{-3}{2}$$

Now for $j = 2 \Rightarrow m$, we have five values that are

$m = -2, -1, 0, +1, +2$.

By the same way,

$$\langle 2,-1|J_x|2,-2\rangle = \frac{\hbar}{2}\sqrt{(2+2)(2-2+1)} = \hbar \qquad m = -2$$

$$\langle 2,0|J_x|2,-1\rangle = \frac{\sqrt{3}}{2}\hbar \qquad\qquad m = -1$$

$$\langle 2,+1|J_x|2,0\rangle = \frac{\hbar}{2}\sqrt{6} = \frac{\sqrt{3}}{2}\hbar \qquad\qquad m = 0$$

$$\langle 2,3|J_x|2,2\rangle = 0 \qquad\qquad m = 0$$

$$\langle 2,2|J_x|2,1\rangle = \hbar \qquad\qquad m = 1$$

By the same way, from (9), we have

$$\langle 2,-3|J_x|2,-2\rangle = 0 \qquad\qquad m = -2$$

$$\langle 2,-2|J_x|2,-1\rangle = \hbar \qquad\qquad m = -2$$

$$\langle 2,-1|J_x|2,0\rangle = \frac{\sqrt{3}}{2}\hbar \qquad\qquad m = 0$$

$$\langle 2,0|J_x|2,1\rangle = \frac{\sqrt{3}}{2}\hbar \qquad\qquad m = 1$$

$$\langle 2,1|J_x|2,2\rangle = \hbar \qquad\qquad m = 2$$

Now for

$$j = \frac{5}{2} \Rightarrow m = \frac{5}{2}, \frac{3}{2}, \frac{1}{2}, \frac{-1}{2}, \frac{-3}{2}, \frac{-5}{2}$$

and by the same way,

$$\left\langle \frac{5}{2},\frac{7}{2}\middle| J_x\middle| \frac{5}{2},\frac{5}{2}\right\rangle = 0 \qquad , m = j = \frac{5}{2}$$

$$\left\langle \frac{5}{2},\frac{5}{2}\middle| J_x\middle| \frac{5}{2},\frac{3}{2}\right\rangle = \frac{\sqrt{5}}{2}\hbar \qquad , m = \frac{3}{2}$$

$$\left\langle \frac{5}{2},\frac{3}{2}\middle| J_x\middle| \frac{5}{2},\frac{1}{2}\right\rangle = \sqrt{2}\hbar \qquad , m = \frac{1}{2}$$

$$\left\langle \frac{5}{2},\frac{-3}{2}\middle| J_x\middle| \frac{5}{2},\frac{-5}{2}\right\rangle = \frac{\sqrt{5}}{2}\hbar \qquad , m = \frac{-5}{2}$$

$$\left\langle \frac{5}{2},\frac{-1}{2}\middle| J_x\middle| \frac{5}{2},\frac{-3}{2}\right\rangle = \sqrt{2}\hbar \qquad , m = \frac{-3}{2}$$

$$\left\langle \frac{5}{2},\frac{1}{2}\middle| J_x\middle| \frac{5}{2},\frac{-1}{2}\right\rangle = \frac{3}{2}\hbar \qquad , m = \frac{-1}{2}$$

Also from (9), we have

$$\left\langle \frac{5}{2},\frac{3}{2}\middle| J_x\middle| \frac{5}{2},\frac{5}{2}\right\rangle = \frac{\sqrt{5}}{2}\hbar \qquad , m = \frac{5}{2}$$

$$\left\langle \frac{5}{2},\frac{1}{2}\middle| J_x\middle| \frac{5}{2},\frac{3}{2}\right\rangle = \sqrt{2}\hbar \qquad , m = \frac{3}{2}$$

$$\left\langle \frac{5}{2},\frac{-1}{2}\middle| J_x\middle| \frac{5}{2},\frac{1}{2}\right\rangle = \frac{3}{2}\hbar \qquad , m = \frac{1}{2}$$

$$\left\langle \frac{5}{2},\frac{-7}{2}\middle| J_x\middle| \frac{5}{2},\frac{-5}{2}\right\rangle = 0 \qquad , m = \frac{-5}{2}$$

$$\left\langle \frac{5}{2}, \frac{-5}{2} \middle| J_x \middle| \frac{5}{2}, \frac{-3}{2} \right\rangle = \frac{\sqrt{5}}{2}\hbar \qquad , m = \frac{-3}{2}$$

$$\left\langle \frac{5}{2}, \frac{-3}{2} \middle| J_x \middle| \frac{5}{2}, \frac{-1}{2} \right\rangle = \sqrt{2}\hbar \qquad , m = \frac{-1}{2}$$

Now,

$$\left\langle \frac{3}{2}m' \middle| J_x \middle| \frac{3}{2}m \right\rangle = \begin{pmatrix} 0 & \frac{\sqrt{3}}{2}\hbar & \hbar & \frac{\sqrt{3}}{2}\hbar \\ 0 & 0 & 0 & 0 \\ 0 & 0 & 0 & 0 \\ \frac{\sqrt{3}}{2}\hbar & \hbar & \frac{\hbar}{\sqrt{2}} & 0 \end{pmatrix}$$

$$\left\langle 2m' \middle| J_x \middle| 2m \right\rangle = \begin{pmatrix} 0 & \hbar & \frac{\sqrt{3}}{2}\hbar & \frac{\sqrt{3}}{2}\hbar & \hbar \\ 0 & 0 & 0 & 0 & 0 \\ 0 & 0 & 0 & 0 & 0 \\ 0 & 0 & 0 & 0 & 0 \\ \hbar & \frac{\sqrt{3}}{2}\hbar & \frac{\sqrt{3}}{2}\hbar & \hbar & 0 \end{pmatrix}$$

$$
\left\langle \frac{5}{2}m'\middle|J_x\middle|\frac{5}{2}m\right\rangle =
\begin{pmatrix}
0 & \dfrac{\sqrt{5}}{2}\hbar & \sqrt{2}\hbar & \dfrac{3}{2}\hbar & \sqrt{2}\hbar & \dfrac{\sqrt{5}}{2}\hbar \\
0 & 0 & 0 & 0 & 0 & 0 \\
0 & 0 & 0 & 0 & 0 & 0 \\
0 & 0 & 0 & 0 & 0 & 0 \\
0 & 0 & 0 & 0 & 0 & 0 \\
\dfrac{\sqrt{5}}{2}\hbar & \sqrt{2}\hbar & \dfrac{3}{2}\hbar & \sqrt{2}\hbar & \dfrac{\sqrt{5}}{2}\hbar & 0
\end{pmatrix}
$$

Also, by the same way, using (10) and (11), we find for J_y for

$$
j = \frac{3}{2} \Rightarrow m = \frac{3}{2}, \frac{1}{2}, , \frac{-3}{2}, \frac{-1}{2} .
$$

$$
\left\langle \frac{3}{2}m'\middle|J_y\middle|\frac{3}{2}m\right\rangle =
\begin{pmatrix}
0 & -\dfrac{i}{2}\hbar\sqrt{3} & -i\hbar & -\dfrac{i}{2}\hbar\sqrt{3} \\
0 & 0 & 0 & 0 \\
0 & 0 & 0 & 0 \\
\dfrac{i}{2}\hbar\sqrt{3} & i\hbar & \dfrac{i}{2}\hbar\sqrt{3} & 0
\end{pmatrix}
$$

$$
\left\langle 2m'\middle|J_y\middle|2m\right\rangle =
\begin{pmatrix}
0 & -i\hbar & -i\hbar\dfrac{\sqrt{3}}{2} & -i\hbar\dfrac{\sqrt{3}}{2} & -i\hbar \\
0 & 0 & 0 & 0 & 0 \\
0 & 0 & 0 & 0 & 0 \\
0 & 0 & 0 & 0 & 0 \\
i\hbar & i\hbar\dfrac{\sqrt{3}}{2} & i\hbar\dfrac{\sqrt{3}}{2} & i\hbar & 0
\end{pmatrix}
$$

$$\left\langle \frac{5}{2}m'\middle|J_y\middle|\frac{5}{2}m\right\rangle = \begin{pmatrix} 0 & -i\hbar\dfrac{\sqrt{5}}{2} & -i\hbar\sqrt{2} & -\dfrac{3}{2}\hbar & i\hbar\sqrt{2} & -i\hbar\dfrac{\sqrt{5}}{2} \\ 0 & 0 & 0 & 0 & 0 & 0 \\ 0 & 0 & 0 & 0 & 0 & 0 \\ 0 & 0 & 0 & 0 & 0 & 0 \\ 0 & 0 & 0 & 0 & 0 & 0 \\ i\hbar\dfrac{\sqrt{5}}{2} & i\hbar\sqrt{2} & \dfrac{3}{2}\hbar & i\hbar\sqrt{2} & i\hbar\dfrac{\sqrt{5}}{2} & 0 \end{pmatrix}$$

From equation (12) $\left\langle jm\middle|J_z\middle|jm\right\rangle = m\hbar$

If we follow the same steps, we find for J_z that

$$\left\langle \frac{3}{2}m\middle|J_z\middle|\frac{3}{2}m\right\rangle = \begin{pmatrix} \dfrac{3}{2} & 0 & 0 & 0 \\ 0 & \dfrac{1}{2}\hbar & 0 & 0 \\ 0 & 0 & \dfrac{1}{2}\hbar & 0 \\ 0 & 0 & 0 & -\dfrac{3}{2}\hbar \end{pmatrix}$$

Here, we have only the diagonal elements.

For $j = 2$, we have

$$\left\langle 2m\middle|J_z\middle|2m\right\rangle = \begin{pmatrix} 2\hbar & 0 & 0 & 0 & 0 \\ 0 & \hbar & 0 & 0 & 0 \\ 0 & 0 & 0 & 0 & 0 \\ 0 & 0 & 0 & -\hbar & 0 \\ 0 & 0 & 0 & 0 & -2\hbar \end{pmatrix}$$

For $j = 5$

$$\left\langle \frac{5}{2} m \middle| J_z \middle| \frac{5}{2} m \right\rangle = \begin{pmatrix} \frac{5}{2}\hbar & 0 & 0 & 0 & 0 & 0 \\ 0 & \frac{3}{2}\hbar & 0 & 0 & 0 & 0 \\ 0 & 0 & \frac{1}{2}\hbar & 0 & 0 & 0 \\ 0 & 0 & 0 & -\frac{1}{2}\hbar & 0 & 0 \\ 0 & 0 & 0 & 0 & -\frac{3}{2}\hbar & 0 \\ 0 & 0 & 0 & 0 & 0 & -\frac{5}{2}\hbar \end{pmatrix}$$

Problem 5

Show that

$$\left\langle \alpha II \middle| \mu_\phi \middle| \alpha II \right\rangle = \frac{\left\langle II \middle| I \middle| II \right\rangle \left\langle \alpha II \middle| \mu_{op} I \middle| \alpha II \right\rangle}{I(I+1)}$$

The Proof

I is the total angular momentum;

μ_{op} is the magnetic moment operator,

where I and μ_{op} have the following relations:

$$I_x \mu_x - \mu_x I_x = 0$$

$$I_x \mu_z - \mu_z I_x = 0 \tag{1}$$

$$I_x \mu_y - \mu_y I_x = i\mu_z$$

$$I_y \mu_y - \mu_y I_y = I_z \mu_z - \mu_z I_z = 0$$

Therefore, we can write the following relation between I and μ_{op}:

$$I^4 \mu_{op} - 2I^2 \mu_{op} I^2 + \mu I^2 = 2\left(I^2 \mu_{op} + \mu_{op} I^2\right) - 4I\left(I \cdot \mu_{op}\right) \tag{2}$$

Equation (2) can be written in matrix elements as the following:

$$\left\langle \alpha II \left| I^2 \mu_{op} - 2I^2 \mu_{op} I^2 + \mu_{op} I^2 - 2\left(I^2 \mu_{op} + \mu_{op} I^2\right) \right| \alpha II \right\rangle$$

$$= -4\left\langle \alpha II \left| I\left(I \cdot \mu_{op}\right) \right| \alpha II \right\rangle \tag{3}$$

Using the Hermitian property of I and μ_{op}, we can write equation (3) as follows:

$$\langle \alpha II | \mu_{op} | \alpha II \rangle \left[I'^2 (I'+1)^2 - 2I'(I'+1)I(I+1) + I^2 (I+1)^2 - 2I'(I'+1) \right.$$
$$\left. -4I(I+1) \right] = -4 \langle \alpha II | I(I \cdot \mu) | \alpha II \rangle \tag{4}$$

For $I' = I$, equation (4) becomes

$$\langle \alpha II | \mu_{op} | \alpha II \rangle = \frac{\langle \alpha II | I \cdot \mu_{op} | \alpha II \rangle}{I(I+1)} \tag{5}$$

I is diagonal in α and I, and $I \cdot \mu_{op}$ is diagonal in (I). Therefore, equation (5) can take the form

$$\langle \alpha II | \mu_{\phi} | \alpha II \rangle = \frac{\langle II | I | II \rangle \langle \alpha II | \mu_{op} \cdot I | \alpha II \rangle}{I(I+1)} \tag{6}$$

QED

Problem 6

Consider a case of square well potential show by explicit calculation that there is no bound state for p-state (l = 1). E<<V$_0$. If the force is short range, the potential well is deep.

The Solution

Let us write down the general expression for the wave equation, which is

$$\frac{d^2 f_l}{dr^2} + \frac{M}{\hbar^2} E f_l - \frac{l(l+1)}{r^2} f_l = (-1)^l \frac{M}{\hbar^2} V(r) f_l \tag{1}$$

For p-state (l = 1); therefore, equation (1) becomes

$$\frac{d^2 f_1}{dr^2} + \frac{M}{\hbar^2} E f_1 - \frac{2}{r^2} f_1 = -\frac{M}{\hbar^2} V_\circ(r) f_1 \qquad\qquad r\langle a$$

$$\frac{d^2 f_1}{dr^2} + \frac{M}{\hbar^2} E f_1 - \frac{2}{r^2} f_1 = 0 \qquad\qquad r>a \tag{2}$$

Now let E=- ω , E<0

Therefore, equation (2) becomes

$$\frac{d^2 f_1}{dr^2} - \frac{2}{r^2} f_1 + \frac{M}{\hbar^2} (V_\circ - \omega) f_1 = o \qquad\qquad r\langle a$$

$$\frac{d^2 f_1}{dr^2} - \frac{2}{r^2} f_1 - \frac{M}{\hbar^2} \omega f_1 = o \qquad\qquad r\rangle a \tag{3}$$

The solution of equation (3) is

$$f_1 = A\left(\frac{\sin \kappa r}{\kappa r} - \cos \kappa r\right) \qquad\qquad \text{for} \quad r\langle a$$

$$f_1 = B \exp\left(-\alpha(r-a)\left(1 + \frac{1}{\alpha r}\right)\right) \qquad \text{for} \quad r\rangle a \tag{4}$$

where

$$\kappa = \sqrt{\frac{M}{\hbar^2}(V_\circ - \omega)}$$

$$\alpha = \sqrt{\frac{M}{\hbar^2}\omega}$$

Now using the joint condition at r = a, we have

$$\left(\frac{f_1'}{f_1}\right)_{r=a} = \left(\frac{f_1'}{f_1}\right)_{r=a}, f' = \frac{df}{dr}$$

for r<a for r>a

So we obtain from equation (4) the following:

$$\frac{\cos \kappa a\!\!\!\bigg/\!\!\kappa a + \sin \kappa a\left[1 - (\kappa a)^{-2}\right]}{\sin \kappa a\!\!\!\bigg/\!\!\kappa a - \cos \kappa a} = \frac{-\alpha}{\kappa}\,\frac{1 + (\alpha a)^{-1} + (\alpha a)^{-2}}{1 + (\alpha a)^{-1}} \tag{5}$$

Now we can find the minimum V_0, which is necessary to give a bound state by putting the binding energy $\omega = 0$, which yields $\alpha = 0$.

Now if $\alpha = 0$, then from equation (5), we find

$$\frac{\cos \kappa a / \kappa a + \sin \kappa a \left[1 - (\kappa a)^{-2}\right]}{\sin \kappa a / \kappa a - \cos \kappa a} = -\frac{1}{\kappa a}$$

Let $\kappa a = \beta$, then we have

$$\cos \beta + \sin \beta \left(\beta - \frac{1}{\beta}\right) = -\left(\sin \beta / \beta - \cos \beta\right)$$

$$\Rightarrow \beta \sin \beta = 0 \qquad \beta = \kappa a \neq 0$$

$$\therefore \quad \sin \beta = 0 \Rightarrow \beta = \pi \Rightarrow \kappa a = \pi$$

But $\kappa = \sqrt{\dfrac{M}{\hbar^2}(V_\circ - \omega)}$ and ω is assumed to be zero.

$$= \sqrt{\frac{M}{\hbar^2} V_\circ} \Rightarrow a\sqrt{\frac{M}{\hbar^2} V_\circ} = \pi \Rightarrow \sqrt{\frac{M}{\hbar^2} V_\circ} = \frac{\pi}{a} \Rightarrow V_\circ = \frac{\pi^2 \hbar^2}{M a^2}$$

$$\tag{6}$$

The value of (V_0) in equation (6) is irreconcilable with the conclusion that $(V_0 a^2)$ is only slightly larger than $\left(\dfrac{\pi^2 \hbar^2}{4M}\right)$, which has been found in the case of the square potential well for the deuteron.

V_0 is large compared to the binding energy of the deuteron. (From the theory of the α-particle decay, a value of $V_0 a^2$ of about $4\hbar/M$ can be deduced.) Therefore, it can be concluded that there is no stable p-state ($l = 1$) of the deuteron. Hence, p-state is unbound state.

QED

Problem 7

Show that $x \pm iy = r \sin \theta \, e^{\pm i\varphi}$

The Solution

$x = r \sin \theta \cos \varphi$

$y = r \sin \theta \sin \varphi$

$z = r \cos \theta$

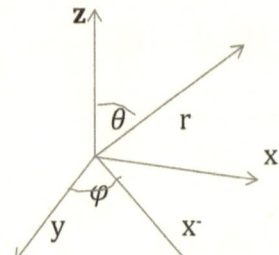

Now take $x + iy = r \sin \theta \cos \varphi + ir \sin \theta \sin \varphi$

$= r \sin \theta [\cos \varphi + i \sin \varphi] = r \sin \theta \, e^{+i\varphi}$ (from complex varibles)

$\therefore x + iy = r \sin \theta \, e^{+i\varphi}$ (1)

Now take $x - iy = r \sin \theta [\cos \varphi - i \sin \varphi] = r \sin \theta \, e^{-i\varphi}$ (2)

So from equations (1) and (2),

$[x \pm iy] = r \sin \theta \, e^{\pm i\varphi}$, as requested

QED

Problem 8:

Verify by direct calculation that

$\vec{J} \times \vec{J} = i\vec{J}$ is satisfied.

The Proof

We have $\vec{J} = \vec{L} + \vec{S}$ and $\vec{L} = -i\vec{r} \times \vec{\nabla}$, $\hbar = 1$, and $\vec{S}$ has the following components:

$$S_x = \begin{pmatrix} 0 & 0 & 0 \\ 0 & 0 & -i \\ 0 & i & 0 \end{pmatrix}; \quad S_y = \begin{pmatrix} 0 & 0 & i \\ 0 & 0 & 0 \\ -i & 0 & 0 \end{pmatrix}; \quad S_z = \begin{pmatrix} 0 & -i & 0 \\ i & 0 & 0 \\ 0 & 0 & 0 \end{pmatrix} \qquad (1)$$

Using unit vectors $\hat{e}_1, \hat{e}_2, \hat{e}_3$, we have

$$\vec{J} \times \vec{J} = \begin{vmatrix} \hat{e}_1 & \hat{e}_2 & \hat{e}_3 \\ J_x & J_y & J_z \\ J_x & J_y & J_z \end{vmatrix} = \begin{vmatrix} \hat{e}_1 & \hat{e}_2 & \hat{e}_3 \\ L_x + S_x & L_y + S_y & L_z + S_z \\ L_x + S_x & L_y + S_y & L_z + S_z \end{vmatrix} \qquad (2)$$

$$\vec{J} \times \vec{J} = \hat{e}_1 \left[\left(L_y + S_y \right)\left(L_z + S_z \right) - \left(L_z + S_z \right)\left(L_y + S_y \right) \right] + \hat{e}_2 \left[\left(L_z + S_z \right) \right.$$

$$\left(L_x + S_x \right) - \left(L_x + S_x \right)\left(L_z + S_z \right) \right] + \hat{e}_3 \left[\left(L_x + S_x \right)\left(L_y + S_y \right) \right.$$

$$\left. - \left(L_y + S_y \right)\left(L_x + S_x \right) \right]$$

$$= \hat{e}_1 \left[L_y + S_y, L_z + S_z \right] + \hat{e}_2 \left[L_z + S_z, L_x + S_x \right] + \hat{e}_3 \left[L_y + S_y, L_x + S_x \right] \quad (3)$$

Using the commutator properties and the relations

$\left[L_i, L_j \right] = iL_\kappa, \quad i,j = x,y,z, \quad \kappa = x,y,z$ and the matrix of S_x, S_y, S_z

in equation (1) and the fact $\left[S_y, S_z \right] = S_y S_z - S_z S_y = iS_x$ and

$\left[S_z, S_x \right] = iS_y, \left[S_x, S_y \right] = iS_z$, then equation (2) becomes

$$\vec{J} \times \vec{J} = \hat{e}_1 \left[iL_x + iS_x \right] + \hat{e}_2 \left[iL_y + iS_y \right] + \hat{e}_3 \left[iL_z + iS_z \right]$$

$$= i\hat{e}_1 J_x + i\hat{e}_2 J_y + i\hat{e}_3 J_z = i \left[\hat{e}_1 J_x + \hat{e}_2 J_y + \hat{e}_3 J_z \right] =$$

$$i\vec{J} \Rightarrow \left[\vec{J} \times \vec{J} = i\vec{J} \right]$$

QED

Problem 9

Show that

$$\left[(J+M)(J-M+1) \right]^{\frac{1}{2}} \left\langle j_1 m_1 j_2 m_2 \middle| JM-1 \right\rangle$$

$$= \left[(j_1 + m_1 + 1)(j_1 - m_1) \right]^{\frac{1}{2}} \left\langle j_1 m_1 + 1 j_2 m_2 \middle| JM \right\rangle$$

$$+ \left[(j_2 + m_2 + 1)(j_2 - m_2) \right]^{\frac{1}{2}} \left\langle j_1 m_1 j_2 m_2 + 1 \middle| JM \right\rangle$$

The Proof

By using $J_- = J_{1-} + J_{2-}$, we can show that

$$J_- \sum_{m_1 m_2} U\left(j_1 m_1 j_2 m_2\right)\left\langle j_1 m_1 j_2 m_2 \mid JM \right\rangle$$

$$= \left\langle JM - 1 \mid J_- \mid JM \right\rangle \sum_{m_1' m_2'} U\left(j_1 m_1' j_2 m_2'\right)\left\langle j_1 m_1' j_2 m_2' \mid JM \right\rangle$$

$$= \sum_{m_1 m_2} \left\langle j_1 m_1 - 1 \mid J_{1-} \mid j_1 m_1 \right\rangle U\left(j_1 m_1 - 1 j_2 m_2\right)\left\langle j_1 m_1 j_2 m_2 \mid JM \right\rangle$$

$$+ \sum_{m_1' m_2'} \left\langle j_2 m_2' - 1 \mid J_{2-} \mid j_2 m_2 \right\rangle U\left(j_1 m_1' j_2 m_2 - 1\right)\left\langle j_1 m_1' j_2 m'_2 \mid JM \right\rangle$$

But $J_- U \left\langle JM \right\rangle = \hbar\left[\left(J+M\right)\left(J-M+1\right)\right]^{1/2} U \left\langle JM - 1 \right\rangle$

And $J_- = J_{1-} + J_{2-}$

$$\therefore \left\langle JM - 1 \mid J_- \mid JM \right\rangle \sum_{m_1' m_2'} U\left(j_1 m_1' j_2 m_2'\right)\left\langle j_1 m_1 j_2 m_2 \mid JM - 1 \right\rangle$$

$$= \hbar\left[\left(J+M\right)\left(J-M+1\right)\right]^{1/2}\left\langle JM - 1 \mid JM - 1 \right\rangle \sum_{m_1' m_2'} U\left(j_1 m_1' j_2 m_2'\right)\left\langle j_1 m_1' j_2 m_2' \mid JM - 1 \right\rangle$$

$$= \hbar\left[\left(J+M\right)\left(J-M+1\right)\right]^{1/2} \sum_{m_1' m_2'} U\left(j_1 m_1' j_2 m_2'\right)\left\langle j_1 m_1' j_2 m_2' \mid JM - 1 \right\rangle$$

Hence,

$$J_- \sum_{m_1 m_2} U\left(j_1 m_1 j_2 m_2\right)\left\langle j_1 m_1 j_2 m_2 \mid JM \right\rangle = \hbar\left[\left(J+M\right)\left(J-M+1\right)\right]^{1/2} \times$$

$$\sum_{m_1' m_2'} U\left(j_1 m_1' j_2 m_2'\right)\left\langle j_1 m_1' j_2 m_2' \mid JM - 1 \right\rangle \tag{1}$$

By the same way,

$$\sum_{m_1 m_2} \langle j_1 m_1 - 1 | J_{1-} | j_1 m_1 \rangle U (j_1 m_1 - 1 j_2 m_2) \langle j_1 m_1 j_2 m_2 | JM \rangle$$

$$+ \sum_{m_1' m_2'} \langle j_2 m_2' - 1 | J_{2-} | j_2 m_2' \rangle U (j_1 m_1' j_2 m_2' - 1) \langle j_1 m_1' j_2 m'_2 | JM \rangle$$

$$= \sum_{m_1 m_2} \hbar \left[(j_1 + m_1)(j_1 - m_1 + 1) \right]^{1/2} U (j_1 m_1 - 1 j_2 m_2) \langle j_1 m_1 j_2 m_2 | JM \rangle$$

$$+ \sum_{m_1' m_2'} \hbar \left[(j_2 + m_2')(j_2 - m_2' - 1) \right]^{1/2} U (j_1 m_1' j_2 m_2' - 1) \langle j_1 m_1' j_2 m'_2 | JM \rangle$$

$$= \sum_{m_1 m_2} \left[\hbar \left[(j_1 - m_1)(j_1 + m_1 + 1) \right]^{1/2} \langle j_1 m_1 + 1 j_2 m_2 | JM \rangle \right.$$

$$\left. + \sum_{m_1' m_2'} \left[(j_2 - m_2')(j_2 + m_2' + 1) \right]^{1/2} (j_1 m_1' + 1 j_2 m_2' + 1) \langle j_1 m_1' + 1 j_2 m'_2 + 1 | JM \rangle \right]$$

$$U (j_1 m_1' j_2 m'_2) \tag{2}$$

Now since the results of equations (1) and (2) have to be identical, therefore, by equating the coefficients of U(...), we get

$$\left[(J + M)(J - M + 1) \right]^{1/2} \langle j_1 m_1 j_2 m_2 | JM - 1 \rangle =$$

$$\left[(j_1 - m_1)(j_1 + m_1 + 1) \right]^{1/2} \langle j_1 m_1 + 1, j_2 m_2 | JM \rangle +$$

$$\left[(j_2 - m_2)(j_2 + m_2 + 1) \right]^{1/2} \langle j_1 m_1 j_2 m_2 + 1 | JM \rangle \tag{3}$$

QED

Problem 10

Find A and B, which are involved in the following symmetry relation of Clebsch-Gordan coefficients, i.e., CGC.

(a) $\langle j_1 m_1 j_2 m_2 | j_3 m_3 \rangle = A \langle j_2 - m_2 j_3 m_3 | j_1 m_1 \rangle$

(b) $\langle j_1 m_1 j_2 m_2 | j_3 m_3 \rangle = B \langle j_1 - m_1 j_2 - m_2 | j_3 - m_3 \rangle$

The Solution

By representing CGC by 3j-symbol, we can find out what A and B are. Now we have

$$\begin{pmatrix} j_1 & j_2 & j_3 \\ m_1 & m_2 & m_3 \end{pmatrix} = (-1)^{j_1 - j_2 - m_3} (2j_3 + 1)^{-\frac{1}{2}} \langle j_1 m_1 j_2 m_2 | j_3 - m_3 \rangle$$

$$\therefore \langle j_1 m_1 j_2 m_2 | j_3 - m_3 \rangle = (2j_3 + 1)^{\frac{1}{2}} (-1)^{j_2 + m_3 - j_1} \begin{pmatrix} j_1 & j_2 & j_3 \\ m_1 & m_2 & -m_3 \end{pmatrix} \quad (1)$$

Now take (a)

$$\langle j_1 m_1 j_2 m_2 | j_3 m_3 \rangle = (2j_3 + 1)^{\frac{1}{2}} (-1)^{j_2 + m_3 - j_1} \begin{pmatrix} j_1 & j_2 & j_3 \\ m_1 & m_2 & -m_3 \end{pmatrix}$$

By symmetry, we have

$$\begin{pmatrix} j_1 & j_2 & j_3 \\ m_1 & m_2 & -m_3 \end{pmatrix} = \begin{pmatrix} j_2 & j_3 & j_1 \\ m_2 & -m_3 & m_1 \end{pmatrix}$$

But

$$\begin{pmatrix} j_2 & j_3 & j_1 \\ m_2 & -m_3 & m_1 \end{pmatrix} = (-1)^{j_3+j_2+j_1} \begin{pmatrix} j_2 & j_3 & j_1 \\ -m_2 & m_3 & -m_1 \end{pmatrix}$$

$$\therefore \begin{pmatrix} j_1 & j_2 & j_3 \\ m_1 & m_2 & -m_3 \end{pmatrix} = (-1)^{j_3+j_2+j_1} \begin{pmatrix} j_2 & j_3 & j_1 \\ -m_2 & m_3 & -m_1 \end{pmatrix}$$

But

$$\begin{pmatrix} j_2 & j_3 & j_1 \\ m_2 & -m_3 & m_1 \end{pmatrix} = (-1)^{j_2-j_3-m_1} (2j_1+1)^{-\frac{1}{2}} \langle j_2 - m_2 j_3 m_3 | j_1 m_1 \rangle$$

$$\therefore \begin{pmatrix} j_1 & j_2 & j_3 \\ m_1 & m_2 & -m_3 \end{pmatrix} = (-1)^{j_1+j_2+j_3-j_2-j_3-m_1} (2j_1+1)^{-\frac{1}{2}} \langle j_2 - m_2 j_3 m_3 | j_1 m_1 \rangle$$

$$= (-1)^{j_1-m_1} (2j_1+1)^{-\frac{1}{2}} \langle j_2 - m_2 j_3 m_3 | j_1 m_1 \rangle$$

Hence,

$$\begin{pmatrix} j_1 & j_2 & j_3 \\ m_1 & m_2 & -m_3 \end{pmatrix} = \frac{(-1)^{j_1-m_1}}{(2j_1+1)^{\frac{1}{2}}} \langle j_2 - m_2 j_3 m_3 | j_1 m_1 \rangle \qquad (2)$$

Now substitute equation (2) into equation (1); you get

$$\langle j_1 m_1 j_2 m_2 | j_3 m_3 \rangle = \left(\frac{2j_3+1}{2j_1+1} \right)^{\frac{1}{2}} (-1)^{j_2+m_3-j_1+j_1-m_1} \langle j_2 - m_2 j_3 m_3 | j_1 m_1 \rangle$$

Also, remember that $m_1 + m_2 = m_3 \Rightarrow m_3 - m_1 = m_2$

Hence,

$$\langle j_1 m_1 j_2 m_2 | j_3 m_3 \rangle = \left(\frac{2j_3 + 1}{2j_1 + 1} \right)^{\!\!1/2} (-1)^{j_2 + m_2} \langle j_2 - m_2 j_3 m_3 | j_1 m_1 \rangle \tag{3}$$

Now compare equation (3) with (a); you find A as

$$\left[A = (-1)^{j_2 + m_2} \left(\frac{2j_3 + 1}{2j_1 + 1} \right)^{\!\!1/2} \right]$$

By the same way, take (b):

$$\langle j_1 m_1 j_2 m_2 | j_3 m_3 \rangle = (-1)^{j_2 + m_3 - j_1} (2j_3 + 1)^{1/2} \begin{pmatrix} j_1 & j_2 & j_3 \\ m_1 & m_2 & -m_3 \end{pmatrix} \tag{4}$$

Now, $\begin{pmatrix} j_1 & j_2 & j_3 \\ m_1 & m_2 & -m_3 \end{pmatrix} = (-1)^{j_1 + j_2 + j_3} \begin{pmatrix} j_1 & j_2 & j_3 \\ -m_1 & -m_2 & m_3 \end{pmatrix}$

But $\begin{pmatrix} j_1 & j_2 & j_3 \\ -m_1 & -m_2 & m_3 \end{pmatrix} = \dfrac{(-1)^{j_1 - j_2 - m_3}}{(2j_3 + 1)^{1/2}} \langle j_1 - m_1 \; j_2 - m_2 | j_3 - m_3 \rangle \tag{5}$

Now substitute equation (5) into equation (4); you find

$$\langle j_1 m_1 j_2 m_2 | j_3 m_3 \rangle = (-1)^{j_1 + j_2 + j_3} \langle j_1 - m_1, \, j_2 - m_2 | j_3 - m_3 \rangle \tag{6}$$

Comparing equation (6) with (b); you find

$\left[B = (-1)^{j_1 + j_2 + j_3} \right]$; therefore,

(a) $\quad \langle j_1 m_1 j_2 m_2 | j_3 m_3 \rangle = (-1)^{j_2 + m_2} \left(\frac{2j_3 + 1}{2j_1 + 1} \right)^{\!\!1/2} \langle j_2 - m_2, j_3 m_3 | j_1 m_1 \rangle$

and

(b) $\langle j_1 m_1 j_2 m_2 | j_3 m_3 \rangle = (-1)^{j_1+j_2+j_3} \langle j_1 - m_1, j_2 - m_2 | j_3 - m_3 \rangle$

QED

Problem 11

Given an electron with $l = 1$, $s = \dfrac{1}{2}$, i.e., p-state, find (a) $\left| \dfrac{1}{2}\dfrac{1}{2} \right\rangle$
, (b) $\left| \dfrac{1}{2}\dfrac{-1}{2} \right\rangle$, (c) $\left| \dfrac{3}{2}\dfrac{3}{2} \right\rangle$, (d) $\left| \dfrac{3}{2}\dfrac{1}{2} \right\rangle$, (e) $\left| \dfrac{3}{2}\dfrac{-3}{2} \right\rangle$, (f) $\left| \dfrac{3}{2}\dfrac{-1}{2} \right\rangle$

The Solution

J takes the value $\dfrac{1}{2}, (l-s), \dfrac{3}{2}, (l+s)$.

Take $j = \dfrac{1}{2}$; there are 2j + 1 values for m, which are $\dfrac{1}{2}, \dfrac{-1}{2}$.

a) Now $\left| \dfrac{1}{2}\dfrac{1}{2} \right\rangle = \left\langle 11\dfrac{1}{2}\dfrac{1}{2} \middle| \dfrac{1}{2}\dfrac{1}{2} \right\rangle \left| 11\dfrac{1}{2}\dfrac{1}{2} \right\rangle$

$= (-1)^{-1+\frac{1}{2}+\frac{1}{2}} (2)^{\frac{1}{2}} \begin{pmatrix} 1 & \dfrac{1}{2} & \dfrac{1}{2} \\ 1 & \dfrac{1}{2} & \dfrac{-1}{2} \end{pmatrix} \left| 11\dfrac{1}{2}\dfrac{1}{2} \right\rangle$

$$= \sqrt{2} \begin{pmatrix} 1 & \dfrac{1}{2} & \dfrac{1}{2} \\ 1 & \dfrac{1}{2} & \dfrac{-1}{2} \end{pmatrix} \left| 11 \dfrac{1}{2} \dfrac{1}{2} \right\rangle = \sqrt{2} \begin{pmatrix} \dfrac{1}{2} & \dfrac{1}{2} & 1 \\ \dfrac{-1}{2} & \dfrac{-1}{2} & 1 \end{pmatrix} |11\rangle \left| \dfrac{1}{2} \dfrac{1}{2} \right\rangle$$

But

$$\begin{pmatrix} \dfrac{1}{2} & \dfrac{1}{2} & 1 \\ \dfrac{-1}{2} & \dfrac{-1}{2} & 1 \end{pmatrix} = (-1)^{+1} \left[\dfrac{(1)(1)(2)}{(3)(2)(1)} \right]^{1/2} = \dfrac{1}{\sqrt{3}}$$

$$\therefore \left| \dfrac{1}{2} \dfrac{1}{2} \right\rangle = \sqrt{\dfrac{2}{3}} |11\rangle \left| \dfrac{1}{2} \dfrac{1}{2} \right\rangle \equiv \sqrt{\dfrac{2}{3}} Y_{11}(\theta, \varphi) \chi_{\frac{1}{2}\frac{1}{2}} = \sqrt{\dfrac{2}{3}} Y_{11}(\theta, \varphi) \begin{pmatrix} 1 \\ 0 \end{pmatrix}$$

Therefore,

$$\left| \dfrac{1}{2} \dfrac{1}{2} \right\rangle = \sqrt{\dfrac{2}{3}} \begin{pmatrix} Y_{11}(\theta, \phi) \\ 0 \end{pmatrix} \tag{1}$$

b) $\left| \dfrac{1}{2} \dfrac{-1}{2} \right\rangle = |11\rangle \left| \dfrac{1}{2} \dfrac{1}{2} \right\rangle \left\langle 11 \dfrac{1}{2} \dfrac{1}{2} \middle| \dfrac{1}{2} \dfrac{-1}{2} \right\rangle$

Now

$$\left\langle 11 \dfrac{1}{2} \dfrac{1}{2} \middle| \dfrac{1}{2} \dfrac{-1}{2} \right\rangle = (-1)^0 \sqrt{2} \begin{pmatrix} 1 & \dfrac{1}{2} & \dfrac{1}{2} \\ 1 & \dfrac{1}{2} & \dfrac{-1}{2} \end{pmatrix} = \sqrt{2} \begin{pmatrix} \dfrac{1}{2} & \dfrac{1}{2} & 1 \\ \dfrac{1}{2} & \dfrac{1}{2} & 1 \end{pmatrix}$$

$$= -\sqrt{2}\begin{pmatrix} \dfrac{1}{2} & \dfrac{1}{2} & 1 \\ \dfrac{-1}{2} & \dfrac{-1}{2} & 1 \end{pmatrix} = \sqrt{2} \cdot \frac{1}{\sqrt{3}} Y_{11}(\theta,\phi)\chi_{\frac{1}{2}\frac{1}{2}}$$

$$\left|\frac{1}{2}\frac{-1}{2}\right\rangle = \sqrt{\frac{2}{3}} Y_{11}(\theta,\phi)\begin{pmatrix} 1 \\ 0 \end{pmatrix} = \sqrt{\frac{2}{3}}\begin{pmatrix} Y_{11}(\theta,\phi) \\ 0 \end{pmatrix} \tag{2}$$

c) For $j = \dfrac{3}{2}$, there are four values for m, i.e., $\dfrac{3}{2}, \dfrac{1}{2}, \dfrac{-3}{2}, \dfrac{-1}{2}$

Hence, $\left|\dfrac{3}{2}\dfrac{3}{2}\right\rangle = \left\langle 11\dfrac{1}{2}\dfrac{1}{2}\middle|\dfrac{3}{2}\dfrac{3}{2}\right\rangle \left|11\right\rangle \left|\dfrac{1}{2}\dfrac{1}{2}\right\rangle = \left|11\right\rangle\left|\dfrac{1}{2}\dfrac{1}{2}\right\rangle = Y_{11}(\theta,\phi)\chi_{\frac{1}{2}\frac{1}{2}}$

$$= Y_{11}(\theta,\phi)\begin{pmatrix} 1 \\ 0 \end{pmatrix} = \begin{pmatrix} Y_{11}(\theta,\phi) \\ 0 \end{pmatrix}$$

Following the same way, you find that

d) $\left|\dfrac{3}{2}\dfrac{1}{2}\right\rangle = \dfrac{1}{\sqrt{3}}\begin{pmatrix} Y_{11}(\theta,\phi) \\ 0 \end{pmatrix}$

e) $\left|\dfrac{3}{2}\dfrac{-3}{2}\right\rangle = Y_{11}(\theta,\phi)\begin{pmatrix} 1 \\ 0 \end{pmatrix} = \begin{pmatrix} Y_{11}(\theta,\phi) \\ 0 \end{pmatrix}$

f) $\left|\dfrac{3}{2}\dfrac{-1}{2}\right\rangle = \dfrac{1}{\sqrt{3}}\begin{pmatrix} Y_{11}(\theta,\phi) \\ 0 \end{pmatrix}$

QED

Problem 12

Prove that $\left(\sum_\mu \partial_\mu \overline{\Psi} \gamma_\mu - \frac{m_\circ c}{\hbar} \overline{\Psi}\right) = 0$, where $\overline{\Psi} = \Psi^+ \gamma_4$ (Dirac ad joint).

The Proof

Take the covariant form of Dirac equation, i.e.,

$$\left(\sum_\mu \gamma_\mu \partial_\mu + \frac{m_\circ c}{\hbar}\right) \Psi^+ = 0 \tag{1}$$

where $\partial_\mu \equiv \dfrac{1}{ic}\dfrac{\partial}{\partial t} - \nabla = -\dfrac{1}{ic}\partial_t - \nabla$

and that $\gamma_\mu \equiv (\gamma_4, \gamma_\kappa)$, where $\gamma_\mu = \sum_\kappa \gamma_\kappa + \gamma_4$

Also, $\gamma_\kappa = -i\beta\alpha_\kappa$

$\gamma_4 = \beta$

and $\partial_\mu = \sum_\kappa \gamma_\kappa \partial_\kappa - \dfrac{i}{c}\gamma_4 \partial_t = \sum_\kappa -i\beta\alpha_\kappa \partial_\kappa - \dfrac{i}{c}\gamma_4 \partial_t \Rightarrow \left(\partial_\mu\right)^+ = -\partial_\mu$

Now the Hermitian conjugate of equation (1) is given by

$$\left(\sum_\mu -\partial_\mu \Psi^+ \gamma_\mu^+ + \frac{m_\circ c}{\hbar} \Psi^+\right) = 0 \tag{2}$$

But $\gamma_\mu^+ = \gamma_\mu$ (Hermitian)

Multiply equation (2) by $\left(\gamma_4\right)$ from the left; you get

$$\left(\sum_\mu -\partial_\mu \Psi^+ \gamma_4 \gamma_\mu + \frac{m_\circ c}{\hbar} \Psi^+ \gamma_4\right) = 0$$

$$(3)$$

Since $\Psi^+ \gamma_4 = \overline{\Psi}$, and multiply by (-1); you find

$$\left(\sum_\mu \partial_\mu \overline{\Psi} \gamma_\mu - \frac{m_\circ c}{\hbar} \overline{\Psi}\right) = 0 \qquad \text{This is the desired result.}$$

QED

Problem 13

Show that $\gamma_5 = \gamma_1 \gamma_2 \gamma_3 \gamma_4$.

The Proof

We have that

$$\gamma_4 = \beta; \qquad \beta^2 = 1; \qquad \gamma_\kappa = -i\alpha_\kappa \beta$$

$$\therefore \gamma_1 = -i\alpha_1 \beta$$

$$\gamma_2 = -i\alpha_2 \beta$$

$$\gamma_3 = -i\alpha_3\beta$$

$$\therefore \gamma_1\gamma_2\gamma_3 = -(i)^3\alpha_1\alpha_2\alpha_3\beta^3 = i\alpha_1\alpha_2\alpha_3\beta^3$$

Therefore, $\gamma_1\gamma_2\gamma_3\gamma_4 = i\alpha_1\alpha_2\alpha_3\beta^4 = i\alpha_1\alpha_2\alpha_3$, since $\beta^4 = 1$.

So $\gamma_1\gamma_2\gamma_3\gamma_4 = i\alpha_1\alpha_2\alpha_3$ (1)

Now $\alpha_\kappa = \begin{pmatrix} 0 & \sigma_\kappa \\ \sigma_\kappa & 0 \end{pmatrix} \Rightarrow \alpha_1 = \begin{pmatrix} 0 & \sigma_1 \\ \sigma_1 & 0 \end{pmatrix}; \alpha_2 = \begin{pmatrix} 0 & \sigma_2 \\ \sigma_2 & 0 \end{pmatrix}$

where $\sigma_1 = \begin{pmatrix} 0 & 1 \\ 1 & 0 \end{pmatrix}; \sigma_2 = \begin{pmatrix} 0 & -i \\ i & 0 \end{pmatrix}; \sigma_3 = \begin{pmatrix} 1 & 0 \\ 0 & -1 \end{pmatrix}$

So $\alpha_1 = \begin{pmatrix} 0 & 0 & 0 & 1 \\ 0 & 0 & 1 & 0 \\ 0 & 1 & 0 & 0 \\ 1 & 0 & 0 & 0 \end{pmatrix}; \alpha_2 = \begin{pmatrix} 0 & 0 & 0 & -i \\ 0 & 0 & i & 0 \\ 0 & -i & 0 & 0 \\ i & 0 & 0 & 0 \end{pmatrix}$ and $\alpha_3 = \begin{pmatrix} 0 & 0 & 1 & 0 \\ 0 & 0 & 0 & -1 \\ 1 & 0 & 0 & 0 \\ 0 & -1 & 0 & 0 \end{pmatrix}$

$$\therefore \alpha_1\alpha_2\alpha_3 = \begin{pmatrix} 0 & 0 & i & 0 \\ 0 & 0 & 0 & i \\ i & 0 & 0 & 0 \\ 0 & i & 0 & 0 \end{pmatrix}$$ (2)

Now substitute equation (2) into equation (1); you get

$$\gamma_1\gamma_2\gamma_3\gamma_4 = i\alpha_1\alpha_2\alpha_3 = i\begin{pmatrix} 0 & 0 & i & 0 \\ 0 & 0 & 0 & i \\ i & 0 & 0 & 0 \\ 0 & i & 0 & 0 \end{pmatrix} = \begin{pmatrix} 0 & 0 & -1 & 0 \\ 0 & 0 & 0 & -1 \\ -1 & 0 & 0 & 0 \\ 0 & -1 & 0 & 0 \end{pmatrix}$$

But

$$\gamma_5 = \begin{pmatrix} 0 & 0 & -1 & 0 \\ 0 & 0 & 0 & -1 \\ -1 & 0 & 0 & 0 \\ 0 & -1 & 0 & 0 \end{pmatrix}$$

Therefore, $\left[\gamma_5 = \gamma_1\gamma_2\gamma_3\gamma_4 \right]$ QED

Problem 14:

Compute $J_{lm}(\vec{\kappa})$, where

$$J_{lm}(\vec{\kappa}) = \int e^{i\kappa.r} e^{-r/a} Y_{lm}(\theta,\phi) r^{-(l+1)} r^2 dr \sin\theta \, d\theta \, d\phi ,$$

where θ, Φ are the polar angles of the vector r (r is the position vector of the electron with respect to the nucleus).

The Solution:

To compute such integral, we assume that

$$(2\pi/\lambda)\rangle\rangle \frac{1}{a} \to e^{-r/a} \to 1$$

i) $e^{-r/a} = 1$

ii)
$$e^{-i\kappa.r} = e^{-i\kappa r \cos\theta} = \sum_{l=0}^{\infty} i^{-l}(2l+1)J_l(\kappa r)P_l(\cos\theta)$$

But $P_l(\cos\theta) = \dfrac{4\pi}{2l+1}\displaystyle\sum_{m=-l}^{l} Y_{lm}^*(\theta,\phi)Y_{lm}(\theta,\phi)$

$\therefore \quad e^{-i\kappa.r} = 4\pi\displaystyle\sum_{l=0}^{\infty}\sum_{m=-l}^{l} i^{-l}J_l(\kappa r)Y_{lm}^*(\theta,\phi)Y_{lm}(\theta,\phi)$ (1)

Now using equation (1) to find

$J_{lm}(\kappa r) = 4\pi\displaystyle\sum_{l=0}^{\infty}\sum_{m=-l}^{l} i^{-l}\iint r^{-(l-1)}dr\, J_l(\kappa r)Y_{lm}(\theta,\phi)\left|Y_{lm}(\theta,\phi)\right|^2 \sin\theta\, d\theta\, d\phi$

$= 4\pi\displaystyle\sum_{l=0}^{\infty}\sum_{m=-l}^{l} i^{-l}\int r^{-(l-1)}J_l(\kappa r)dr\, Y_{lm}(\theta,\phi)\int\left|Y_{lm}(\theta,\phi)\right|^2 \sin\theta\, d\theta\, d\phi$

$= 4\pi\displaystyle\sum_{l=0}^{\infty}\sum_{m=-l}^{l} i^{-l}Y_{lm}(\theta,\phi)\int r^{-l+1}J_l(\kappa r)dr$

Now let us take it for (l) and (m), then,

$J_{lm}(\kappa r) = 4\pi i^{-l}Y_{lm}(\theta,\phi)\int r^{1-l}J_l(\kappa r)dr$

Using the table of Bessel functions, the integral is given as

$\displaystyle\int\dfrac{J_l(\kappa r)}{r^{1-l}}dr = \dfrac{\kappa^{l-2}}{(2l-1)!!}$; therefore,

$\left[J_{lm}(\kappa r) = 4\pi i^{-l}\dfrac{\kappa^{l-2}}{(2l-1)!!}Y_{lm}(\theta,\phi)\right]$,where $(2l-1)!! = 1\cdot3\cdot5\cdot7\cdot\ldots\ldots(2l-1)$

QED

Problem 15

Show that $D_{m'm}^{(l)}(-\alpha,\beta,-\gamma) = (-1)^{m-m'} D_{-m'-m}^{(l)}(\alpha,\beta,\gamma)$.

The Proof

We have

$$Y_{lm}(\theta',\phi') = \sum_{m'=-l}^{l} Y_{lm'}(\theta,\phi) D_{m'm}^{(l)}(\omega) \tag{1}$$

where $D_{m'm}^{(l)}(\omega)$ is the matrix element of $D(\alpha\beta\gamma)$

and $D(\alpha\beta\gamma) = e^{\frac{i\alpha}{\hbar}L_z} e^{\frac{i\beta}{\hbar}L_y} e^{\frac{i\gamma}{\hbar}L_z}$ (Edmonds, p. 55, ex. 4.18)

$\therefore D_{-m'-m}^{(l)}(\alpha\beta\gamma) = \langle lm'|D(\alpha\beta\gamma)|lm\rangle$

Since we are dealing with a representation in which the matrix of L_z are diagonal, therefore, $D_{m'm}^{(l)}(\alpha\beta\gamma) = e^{im'\gamma} d_{mm'}^{(l)}(\beta) e^{im\alpha}$ \qquad (2)

where $\left[d_{mm'}^{(l)}(\beta) = \langle lm'|exp\frac{i\beta}{\hbar}L_z|lm\rangle \right]$ \qquad (3)

Equation (3) is called Wigner small d-matrix.

(For a special choice of arguments, see Edmonds, p. 59, eq. 4.1.27 and p. 57, eq. 4.1.15.)

We can write that

$$d_{mm'}^{(l)}(\beta) = (-1)^{l-m} \left[\frac{(2l)!}{(l+m)!(l-m)!} \right]^{1/2} \left(\cos \frac{\beta}{2}\right)^{l+m} \left(\sin \frac{\beta}{2}\right)^{l-m} \tag{4}$$

Now equation (2) becomes

$$D_{m'm}^{(l)}(\alpha\beta\gamma) = e^{im'\gamma} \left[(-1)^{l-m} \left[\frac{(2l)!}{(l+m)!(l-m)!} \right]^{1/2} \left(\cos \frac{\beta}{2}\right)^{l+m} \left(\sin \frac{\beta}{2}\right)^{l-m} \right] e^{im\alpha} \tag{5}$$

Now $d_{mm'}^{(l)}$ is real; therefore,

$$d_{m'm}^{(l)}(\beta) = d_{mm'}^{(l)}(\beta) \tag{6}$$

By making successive application of rotation to a frame of coordinates corresponding to multiplication of the appropriate matrices that represent the rotations, we, therefore, have

$$d_{m'm}^{(l)}(\beta + \pi) = \sum_{m''} d_{m'm''}^{(l)}(\pi) d_{m''m}^{(l)}(\beta) = (-1)^{l-m'} d_{-m'm}^{(l)}(\beta) \tag{7}$$

By the same way, we have

$$d_{m'm}^{(l)}(\beta - \pi) = (-1)^{l-m'} d_{-m'm}^{(l)}(-\beta) = (-1)^{l-m'} d_{m',-m}^{(l)}(\beta) \tag{8}$$

Now we have

$$d_{m'm}^{(l)}(\beta) = \sum_{m''} d_{m'm''}^{(l)}(\beta + \pi) d_{m''m}^{(l)}(-\pi) = (-1)^{l-m} d_{m',-m}^{(l)}(\beta + \pi)$$

Using equation (6), one gets

$$d_{m'm}^{(l)}(\beta) = (-1)^{m'-m} d_{-m'-m}^{(l)}(\beta) \tag{9}$$

And similarly,

$$d^{(l)}_{m'm}(\beta) = (-1)^{m'-m} d^{(l)}_{m'm}(\beta) \tag{10}$$

$\therefore$ Equation (2) becomes

$$D^{(l)}_{m'm}(\alpha\beta\gamma) = e^{im'\gamma}(-1)^{m'-m} d^{(l)}_{-m'-m}(\beta) e^{im\alpha} \tag{11}$$

or $D^{(l)}_{m'm}(\alpha\beta\gamma) = e^{im'\gamma}(-1)^{m'-m} d^{(l)}_{m'm}(\beta) e^{im\alpha}$ (12)

Now from equation (11), we have

$$D^{(l)}_{m'm}(-\alpha\ \beta - \gamma) = e^{-im'\gamma}(-1)^{m'-m} d^{(l)}_{-m'-m}(\beta) e^{-im\alpha} = D^{(l)*}_{m'm}(\alpha\beta\gamma) \tag{13}$$

From equation (12) $D^{(l)}_{-m'-m}(\alpha\beta\gamma) = e^{-im'\gamma}(-1)^{m-m'} d^{(l)}_{-m'-m}(\beta) e^{-im\alpha}$ (14)

Comparing equation (13) and equation (14), one gets

$$D^{(l)}_{m'm}(-\alpha\ \beta - \gamma) = (-1)^{m-m'} D^{(l)}_{-m'-m}(\alpha\beta\gamma)$$ as required.

QED

Problem 16

Find the spherical components of second rank tensor symmetrical.

The Solution

We have

$$S_{i\kappa} = \frac{1}{2}\left(x_i x_\kappa + x_\kappa x_i\right) - \frac{1}{3}\tau_{i\kappa} \tag{1}$$

But

$$\tau_{i\kappa} = \frac{1}{3}\delta_{i\kappa}T \tag{2}$$

where T is scalar tensor, i.e., of rank zero $\left(\lambda = 0\right)$

Hence, $T = Y_\infty\left(\theta,\phi\right), \quad l = 0, \quad m = 0$

Therefore, $S_{i\kappa}$ can be written as

$$S_{i\kappa} = \frac{1}{2}\left(x_i x_\kappa + x_\kappa x_i\right) - \frac{1}{9}\delta_{i\kappa}T \tag{3}$$

Now let us find the Cartesian components, and then we transform to the spherical components.

So

$$S_{11} = x^2 - \frac{1}{9}T$$
$$S_{12} = xy$$
$$S_{13} = xz$$
$$S_{21} = yx$$
$$S_{22} = y^2 - \frac{1}{9}T$$
$$S_{23} = yz$$
$$S_{31} = zx$$
$$S_{32} = zy$$
$$S_{33} = z^2 - \frac{1}{9}T$$

$$\Rightarrow S_{i\kappa} = \begin{pmatrix} x^2 - \frac{1}{9}T & xy & xz \\ yx & y^2 - \frac{1}{9}T & yz \\ zx & zy & z^2 - \frac{1}{9}T \end{pmatrix}$$

(4)

For a tensor of second rank $l = 2, \Rightarrow m = 2,1,0,-1,-2$, therefore, this tensor has $(2l+1) = 5$ spherical components, which are $r^2 Y_{22}(\theta,\phi), r^2 Y_{21}(\theta,\phi), r^2 Y_{20}(\theta,\phi), r^2 Y_{2,-2}(\theta,\phi), r^2 Y_{2,-1}(\theta,\phi)$, which corresponds to x², xy, z², y², yx; therefore, $S_{i\kappa}$ can take the following matrices:

$$S_{i\kappa} = \begin{pmatrix} r^2 Y_{22} - \frac{1}{9}T & r^2 Y_{21} & 0 \\ r^2 Y_{2,-1} & r^2 Y_{2,-2} - \frac{1}{9}T & 0 \\ 0 & 0 & r^2 Y_{20} - \frac{1}{9}T \end{pmatrix}$$

But T is a tensor of rank zero, i.e., l = m = 0

$$\therefore T = Y_{oo}(\theta,\phi) = \frac{1}{\sqrt{4\pi}}$$

Therefore,

$$S_{i\kappa} = \begin{pmatrix} r^2 Y_{22} - \dfrac{1}{9} Y_{00} & r^2 Y_{21} & 0 \\ r^2 Y_{2,-1} & r^2 Y_{2,-2} - \dfrac{1}{9} Y_{00} & 0 \\ 0 & 0 & r^2 Y_{20} - \dfrac{1}{9} Y_{00} \end{pmatrix} \tag{5}$$

Now we want to find these spherical components, remembering that

$(x \pm iy) = re^{\pm i\phi} \sin\theta$ (problem 7).

Or one can find these from a special table.

Hence, from the table, we find

$$
\left.
\begin{aligned}
r^2 Y_{2,2} &= \frac{1}{4}\sqrt{\frac{15}{2\pi}}(x+iy)^2 \\
r^2 Y_{2,1} &= -\frac{1}{2}\sqrt{\frac{15}{2\pi}}z(x+iy) \\
r^2 Y_{2,-1} &= \frac{1}{2}\sqrt{\frac{15}{2\pi}}z(x-iy) \\
r^2 Y_{2,-2} &= \frac{1}{4}\sqrt{\frac{15}{2\pi}}(x-iy)^2 \\
r^2 Y_{2,0} &= \frac{1}{4}\sqrt{\frac{5}{\pi}}(2z^2 - x^2 - y^2) \\
Y_{0,0} &= \frac{1}{\sqrt{4\pi}}
\end{aligned}
\right\} \tag{6}
$$

Now substitute equation (6) in equation (5); you get the desired spherical components of the given tensor, which are

$$
S_{i\kappa} = \begin{pmatrix} \frac{1}{4}\sqrt{\frac{15}{2\pi}}(x+iy)^2 - \frac{1}{18\sqrt{\pi}} & -\frac{1}{2}\sqrt{\frac{15}{2\pi}}z(x+iy) & 0 \\ \frac{1}{2}\sqrt{\frac{15}{2\pi}}z(x-iy) & \frac{1}{4}\sqrt{\frac{15}{2\pi}}(x-iy)^2 - \frac{1}{18\sqrt{\pi}} & 0 \\ 0 & 0 & \frac{1}{4}\sqrt{\frac{5}{\pi}}(2z^2 - x^2 - y^2) - \frac{1}{18\sqrt{\pi}} \end{pmatrix}
$$

QED

Problem 17

Prove that

$$
T_\mu^{(\lambda)} = \sum_{\mu_1 \mu_2} \langle \lambda_1 \mu_1 \lambda_2 \mu_2 | \lambda \mu \rangle T_{\mu_1}^{(\lambda_1)} T_{\mu_2}^{(\lambda_2)}
$$

The Proof

Since $T_{\mu_1}^{(\lambda_1)}$, $T_{\mu_2}^{(\lambda_2)}$ and $T_\mu^{(\lambda)}$ are spherical tensors, one can represent them by the spherical harmonic functions, which are tensors too.

Now if we represent $T_{\mu_1}^{(\lambda_1)}$ by Y_{jm} and $T_{\mu_2}^{(\lambda_2)}$ by $Y_{j'm'}$ and $T_\mu^{(\lambda)}$ by y_{JM}, then we can use Clebsch-Gordan coefficients to prove (1). If Y_{jm} represents an eigenfunction of J_1^2, whose eigenvalue is $j(j+1)$ in unit of $\hbar$, it is also eigenfunction of J_{1z}, whose eigenvalue is (m) in unit of $\hbar$. And that $Y_{j'm'}$ is eigenfunction of J_2^2 with eigenvalue $j'(j'+1)$ in unit of $\hbar$ and also an eigenfunction of J_{2z} with eigenvalue m' in unit of $\hbar$.

The product of Y_{jm} and $Y_{j'm'}$ is also an eigenfunction of J_z, the z-component of the total angular momentum J, where $J_z = J_{1z} + J_{2z}$ with eigenvalue $M = m + m'$.

But in general, the product of Y_{jm} and $Y_{j'm'}$ is not an eigenfunction to $J^2 = J_1^2 + J_2^2 + J_3^2$. But the linear combinations of the products Y_{jm}, $Y_{j'm'}$ can be formed, which are simultaneous eigenfunctions of J^2 with eigenvalue $J(J+1)$ and J_z with eigenvalue M.

Now these eigenfunctions of J^2 is, let us say, y_{JM}. Therefore, by using Clebsch-Gordan series, we can find y_{JM} as

$$y_{JM} = \sum_{m=-j}^{j} \sum_{m'=-j'}^{j'} \langle jmj'm' | JM \rangle Y_{jm} Y_{j'm'}$$

(1)

Now if we go back to our assumption, we find that

$$T_\mu^{(\lambda)} = \sum_{\mu_1=-\lambda_1}^{\lambda_1} \sum_{\mu_2=-\lambda_2}^{\lambda_2} \langle \lambda_1 \mu_1 \lambda_2 \mu_2 | \lambda \mu \rangle T_{\mu_1}^{(\lambda_1)} T_{\mu_2}^{(\lambda_2)}$$

where $\mu = \mu_1 + \mu_2$

and $\lambda = \lambda_1 + \lambda_2, \lambda_1 + \lambda_2 - 1, \ldots \ldots |\lambda_1 - \lambda_2|$

Hence,

$$\left[T_{\mu}^{(\lambda)} = \sum_{\mu_1 \mu_2} \langle \lambda_1 \mu_1 \lambda_2 \mu_2 | \lambda \mu \rangle T_{\mu_1}^{(\lambda_1)} T_{\mu_2}^{(\lambda_2)} \right]$$

where $\langle \lambda_1 \mu_1 \lambda_2 \mu_2 | \lambda \mu \rangle$ are CGC

QED

Problem 18

Construct $T^{(2)}$ from two $T^{(1)} = \vec{r}_s$, then compute $T_{\pm 2}^{(2)}$, $T_{\pm 1}^{(2)}$, $T_0^{(2)}$ and compare these with problem (16).

The Solution

Let us use $T_{\mu}^{(\lambda)} = A r^{\lambda} y_{\lambda \mu} (\theta, \phi)$. We can then find those components required above, where (λ) is the rank order of the tensor. A is a constant for each value of (μ), where the number of the values of μ is

$(2\lambda + 1)$.

∴ For

$\lambda = 2$, $\mu = 2, 1, 0, -1, -2$.

Hence,

$T_2^{(2)} = A r^2 y_{2,2} (\theta, \phi)$

$$T_{+1}^{(2)} = Ar^2 y_{2,1}(\theta,\phi)$$

$$T_{-2}^{(2)} = Ar^2 y_{2,-2}(\theta,\phi) \tag{1}$$

$$T_{-1}^{(2)} = Ar^2 y_{2,-1}(\theta,\phi)$$

$$T_0^{(2)} = Ar^2 y_{2,0}(\theta,\phi)$$

As in problem (16), we find from the table

$$r^2 y_{2,2}(\theta,\phi) = \frac{1}{4}\sqrt{\frac{15}{2\pi}}(x+iy)^2 = \frac{1}{A}\cdot T_2^{(2)}$$

$$r^2 y_{2,-2}(\theta,\phi) = \frac{1}{4}\sqrt{\frac{15}{2\pi}}(x-iy)^2 = \frac{1}{A}\cdot T_{-2}^{(2)}$$

$$r^2 y_{2,0}(\theta,\phi) = \frac{1}{4}\sqrt{\frac{5}{\pi}}(2z^2 - x^2 - y^2) = \frac{1}{A}\cdot T_0^{(2)} \tag{2}$$

$$r^2 y_{2,1}(\theta,\phi) = -\frac{1}{2}\sqrt{\frac{15}{2\pi}}z(x+iy) = \frac{1}{A}\cdot T_1^{(2)}$$

$$r^2 y_{2,-1}(\theta,\phi) = \frac{1}{2}\sqrt{\frac{15}{2\pi}}z(x-iy) = \frac{1}{A}\cdot T_{-1}^{(2)}$$

Comparing these with those of problem (16), we find that A = 1, so it follows that

$$T_2^{(2)} = \frac{1}{4}\sqrt{\frac{15}{2\pi}}(x+iy)^2$$

$$T_{-2}^{(2)} = \frac{1}{4}\sqrt{\frac{15}{2\pi}}(x-iy)^2$$

$$T_0^{(2)} = \frac{1}{4}\sqrt{\frac{5}{\pi}}\left(2z^2 - x^2 - y^2\right)$$

$$T_1^{(2)} = -\frac{1}{2}\sqrt{\frac{15}{2\pi}}z\left(x+iy\right)$$

$$T_{-1}^{(2)} = \frac{1}{2}\sqrt{\frac{15}{2\pi}}z\left(x-iy\right)$$

QED

Problem 19

Given a nucleus of spin $I = 1$, $\mu \neq 0$, $Q = 0$ under a constant magnetic field $\left(\vec{B}\right)$ parallel to the z-axis in s-system and making an angle $\left(\beta\right)$ with the electric field gradient, which is along $\bar{z} - axis$ in $\bar{s} - system$, as shown in the figure below, where $\cos \beta = 0.1$, discuss the effects of $\left(\vec{B}\right)$ and the electric gradient $\Phi'_{z'z'}$ on such system.

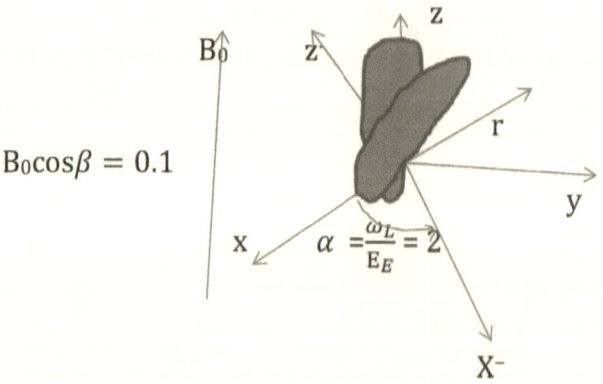

In this problem, the nucleus is under the magnetic interaction (due to the interaction between the external magnetic field and the magnetic dipole moment of the nucleus) and the electric interaction, which is due to the electric field from the source outside the nucleus (electrons). Therefore, the Hamiltonian of the system is given by

$$H = H(\mu) + H_s(E_s) = -\mu \cdot \vec{B} + \frac{4\pi}{5} \sum (-1)^\mu M_\mu^{(2)} \cdot V_\mu^{(2)} \tag{1}$$

The first term is due to magnetic interaction. The second term is due to quadruple interaction. This term (QI) is the only term considered, due to the fact that $M^{(1)}$ (electric dipole moment) and the electric octapole moment $M^{(3)}$ vanish ,because of the parity conservation and $M^{(4)}$ is too small to be considered.

Now let us take the effect of $(\vec{B})$, i.e., H(μ). Hence,

$$H(\mu) = -\mu \cdot B = -\mu_z B_\circ \tag{2}$$

where $\mu = \gamma I$ (magnetic dipole moment, magnetic dipole moment operator).

So the energy matrix elements are given by

$$\langle \mathrm{Im}| H(\mu) | \mathrm{Im}' \rangle = -B_\circ \langle \mathrm{Im}| \mu_z | \mathrm{Im}' \rangle$$

And since $\mu = \gamma I$, the energy matrix is diagonal.

$$\langle \mathrm{Im}| H(\mu) | \mathrm{Im} \rangle = -B_\circ \langle \mathrm{Im}| \mu_z | \mathrm{Im} \rangle$$

$$= -B_{\circ}(-1)^{I-m} \begin{pmatrix} I & 1 & I \\ -m & 0 & m \end{pmatrix} \langle I \| \mu \| \rangle \qquad \text{—(using Wigner-Eckart}$$

theorem)

Now by looking at the table of 3j-symbol, we find that

$$\langle \mathrm{Im} | H(\mu) | \mathrm{Im} \rangle = -B_{\circ} \frac{m}{[(2I+1)(I+1)I]^{\frac{1}{2}}} \langle I \| \mu \| \rangle \tag{3}$$

But the magnetic moment is defined as

$$\mu = \langle II | \mu_z | II \rangle = \frac{m}{[(2I+1)(I+1)I]^{\frac{1}{2}}} \langle I \| \mu \| \rangle \tag{4}$$

From equations (3) and (4), the energy eigenvalues are

$$E_m = \langle \mathrm{Im} | H(\mu) | \mathrm{Im} \rangle = -\frac{m\mu B_{\circ}}{I}$$

If we introduce $\omega_L = \dfrac{\mu B_{\circ}}{I\hbar}$

$$E_m = -m\omega_L \hbar \quad \text{or, in general,}$$

$$E_{mm'} = \langle \mathrm{Im} | H(\mu) | \mathrm{Im'} \rangle = -m\omega_L \hbar \delta_{mm'} \tag{4'}$$

where

$$\delta_{mm'} = \begin{cases} 1 & m = m' \\ 0 & m \neq m' \end{cases}$$

In this problem,

$$I = 1 \Rightarrow m = +1,0,-1.$$

So we have

$$\left. \begin{aligned} E_{+1} &= -\omega_L \hbar \\ E_0 &= 0 \\ E_{-1} &= \omega_L \hbar \end{aligned} \right\}$$ This energy splitting is due to the external field $\left(\vec{B}\right)$. It removes the degeneracy.

Now we treat the electric interaction as follows:

$$H_s(E_s) = \frac{4\pi}{5} \sum (-1)^\mu M_\mu^{(2)} \cdot V_{-\mu}^{(2)} \qquad (5)$$

The quadruple interaction matrix elements, with respect to the system s, are given by

$$\langle \mathrm{Im}|H(E_s)|\mathrm{Im'}\rangle = \frac{4\pi}{5} \sum (-1)^\mu \langle \mathrm{Im}|M_\mu^{(2)}|\mathrm{Im'}\rangle V_{-\mu}^{(2)} \qquad (6)$$

We assume in equation (6) that the classical treatment of the electric field is a good approximation. So by applying the Wigner-Eckart theorem, we find that

$$\langle \mathrm{Im} | H(E_s) | \mathrm{Im'} \rangle = \frac{4\pi}{5} \sum_{\mu} (-1)^{I-m+\mu} \begin{pmatrix} I & 2 & I \\ -m & \mu & m' \end{pmatrix} \langle I \| M^{(2)} \| I \rangle V^{(2)}_{-\mu}$$

where μ takes $(2I + 1)$ values.

Since 3j-symbol = 0 unless $-m + \mu + m' = 0$, there exists only one μ for given values m and m', which implies that the sum over μ can be omitted. Hence,

$$\langle \mathrm{Im} | H(E_s) | \mathrm{Im'} \rangle = \frac{4\pi}{5} (-1)^{I-m'} \begin{pmatrix} I & 2 & I \\ -m & \mu & m' \end{pmatrix} \langle I \| M^{(2)} \| I \rangle V^{(2)}_{-\mu} \tag{7}$$

Now the electrostatic field is known in $\bar{s} - system$, so $V^{(2)}_{-\mu}$ in equation (7) must be expanded in terms of $V'^{(2)}_{-\mu}$ in $\bar{s} - system$,

i.e., $$\left[V^{(2)}_{-\mu} = \sum_{\mu'=-2}^{+2} V'^{(2)}_{-\mu'} D^{(2)}_{\mu,-\mu'}(\omega) \right] \tag{8}$$

By symmetry, $V^{(2)}_{\pm 1} = V^{(2)}_{\pm 2} = 0$ and $V'^{(2)}_{0} \neq 0$

But $V'^{(2)}_{0}$ is a tensor field operator, which is given by

$$V'^{(2)}_{0} = \sum_{\wedge} \frac{e}{r'^3_{\wedge}} Y_{20}(\theta', \phi') = \sum_{\wedge} \frac{1}{4}\sqrt{\frac{5}{\pi}} \frac{e}{r'^3_{\wedge}} (3\cos^2 \phi_{\wedge} - 1)$$

But

$$\sum_{\wedge} \frac{e_{\wedge}}{r_{\wedge}'^3}\left(3\cos^2\phi'-1\right)=\Phi'_{z'z'}$$

$$\therefore \quad V_0'^{(2)}=\frac{1}{4}\sqrt{\frac{5}{\pi}}\Phi'_{z'z'} \tag{9}$$

Now using equation (8),

$$V_{-\mu}^{(2)}=\sum_{q=-2}^{+2}V_q'^{(2)}D_{q,-\mu}^{(2)}(\omega)=V_0'^{(2)}D_{0,-\mu}^{(2)}(\omega)=V_0'^{(2)}\left(\frac{4}{5}\pi\right)^{1/2}Y_{2,-\mu}(\beta,\lambda)$$

From equation (9), we have

$$V_{-\mu}^{(2)}=\frac{1}{2}\Phi'_{z'z'}Y_{2,-\mu}(\beta,\lambda) \tag{10}$$

$\therefore$ The quadruple interaction matrix elements are

$$\langle \mathrm{Im}|H(E_2)|\mathrm{Im}'\rangle=\frac{2}{5}\pi(-1)^{I-m'}\begin{pmatrix}I & 2 & I \\ -m & \mu & m'\end{pmatrix}\langle I\|M^{(2)}\|I\rangle\Phi'_{z'z'}Y_{2,-\mu}(\beta,\lambda)$$

$$=\frac{2}{5}\pi(-1)^{I-m'}\begin{pmatrix}I & 2 & I \\ -m & \mu & m'\end{pmatrix}\langle I\|M^{(2)}\|I\rangle\Phi'_{z'z'}Y_{2,-\mu}(\beta,\lambda) \tag{11}$$

Now we have

$$M_\mu^{(2)}=\sum e_n r_n^2 Y_{2\mu}(\theta_n\phi_n); \text{ n means nucleus.}$$

So $\quad M_0^{(2)}=\sum e_n r_n^2 Y_{20}(\theta_n\varphi_n)=\frac{1}{4}\sqrt{\frac{5}{\pi^2}}\sum e_n\left(3z^2 r_n^2\right)$

By conventional definition, the electric quadruple moment of the nucleus is given by

$$eQ = \left\langle II \left\| \sum e_n \left(3z^2 r_n^2\right) \right\| II \right\rangle = 4\sqrt{\frac{\pi}{5}} \left\langle II \left\| M_0^{(2)} \right\| II \right\rangle$$

$$\therefore \quad eQ = 4\sqrt{\frac{\pi}{5}} \begin{pmatrix} I & 2 & I \\ -I & 0 & I \end{pmatrix} \left\langle II \left\| M_0^{(2)} \right\| II \right\rangle, \text{ using W-E theorem}$$

$$= 4 \left(\frac{\pi}{5}\right)^{1/2} \frac{2I(2I-1)}{\left[(2I+3)(2I+2)(2I+1)(2I)(2I-1)\right]^{1/2}} \left\langle II \left\| M^{(2)} \right\| II \right\rangle \tag{12}$$

Now using equations (11) and (12), one finds

$$\langle \text{Im} | H_s(E_2) | \text{Im}' \rangle = \frac{1}{2} \left(\frac{\pi}{5}\right)^{1/2} eQ\Phi'_{z'z'} (-1)^{I-m'} \frac{\begin{pmatrix} I & 2 & I \\ -m & \mu & m' \end{pmatrix}}{\begin{pmatrix} I & 2 & I \\ -I & 0 & I \end{pmatrix}} Y_{2,-\mu}(\beta,\lambda) \tag{13}$$

Now the matrix element of the total interaction is given by [see equation (1)]

$$Hmm' = \langle \text{Im} | H | \text{Im}' \rangle = \langle \text{Im} | H(\mu) | \text{Im}' \rangle + \langle \text{Im} | H_s(E_2) | \text{Im}' \rangle$$

So from equations (4`) and (13), we find that (from the table of 3j)

$$Hmm' = -\hbar\omega_L m\delta_{mm'} + \frac{1}{2} eQ\Phi'_{z'z'} \left(\frac{\pi}{5}\right)^{1/2} (-1)^{I-m'} \frac{\left[(2I+3)(2I+2)(2I+1)(2I)(2I-1)\right]^{1/2}}{2I(2I-1)}$$

$$\times \begin{pmatrix} I & 2 & I \\ -m & (m-m') & m' \end{pmatrix} Y_{2,m'-m}(\beta,\lambda) \tag{14}$$

where

$$m - m' = \mu.$$

Let us introduce

$$\omega_E = \frac{1}{2}\frac{eQ}{\hbar}\Phi'_{z'z'}\frac{1}{2I(2I-1)},$$

then equation (14) becomes

$$Hmm' = -\hbar\omega_L m\delta_{mm'} + \hbar\omega_E(-1)^{I-m'}\left(\frac{\pi}{5}\right)^{1/2}\begin{pmatrix} I & 2 & I \\ -m & (m-m') & m' \end{pmatrix} Y_{2,m'-m}(\beta,\lambda)\times[\quad]$$

(15)

where $[\quad] = \left[(2I+3)(2I+2)(2I+1)(2I)(2I-1)\right]^{1/2}$

Now we diagonalize $H(\beta,0)$ as

$$H(\beta,0) = D(\gamma)H(\beta,\gamma)D^{-1}(\gamma), \text{ where}$$

$$D(\gamma) = \begin{pmatrix} e^{il\gamma} & & 0 \\ & e^{im\gamma} & \\ 0 & & e^{-il\gamma} \end{pmatrix}$$

This implies that the eigenvalues of $H(\beta,\gamma)$ are independent of γ, i.e.,

$$E(\beta,\gamma) = E(\beta,0)$$

∴ The unitary matrix $U(\beta,\gamma)$, which diagonalizes $H(\beta,\gamma)$, is then given by

$$U(\beta,\gamma) = U(\beta,0)D(\lambda).$$

$\therefore$ The eigenvalues of $E(\beta,0) = E(\beta)$ is given by

$$E(\beta) = U(\beta,0)H(\beta,0)U^{-1}(\beta,0) \tag{16}$$

For simplification, let us introduce the matrix (k) with the elements

$$k_{mm'} = \frac{1}{\hbar\omega_E} H_{mm'}(\beta,0) \text{ , then}$$

$$\frac{H_{mm'}(\beta)}{\hbar\omega_E} = -\frac{\omega_L}{\omega_E}m + (-1)^{I-m'}\left(\frac{\pi}{5}\right)^{1/2}\begin{pmatrix} I & 2 & I \\ -m & 0 & m \end{pmatrix} Y_{2,0}(\beta,0) \times [\quad] \tag{17}$$

where $[\quad] = \left[(2I+3)(2I+2)(2I+1)(2I)(2I-1)\right]^{1/2}$

but

$$Y_{2,0}(\beta,0) = \frac{1}{4}\sqrt{\frac{5}{\pi}}(3\cos^2\beta - 1)$$

and

$$\begin{pmatrix} I & 2 & I \\ -m & 0 & m \end{pmatrix} = (-1)^{I-m} \frac{2\left|3m^2 - I(2I-1)\right|}{\left[(2I+3)(2I+2)(2I+1)(2I)(2I-1)\right]^{1/2}}$$

$$\therefore \frac{H_{mm'}(\beta,0)}{\hbar\omega_E} = -\frac{\omega_L}{\omega_E}m + \frac{1}{2}2\left[3m^2 - I(2I-1)\right]\left(3\cos^2\beta - 1\right)$$

$$E_m(\beta) = \hbar\omega_E\left[\frac{-\omega_L}{\omega_E}m + \frac{1}{2}\left[3m^2 - I(I+1)\right]\right]\left(3\cos^2\beta - 1\right) \tag{17}$$

In our problem,

$$\frac{\omega_L}{\omega_E} = 2,\ \cos\beta = 0.1$$

Also, there are other two non-vanishing matrix elements; they are

$$\frac{E_{m,m-1}}{\hbar\omega_E} = -\frac{3}{2}\cos\beta\sin\beta(I-2m)\left[(I-m+1)(I+m)\right]^{1/2}$$

$$\frac{E_{m,m-2}}{\hbar\omega_E} = \frac{3}{4}\sin^2\beta(I-2m)\left[(I+m-1)(I+m)(I-m+1)(I-m+2)\right]^{1/2}$$

Now go back to equation (17), where

$$\frac{\omega_L}{\omega_E} = 2\ and\ \cos\beta = 0.1$$

So,

$$E_m(\beta) = \hbar\omega_E\left[-2m + \frac{1}{2}\left[3m^2 - I(I+1)\right]\right]\left(3\cos^2\beta - 1\right)$$

Since I = 1, m = 1, 0, –1, $\cos\beta = 0.1$

Take

$$\frac{E_m(\beta)}{\hbar\omega_E} = K_m(\beta)$$

then,

$$K_m(\beta) = -2m + \frac{1}{2}\left[3m^2 - I(I+1)(3\cos^2\beta - 1)\right]$$

Now we can calculate for $K_m(\beta)$, where m = 1, 0, –1.

The table below gives those values:

$\cos\beta$	$\dfrac{\omega_L}{\omega_E}$	m	K	$\dfrac{E}{\hbar\omega_E}$
0.1	2	+1	-2.99	$E'_o - 2.99$
		0	0.97	$E'_o + 0.97$
		-1	2.02	$E'_o + 2.02$

where E_o is the ground state energy. We see here the effect of the external electric field that removes the level degeneracy.

QED

Problem 20

Show that if $P(m_i)$ is a constant then, $P(m_i) = \dfrac{1}{2I_i + 1}$, and $I(\theta)$ are constants.

The Solution

Let us write down $I(\theta)$ as

$$I(\theta) = const. \sum_{m_i,m,\kappa} P(m_i) \begin{pmatrix} I & L & I_i \\ -m & M & m_i \end{pmatrix}^2 \begin{pmatrix} L & L & \kappa \\ 1 & -1 & 0 \end{pmatrix} \begin{pmatrix} L & L & \kappa \\ M & -M & 0 \end{pmatrix} \times$$

$$(2\kappa+1)(-1)^M P_\kappa(\cos\theta) \tag{1}$$

Now if $P(m_i)$ is constant, that means the population of the substates is constant, and since for an eigen angular momentum (I_i), there are $(2I_i+1)$ substates, which imply that $(2I_i+1)P(m_i)$ is the population of these substates, therefore, $P(m_i)$ is given by

$$P(m_i) = \frac{1}{2I_i+1}$$

which yields

$$I(\theta) = \frac{const.}{2I_i+1} \sum_{m_i,m,\kappa} \begin{pmatrix} I & L & I_i \\ -m & M & m_i \end{pmatrix}^2 \begin{pmatrix} L & L & \kappa \\ 1 & -1 & 0 \end{pmatrix} \begin{pmatrix} L & L & \kappa \\ M & -M & 0 \end{pmatrix} \times$$

$$(2\kappa+1)(-1)^M P_\kappa(\cos\theta)$$

Now since $P(m_i)$ is constant, that means the population of the substates is the same, i.e., it is independent of the orientation, so we consider

$$\theta = 0 \Rightarrow P_\kappa = 1.$$

$$\therefore I(\theta = 0) = \frac{const.}{2I_i+1} \sum_{m_i,m,\kappa} \begin{pmatrix} I & L & I_i \\ -m & M & m_i \end{pmatrix}^2 \begin{pmatrix} L & L & \kappa \\ 1 & -1 & 0 \end{pmatrix} \begin{pmatrix} L & L & \kappa \\ M & -M & 0 \end{pmatrix} (2\kappa+1)^M$$

$$\tag{2}$$

Using W-E theorem,

$$\begin{pmatrix} I & L & I_i \\ -m & M & m_i \end{pmatrix} = (-1)^{I-L-m_i} \frac{\langle Im\,LM\,|\,I_i m_i \rangle}{\sqrt{2I_i+1}} = (-1)^{I_i-L+m_i} \frac{\langle I_i m_i\,LM\,|\,I_i m_i \rangle}{\sqrt{2I+1}}$$

$$\therefore \begin{pmatrix} I & L & I_i \\ -m & M & m_i \end{pmatrix}^2 = (-1)^{2(I-L-m_i)} \frac{\left[\langle Im\,LM\,|\,I_i m_i \rangle\right]^2}{2I+1}$$

Also, $$\begin{pmatrix} L & L & \kappa \\ 1 & -1 & 0 \end{pmatrix} = \frac{1}{[(2L+1)(2L+1)L]^{1/2}}$$

And $$\begin{pmatrix} L & L & \kappa \\ M & -M & 0 \end{pmatrix} = \frac{M}{[(2L+1)(2L+1)L]^{1/2}}$$

$$\therefore I(\theta=0) = \frac{const.}{(2I_i+1)(2I+1)} \sum_{m_i,m,\kappa} \frac{M(2\kappa+1)(-1)^M}{(2L+1)(2L+1)L}(-1)^{I-L-m_i}\left[\langle Im\,LM\,|\,I_i m_i \rangle\right]^2$$

$$= \frac{const.}{L(2L+1)^2(2I+1)(2I_i+1)} \sum_{m_i,m,\kappa} M(2\kappa+1)(-1)^{I-L+m}\left[\langle Im\,LM\,|\,I_i m_i \rangle\right]^2 \quad (4)$$

The RHS of equation (4) is constant, which implies that $I(\theta=0)$ is a constant.

QED

Problem 21

Find $U_{LM}(m,\Omega) = \frac{c}{2\pi\kappa^2} Z_{LM}(\theta,\phi)|a_{LM}(m)|^2$ for the energy in pure magnetic dipole radiation.

The Solution

The energy emitted per second is the flow of energy in the multipole radiation, which is determined by the well-known pointing vector (S), which is defined as

$$S = \frac{c}{4\pi} \left[E \times H \right]$$

(1)

where $\vec{E}$ and $\vec{H}$ are electric and magnetic field intensities, respectively. Far away from the source of the radiation, in the so-called wave zone, $\vec{E}$ and $\vec{H}$ are perpendicular to each other and to the radial direction r, and they are equal in magnitude. Hence, in the wave zone, we have

$$S \cong \frac{c}{4\pi} E^2 \cong \frac{c}{4\pi} H^2$$; from equation (1), we have

$$\langle S \rangle = \frac{c}{2\pi} \left(\vec{E} \times \vec{H} \right) \Rightarrow \left| \langle S \rangle \right| = \frac{c}{2\pi} \left| \vec{E} \right|^2 = \frac{c}{2\pi} \left| \vec{H} \right|^2$$

where

$$\vec{E}(r,t) = \vec{E} e^{i\omega t} + \vec{E}^* e^{-i\omega t}$$

$$\vec{H}(r,t) = \vec{H} e^{i\omega t} + \vec{H}^* e^{-i\omega t}$$

(2)

Now let us consider the pure multipole radiation with amplitude a_{LM}. Energy emitted into solid angle $d\,\Omega$ is given as

$$dE_{\Omega} = \left| \left\langle \vec{S} \right\rangle \right| \cdot R^2 d\Omega = U(\Omega) d\Omega$$

$$U_{LM}(m,\Omega) = \frac{c}{2\pi} \left| \vec{H} \right|^2 R^2 = \frac{c}{2\pi} \left| \vec{H}(R) \right|^2 R^2$$

$$\therefore \left[U_{LM} = \frac{c}{2\pi\kappa^2} \left| H_{LM}(m) \right|^2 \left| a_{LM}(m) \right|^2 R^2 \right]$$

(3)

Equation (3) can be written as

$$U(m,\Omega) = \frac{c}{2\pi} \left| \vec{A}_{LM}(m) \right|^2 \left| a_{LM}(m) \right|^2 R^2$$

(4)

But $\vec{A}_{LM}(m) = h_L^{(1)}(\kappa r) \vec{Y}_{LL}^{M}$

We take the asymptotic limit for $h^{(1)}(\kappa r)$ (Hankel function) $\rightarrow \dfrac{e^{i\kappa r}}{\kappa r}$

$$\therefore \vec{A}_{LM}(m) \rightarrow \frac{e^{i\kappa r}}{\kappa r} \vec{Y}_{LL}^{M}$$

Substituting this into equation (4), you get

$$U_{LM}(m,\Omega) = \frac{c}{2\pi} R^2 h_L^{(1)}(\kappa r) \vec{Y}_{LL}^{M}(\theta,\phi) h_L^{(1)*}(\kappa r) \vec{Y}_{LL}^{M*}(\theta,\phi) \times \left| a_{LM}(m) \right|^2$$

$$= \frac{c}{2\pi\kappa^2} \left| Y_{LL}^{M} \right|^2 \left| a_{LM}(m) \right|^2$$

Since $Z_{LM}(\theta,\phi) = \vec{Y}_{LL}^{M} \cdot \vec{Y}_{LL}^{M+}$, then,

$$\left[U_{LM}(m,\Omega) = \frac{c}{2\pi\kappa^2} Z_{LM}(\theta,\phi) |a_{LM}(m)|^2 \right], \text{ as required.}$$

It is the same as that for electric multiple radiation. The only difference is in the amplitude of radiation, i.e., $a_{LM}(\Omega)$ and $a_{LM}(m)$.

QED

Problem 22

Prove that

$$Z_{LM}(\theta,\phi) = \vec{Y}_{LL}^M \cdot \vec{Y}_{LL}^{M+} = \frac{1}{2}\left[1 - \frac{M(M+1)}{L(L+1)}\right]|Y_L^{M+1}|^2 + \frac{1}{2}\left[1 - \frac{M(M-1)}{L(L+1)}\right]|Y_L^{M-1}|^2$$

The Proof

We have that

$$Y_{LL}^M = -\frac{\vec{L}Y_L^M}{\sqrt{L(L+1)}} \tag{1}$$

From the commutation relations, we have that

$$[L_z, L_+] = \hbar L_+$$

$$[L_z, L_-] = -\hbar L_-$$

$$[L_+, L_-] = 2\hbar L_z; \; L_\pm = L_x \pm iL_y \; L$$

$$[L_\pm, L_z^2] = 0$$

These give

$$L^2 = \frac{1}{2}\left[L_+L_- + L_-L_+\right] + L_z^2 \tag{2}$$

From the definition of $L_\pm$, we see that $\left(L_+\right)^* = L_-$ and $\left(L_-\right)^* = L_+$.

$\therefore L_+L_-$ can be written as $|L_+|^2 = |L_+|^*L_+$

and

L_-L_- can be written as

$$\left|L_-\right|^2 = \left(L_-\right)^* L_-$$

Hence,

$$\left|LY_L^M\right|^2 = \left(\frac{\left(L_+Y_L^M\right)}{\sqrt{2}}\right)^* \cdot \left(\frac{\left(L_+Y_L^M\right)}{\sqrt{2}}\right) + \left(\frac{\left(L_-Y_L^M\right)}{\sqrt{2}}\right)^* \left(\frac{\left(L_-Y_L^M\right)}{\sqrt{2}}\right) + \left(L_zY_L^M\right)^* \left(L_zY_L^M\right) \tag{3}$$

But it is known that

$$L_+Y_L^M = \sqrt{L(L+1) - M(M+1)}Y_L^{M+1}$$

$$L_-Y_L^M = \sqrt{L(L+1) - M(M-1)}Y_L^{M-1} \tag{4}$$

$$L_zY_L^M = MY_L^M$$

From equations (3) and (4), we have

$$\left|LY_L^M\right|^2 = \frac{1}{2}\left[L(L+1) - M(M+1)\right]\left|LY_L^{M+1}\right|^2 + \frac{1}{2}\left[L(L+1) - M(M-1)\right]\left|LY_L^{M-1}\right|^2$$

$$+ M^2\left|Y_L^M\right|^2 \tag{5}$$

Now, $Z_{LM}(\theta,\phi) = \left|\vec{Y}_{LL}^{M}\right|^2$ (6)

Substitute equation (1) into equation (6); you get

$$Z_{LM}(\theta,\phi) = \frac{\left|L\vec{Y}_{L}^{M}\right|^2}{L(L+1)}$$ (7)

and substitute equation (5) into equation (6); you get

$$\left[\left[Z_{LM}(\theta,\phi) = \frac{1}{2}\left[1 - \frac{M(M+1)}{L(L+1)}\right]\left|Y_{L}^{M+1}\right|^2 + \frac{1}{2}\left[1 - \frac{M(M-1)}{L(L+1)}\right]\left|Y_{L}^{M-1}\right|^2\right.\right.$$

$$\left.\left.+ \frac{M^2}{L(L+1)}\left|Y_{L}^{M}\right|^2\right]\right]$$ (8)

This is the desired result.

QED

Problem 23

Compute $Z_{LM}(\theta,\phi)$ for L = 2, presenting the result in diagrams.

The Computation

To compute $Z_{LM}(\theta,\phi)$, we use equation (8) in problem (22). For L = 2, M takes the values $0,\pm1,\pm2$. Now we compute Z for each value of M.

1. For M = 0 then from equation (8), we find that

$$Z_{20} = \frac{1}{2}\left|Y_2^1\right|^2 + \frac{1}{2}\left|Y_2^{-1}\right|^2$$

where $\quad Y_2^1 = -\sqrt{\frac{15}{8\pi}}\sin\theta\cos\theta\, e^{i\psi}$

$$Y_2^{-1} = -\sqrt{\frac{15}{8\pi}}\sin\theta\cos\theta\, e^{-i\psi} \quad\Rightarrow$$

$$\left[Z_{2,0} = \frac{15}{8\pi}\sin^2\theta\cos^2\theta\right] \tag{1}$$

2. For $M = \pm 1$

$$Z_{2,\pm 1} = \frac{1}{2}\left|Y_2^0\right|^2 + \left|Y_2^{-2}\right|^2 + \frac{1}{3}\left|Y_2^{\pm 1}\right|^2$$

But

$$Y_2^0 = -\sqrt{\frac{5}{4\pi}}\left(\frac{3}{2}\cos^2\theta - \frac{1}{2}\right), \quad Y_2^{-2} = \frac{1}{4}\sqrt{\frac{15}{2\pi}}\sin^2\theta\, e^{-2i\psi}$$

And

$$Y_2^{\pm 1} = \pm\sqrt{\frac{15}{8\pi}}\sin\theta\cos\theta\, e^{\pm i\psi} \quad\Rightarrow$$

$$Z_{2,\pm1} = \frac{5}{16\pi}\left(1 - 3\cos^2\theta + 4\cos^4\theta\right) \qquad (2)$$

3. For $M = \pm 2$, following the same way, substituting for $M = 2,-2$ into equation (8), problem (22) and for both values, we find

$$Z_{2,\pm2} = \frac{1}{3}\left[\left|Y_2^1\right|^2 + 2\left|Y_2^2\right|^2\right]$$

$$Y_2^1 = -\sqrt{\frac{15}{8\pi}}\sin\theta\cos\theta\, e^{i\psi}$$

$$Y_2^2 = \frac{1}{4}\sqrt{\frac{15}{2\pi}}\sin^2\theta\, e^{2i\psi}$$

$$Z_{2,\pm2}(\theta,\phi) = \frac{5}{16\pi}\left(1 - \cos^4\theta\right) \qquad (3)$$

These results can be tabulated as

$$Z_{LM}(\theta,\phi) = \left|Y_{LL}^M(\theta,\phi)\right|^2$$

	M		
L	0	±1	±2
2	$\dfrac{15}{8\pi}\sin^2\theta\cos^2\theta$	$\dfrac{5}{16\pi}\left(1-3\cos^2\theta+4\cos^4\theta\right)$	$\dfrac{5}{16\pi}\left(1-\cos^4\theta\right)$

These can be represented diagrammatically as follows:

1. For $Z_{2,0} = \dfrac{15}{8\pi}\sin^2\theta\cos^2\theta$, L = 2, M = 0

L = 2, M = 0

2. For $Z_{2,\pm1} = \dfrac{5}{16\pi}\left(1-3\cos^2\theta+4\cos^4\theta\right)$, L = 2, M = ±1

L = 2, M = ±1

3. For $Z_{2,\pm2}(\theta,\phi) = \dfrac{5}{16\pi}\left(1 - \cos^4\theta\right)$, L = 2, M = ±2

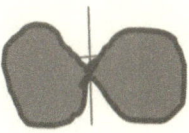

L = 2, M = ±2

The References

1. A. de-Shalit and Feshbach, *Theoretical Nuclear Physics*, vol. 1, John Wiley & Sons Inc., 1974.

2. A. M. Lane, *Nuclear Shell Theory*, W. A. Benjamin Inc., 1964.

3. de-Shalit and Talmi, *Nuclear Shell Theory*, Academic Press, 1963.

4. Robert B. Lighton, *Principles of Modern Physics*, McGraw-Hill Book Company Inc., 1959.

5. David Halliday, *Introductory Nuclear Physics*, 2nd edition, John Wiley & Sons Inc.1955.

6. R. Martin Eisberg, *Fundamentals of Modern Physics*, John Wiley & Sons Inc., 1961.

7. R. D. Evans, *The Atomic Nucleus*, McGraw-Hill Book Company Inc., 1955.

8. *Experimental Nuclear Physics*, vol. 1, Segre, editor, 2nd edition, John Wiley & Sons Inc., 1960.

9. *Introductory Nuclear Physics*, John Wiley & Sons, New York, 1987.

10. M. Pfitzner et al., *European Journal of Physics* 14, 279, 2002.

11. J. Giovinazzo et al., *Physical Review Letters* 89, 102501, 2002.

12. P. Woods and Davis, *Annual Review of Nuclear and Particle Science* 47, 541, 1997.

13. A. A. Oglobin et al., *Physical Review* 61, 034301, 2000.

14. G. Ardisson and M. Hussonnios, Radio Chemist, Acta, 70/71, 123, 1995.

15. B. A. Brown, *Physical Review Letters* 69, 1034, 1992.

16. D. H. Wilkinson, A. Gallman, and D. E. Alburger, *Physical Review Letters* c18, 401, 1978.

17. D. H. Wilkinson, *Nuclear Physics* A 377, 474, 1982.

18. Niels Bohr, *Science* 86, 161, 1937.

19. Maria G. Mayer, *Physical Review* 74, 235, 1948.

20. Maria G. Mayer, *Physical Review* 75, 1969, 1949.

21. O. Haxel, J. H. D. Jensen, and H. E. Suess, *Physical Review* 75, 1766, 1949.

22. M. Honma, Otsuka, B. A. Brown and Mistake, *Physical Review* C65061301, 2002.

23. K. L. G. Heyde, *The Nuclear Shell Model*, Spring Verilog, 1994.

24. R. D. Lawson, *Theory of the Nuclear Shell Model*, Clarendon Press, 1980.

25. A. Bohr and B. R. Mottelson, *Nuclear Structure*, vol. 1, W. A. Benjamin, 1969.

26. A. Bohr and B. R. Mottelson, *Nuclear Structure*, vol. 11, W. A. Benjamin, 1975.

27. R. M. Dreigler and E. K. U. Gross, *Density Functional Theory: An Approach to the Quantum Many-Body Problems*, Springer Berlin, 1990.

28. T. Suzuki, H. Sagawa, and A. Arima, *Nuclear Physics* A536, 141, 1998.

29. B. A. Brown, *Physical Review* 43,R 1513, 1991, C44, 9 24, 1991.

30. B. J. Cole, *Physical Review* 54, 1240, 1996.

31. W. E. Ormand, *Physical Review* C53, 214, 1997.

32. B. J. Cole, *Physical Review* C58, 2831, 1998.

33. B. Blank et al., *Physical Review Letters* 84, 1116, 2000.

34. A. Bulgac and V. R. Shaginyan, *Nuclear Physics* A 601, 103, 1996.

35. A. Bulgac and V. R. Shaginyan, *European Physics Journal* A5, 247, 1995.

36. C. J. Horowitz and J. Piekarewitz, *Physical Review* C63, 001330 R, 2000.

37. H. T. Furtune, R. Sherr, and B. A. Brown, *Physical Review* C61, 057303, 2000.

38. H. Schatz et al., *Physical Review* 294, 167, 1998.

39. A. R. Edmonds, *Angular Momentum in Quantum Mechanics*, Princeton University Press, 1957.

40. D. A. Varshalovich, A. N. Moskalev, and V. K. Khersonskii, *Quantum Theory of Angular Momentum*, World Scientific, 1988.

41. W. J. Thompson, *Angular Momentum*, John Wiley & Sons Inc., 1994.

42. I. Talmi, *Simple Methods of Complex Nuclei*, Harwood Academic Publisher, 1993.

43. P. J. Brussard and P. W. M. Gloudemans, *Shell Model Applications in Nuclear Spectroscopy*, North Holland, 1977.

44. D. M. Brink and G. R. Satchler, *Angular Momentum*, Clarendon, 1975

45. A. Messiah, *Quantum Mechanics*, vol. 1 and vol. 2, North Holland, Amsterdam, 1975.

46. I. S. Townler, *Shell Model Description of Light Nuclei*, Clarendon Press, 1977.

47. M. Rotenberg, R. Bivins, Metropolis and K. R. Wooten Jr., *The 3j and 6j Symbols*, Cambridge: Technology Press MIT, 1959.

48. A. B. Majdal, *Nuclear Theory: The Quasi Particle Method*, Benjamin, New York, 1968.

49. A. B. Majdal, *Theory of Finite Fermi Systems and Applications to Atomic Nuclei*, New York: John Wiley & Sons, 1967.

50. S. Bogner et al., *Physical Review* 65, 051130®, 2000.

51. M. Radici et al., *Physical Review* C50, 3010, 1994.

52. D. Van Neck et al., *Physical Review* C57, 2308, 1998.

53. P. G. Hansen et al., *Annual Review of Nuclear Physics and Particle Science*.

54. G. J. Kramer et al., *Nuclear Physics* A679, 267, 2001.

55. J. Vernote et al., *Nuclear Physics* A571, 1, 1994.

56. B. A. Brown, *Physical Review* C58, 220, 1998.

57. M. Leuschner et al., *Physical Review* C49, 955, 1994.

58. J. Gao et al., *Physical Review Letters* 84, 3265, 2000.

59. D. Branford et al., *Physical Review* C63, 014310, 2000.

60. M. Radici et al., *Physical Review* C66, 014613, 2002.

61. L. Lapihas et al., *Physical Review* C61, 064325, 2000.

62. L. Frankfurt et al., *Physics Letter* B503, 73, 2001.

63. A. Navin et al., *Physical Review Letters* 81, 5089, 1998.

64. T. Aumann et al., *Physical Review Letters* 84, 35, 2000.

65. A. Navin et al., *Physical Review Letters* 85, 266, 2000.

66. V. Guimaraes et al., *Physical Review* C61, 06409, 2000.

67. V. Maddalene et al., *Physical Review* C 63, 024613, 2001.

68. B. A. Brown, *Physical Review Letters* 85, 5300, 2000.

69. D. H. Wilkinson et al., *Physical Review* C18, 401, 1978.

70. D. H. Wilkinson et al., *Nuclear Physics* A377, 474, 1982.

71. B. A. Brown et al., *Atomic Data and Nuclear Data Tables* 93, 347, 1985.

72. E. C. Adelbecger et al., *Nuclear Physics* A417, 269, 1984.

73. Donald H. Perkins, *Introduction to High Energy Physics*, 2nd edition, Reading, Massachusetts: Addison-Wesley, 1982.

74. Hermann Weyl, *The Theory of Groups and Quantum Mechanics*, transl. H. P. Robertson, Dover Publication, 1931.

75. Y. Neeman, *Nuclear Physics* 26, 222, 1961.

76. M. Gell-Mann, "CalTech Report" CTSL-20 (unpublished 1961).

77. M. Gell-Mann, *Physical Review* 125, 1069, 1962.

78. S. Okubo, *Physics Letter, 4,* 14, 1963.

79. S. Okubo, *Progress Theoretical Physics* 27, 946, 1962.

80. S. Okubo, Progress Theoretical Physics , *28, 24,* 1962.

81. C. A. Levinson et al., *Physics Letter* 1, 144, 1962.

82. C. A. Levinson et al., *Nuovo cimento, 23, 226, 1962.*

83. S. Meshkov et al., *Physical Review Letters* 10, 361, 1963.

84. S. P. Rosen, *Physical Review Letters* 11, 100, 1963.

85. S. P. Rosen, "Lectures on Weak Interactions" (unpublished, Physics Department, Purdue University, 1964).

86. H. Pitchman, *Studies in Particles and Field Physics and Weak Interactions*, edited by H. Aly, Edwardsville, USA: Southern Illinois University.

87. A. P. Zuker et al., *Physical Review Letters* 89, 14502 (2002)^{42}Ti,^{42}Sc and ^{42}Ca.

Curriculum Vitae of the Author

Full name: Ali A. Abdulla

Birth date: 7/1/1939

Place of birth: Iraq

Scientific degrees:

1. PhD in Physics, OSU, USA, 1968

2. MS in Physics, NDU, USA, 1964

3. BSc in Physics (honor degree), Baghdad University, Iraq, 1961

Jobs Held

1. **Scientific Managements**

   A. Director general, Arab Atomic Energy Agency (AAEA), 1989–1993

   B. Director manager, Sciences Management, Arab Bureau for Education, 1985–1987, Riyadh, KS

   C. Director, R and Ministry of Industry, 1973–2/15/1975, Iraq

   D. Chairman, Physics Department, College of Science, BU, Iraq, 9/15/1975–6/20/1981

   E. Chairman, Physics Department, College of Science, BU, Iraq, 9/7/1987–2/14/1989

   F. Director, Nuclear Research Institute, Iraqi Atomic Energy Committee, 1969–1972, Iraq

   G. Scientific researcher, Nuclear Institute, IAEC, 1968–1969

2. **Scientific Status**

    A. Scientific researcher, Iraqi Atomic Energy Committee (IAEC), 12/1968–7/1969

    B. Instructor, 1972–1975

    C. Assistant professor, 1975

    D. Professor, 1984

3. **Teaching Experiences**

    A. Teaching the following courses for graduate studies: advanced nuclear physics, advanced quantum mechanics, quantum field theory, particles physics, 1975–1985, 1987–1989, 1993–2009, Physics Department, College of Sciences, Baghdad University, Iraq

    B. For undergraduate, the following courses were taught: quantum mechanics, classical mechanics, electromagnetics theory, optics, nuclear physics, thermodynamics, modern physics, labs

    C. These courses were taught at (1) Oklahoma State University, USA, 1967–1968; (2) Al-Mustansiriya University, Iraq, 1969–1974; (3) Al Basrahok University, Iraq, 1972–1974; (4) Baghdad University, 1975–1985

4. **Supervising MS and PhD Students' Theses**

Supervised thirty MS students' theses and twelve PhD students during the periods 1978–1985, 1993–2009, University of Baghdad, Iraq

5. **Other Scientific Activities**

    A. Chairman of the Iraqi Society of Physicists and Mathematicians, 1976–1985, Iraq

    B. Secretary general of Arab Union for Physicists and Mathematicians, 1976–1985, Iraq

    C. Full member of the Iraqi Academy of Sciences, 1979–1996, Iraq

    D. Honor member of the Iraqi Academy of Sciences, 1996 till now, Iraq

    E. Member of Sigma Pi Sigma, Oklahoma State University Chapter, USA, 1967

6. **Books Authored**

    A. *Nuclear Physics*, text for third year physics (in Arabic), issued by the Iraqi Ministry of Higher Education and Scientific Research (MHESR), 1982

    B. *Properties of Matter and Wave Motion*, text for second year physics (in Arabic), issued by MHESR, 1978

    C. *General Physics for Teachers Institutes*, text (in Arabic), issued by Ministry of Education, 1978, Iraq

    D. *Experimental Nuclear Physics for Fourth Year Students*, text (in Arabic), issued by the University of Baghdad, 1991, Iraq

    E. *Advance Nuclear Physics* (under printing)

    F. *Advance Quantum Mechanics* (under printing)

G. *Philosophical and Thoughts Approaches in Physics* (in Arabic), issued by the Iraqi Academy of Sciences, 2010

H. *Survey Study in the Physics of Nuclear Particles*, issued by the Physics Department Press, 2009

I. *Nuclear Energy and Its Future in the World*, published by the Arab Atomic Energy Agency, 1991

J. There are four books related to a non-physics subject, published in 2005–2009, total of twelve books

K. There are fifteen books that are supervised, edited for, or introduced too.

7. Conferences Attended

(63) local and Arabic planning committees, during the period 1969–2002, in different establishments in Iraq and the Arab organizations,

A. Iraqi Atomic Energy Commission (IAEC), member, 1970–1972

B. Central committee to supervise the execution of the IAEC projects, president, 1971–1972

C. Administration board of the Regional Center of Radio Isotopes (RCRI) for Arab states, Cairo, president, 1971–1972

D. Committee of scientific research projects assessment of RCRI, president, 1970–1972

E. Committee of the development of the College of Sciences, University of Baghdad, member, 1976

F. Higher committee of developing the curriculum and the teaching means, ministry of education, Iraq, member, 1975–1982

G. Committee of physics programs at the high school level, ministry of education, president, 1976–1982

H. Committee to study the establishment of research and development center in the industrial companies in Iraq, member, 1974

I. Consultant committee for the center of astronomy, scientific research establishment, Iraq, member, 1970–1978

J. Thirty committees of assessment of MS and PhD students' theses, chairman, BU and others, 1978–2009

K. The assessment of the scientific degrees obtained from international institutes in different scientific specialties committee, chairman, 1975–1982

8. **Scientific Journals Issued**

A. *Arab Journal for Physics*, editor in chief, 1977–1985, issued by the Arab Union for Arab Physicists and Mathematicians, Iraq

B. *Mathematics and Physics Journal*, editor in chief, 1977–1985, issued by the Iraqi Society for Physics and Mathematics, Iraq

C. *Arab Gulf Journal of Scientific Research*, A, editor in chief, 1985–1987, issued by the Arab Bureau for Education for Arab Gulf States, Riyadh, KS

D. *Arab Gulf Journal of Scientific Research*, B, agricultural and biological sciences, issued by the Arab Bureau for Education for Arab Gulf States, Riyadh, KS

E. *The Atom and Arab Society Development*, popular journal, editor in chief, 1989–1993, issued by AAEA

9. **Selected Published Papers**

The following are some selected published papers out of fifty published papers in many different national and international journals:

Paper Title	Journal
1. Inelastic Magnetic Electron Scattering M1 Form Factors in Ca-48 (M3Y Fitting Parameters Consideration)	*Proceeding of the Iraqi National Scientific Conference* (2010)
2. "Calculation of Elastic and Inelastic Electron Scattering on F19, Using Large Basis No No-Core Shell Wave Function"	*Nuclear Physics A*, 798, 16–28 (2008), North Holland
3. "Role of the Excitation Model Space with M3Y Core Polarization Interaction on the Electron Scattering Form Factors for 19 F"	*Nuclear Physics A*, 2009, North Holland
4. "Isovector Transverse Nuclear Excitation in B10	*Journal of Al-Nahrain University*, vol. 7, 125, 2004
5. "Isoscaler and Isovector Longitudinal Quadruple Nuclear Excitation in C12"	*Dirasat: Pure Science*, vol. 29, no. 1, 40, 2002
6. "Mixing of Te/In Bilayer System Using Ionic Radiation"	*Dirasat: Pure Science*, vol. 29, no. 2, 2002, *Jordan*

7. "Study of the Gamma Irradiation Effect on the Fine Structure of PE of High Density Using PALS Method"	*Iraqi Journal for Physics*, vol. 1, no. 1, 2002
8. "Study the Effect of Beta Irradiation on the Fine Structure of PE of High Density, Using PALS"	*Iraqi Journal for Physics*, vol. 1, no. 2, 2002
9. "Studying the Effect of Irradiation by Gamma on PMMA Using PALT Method"	*Iraqi Journal for Physics*, vol. 1, no. 4, 2002
10. "Studying the Gamma Irradiation Effect on the Fine Structure of PE of Low Density LDPE"	*Iraqi Journal for Physics*, vol. 2, no. 1, 2002
11. "Studying the Effect of Beta Low Dose on PMMA Using PALT Method"	*Indian Journal for Physics*, vol. 75A, no. 5, pp. 457–462, 2002
12. "Studying the Effect of Beta Irradiation on the Fine Structure of PE of Low Density (LDPE) Using PALT Method"	*Journal of Solid State Physics*, vol. 95, no. 12, 2002

13. "Studying the Changes in the Fine Structure of PMMA Due to Gamma Irradiation by Positron Annihilation Method"	*Saddam Journal for Science*, vol. 1, no. 2, 2002
14. "Inelastic (1^+, 02^+, and 3^+) Electron Scattering Form Factor of Li (6)"	*Iraqi Journal for Science*, vol. 43, no. 2, 2002
15. "Positron Lifetime Measurement in Distorted Cupper"	*Al-Mustansiriya Journal of Science*, vol. 12, no. 4, 2002
16. "Core Polarization Effect on the Form Factors for C2 of Nuclei of p Shell"	*Nuclear Physics A*, no. 696, pp. 442–452, 2001, North Holland
17. "Mixing of the Binary System Se/In by Ionizing Radiation"	*Scientific Journal of Iraqi Atomic Energy Commission*, vol. 3, no. 1, p. 5, 2001
18. "Using PALS to Study Structure Changes of Polystyrene (PS) Due to Irradiation"	*Iraqi Journal of Sciences*, vol. 42C, no. 1, 2001 *Indian Journal of Physics*, vol. 75A, no. 5, p. 579, 2001
19. "Form Factor of Magnetic Elastic Scattering for C12"	*Nuclear Physics A*, 696, 2001

20. "Microscopic Calculations of C2 and C4 Form Factors in sd-Shell Nuclei"	Proceeding of the fifth Arab conference for peaceful uses of atomic energy, 2000
21. "Inelastic Electron Scattering for Dominating Negative Parity States in C13"	*Iraqi Journal of Sciences*, vol. 1c, no. 1, 2000
22. "Mixing of the Binary System S/In by Ionizing Radiation"	*Education College Journal*, vol. 3, p. 57, 2000, Al-Mustansiriya
23. "Study of the Properties of the Positron Annihilation Spectrum in Distorted Zinc"	*Dirasat: Pure Sciences*, vol. 46, no. 2, 1999
24. "Elastic Form Factor M1 of 3.562 MeV for Li (6)"	*Nuclear Physics A*, 458, pp. 51–76, 1986, North Holland
25. "Angular Distribution Study in Yb (172, 174)"	*Radiation Research*, 000-00092, 1982
26. "Measurement of the Changes in the Porcelain Electrical Resistance Due to the Effect of the Neutron Flux"	Proceeding of the second Arab scientific conference, Jordan, 1981
27. "Measurement of the Fission Neutron Flow through the Changes in the Electrical Resistance of Cellulose Nitrates"	*Nuclear Instruments and Methods*, vol. 17, p. 505–512, 1981

28. "Degradation of the Polycarbonate Material under the Effect of the Fast Neutrons and Gamma Radiation, Its Application in the Radiation Detection Apparatuses"	Proceeding of the second Arab scientific conference, Jordan, 1981
29. "About the Minimum Cell in the Crystal Space and the Uncertainty Relation between Energy and Time"	*Nuclear Instruments and Methods*, 000 (pp. 71–74) 1979, North Holland
30. "Method to Measure Neutron Doses by the Changes Occurs in the Average Molecular Weight for Cellulose Nitrates"	*Al-Mustansiriya Journal for Sciences*, vol. 2, 1977
31. "Manufacturing a Thin Silicon Radiation Detector"	*Indian Journal for Applied Physics and Mathematics*, vol. 12, pp. 290–295, 1974
32. "Studying the Decay Scheme of Rh (104)"	*Oil and the World Journal*, vol. 6, Iraq and Naft al-Arab, Beirut, 1973
33. "The Role of Nuclear Energy in the Future of the Energy in the World"	*Russian Nuclear Physics*, vol. 6, no. 2, 1972

34. "Study of Gamma Radiation Produced Due to Thermal Neutron from the Isotopes Si30 and Si31"	*The Physical Review C*, vol. 1, no. 3, pp. 1093–1099, March 1970, USA
35. "Search for ekk Emission from the Element Xe (131 m)"	Proceeding the sixth Arab scientific conference, vol. 3c1, 1969, Damascus, Syria
36. "Studying the Reaction Cl (35) (n, Gamma) Cl (36), Using Anti-Compton with Three Crystal Angles"	Proceeding the sixth Arab scientific conference, vol. 3C1, 1969, Damascus, Syria
37. "Using the Neutron Activation Analysis to Determine the Content of Uranium in Some Geological Samples"	

Regional and international are forty-two, 1969–2009, such as the selected examples:

a. The fourth conference on the peaceful uses of the atomic energy, Geneva September 1970

b. The general conference of the International Atomic Energy Agency, 1970, 1991, Vienna, Austria

c.   The conference of the Third World Academy of Science, Trieste, Italy, August, 1985

d.   The scientific conference of the Islamic Organization for Sciences and Technology, Istanbul, Turkey, October 1986

e.   The second Arab scientific conference for physics and mathematics, chairman, Amman, Jordan, 1981

f.   The third Arab scientific conference for physics and mathematics, chairman, Tunis, 1983

g.   The first Arab scientific conference for peaceful uses of the atomic energy, Tripoli, Libya, 1992

h.   The first scientific conference for Arab Scientists and Technologists Abroad (ASTA) and in the Arab world, chairman, Amman, Jordan, 1992

i.   The second scientific conference for ASTA, Amman, Jordan, 1994

j.   The scientific research directors' conference in the Arab Gulf States, organizer, King Fahd University, KS, September 1985

k.   The international conference on physics and mathematics education, Brazil, 2002

l.   The scientific conference of the Iraqi Society for Physics and Mathematics, chairman, Baghdad, 1977

m.   The first Arab scientific conference for physics and mathematics, chairman, 1978

n.   The scientific conference of the College of Science, BU, 1993

o.   The scientific conference of the College of Science, BU, 2009, chairman or a member of these confrences

Index

www.ingramcontent.com/pod-product-compliance
Lightning Source LLC
Chambersburg PA
CBHW020719180526
45163CB00001B/29